Mathematics of Wave Propagation

Mathematics of Wave Propagation

Julian L. Davis

PRINCETON UNIVERSITY PRESS
PRINCETON, NEW JERSEY

Copyright © 2000 by Princeton University Press
Published by Princeton University Press, 41 William Street,
Princeton, New Jersey 08540
In the United Kingdom: Princeton University Press,
Chichester, West Sussex

All Rights Reserved

Library of Congress Cataloging-in-Publication Data

Davis, Julian L.
Mathematics of wave propagation/Julian L. Davis.
p. cm.
Includes bibliographical references and index.
ISBN 0-691-02643-2 (cl : alk. paper)
1. Wave-motion, Theory of. I. Title.

QA927 .D32 2000
530. 12′4 21—dc21   99-044938

This book has been composed in Times Roman

The paper used in this publication meets the minimum requirements of
ANSI/NISO Z39.48-1992 (R1997) (*Permanence of Paper*)

www.pup.princeton.edu

Printed in the United States of America

10 9 8 7 6 5 4 3 2 1

Science is based on facts, as a house is with stones. But a collection of facts is no more a science than a heap of stones is a house.... Scientists believe that there is a hierarchy of facts and that a judicious selection can be made. One has only to open one's eyes to see that the triumphs of industry, which have enriched so many practical men, would never have seen the light of day if they had not been preceded by disinterested fools [the scientists] who sometimes died poor, who never thought of the useful, and yet had a guide that was based on the harmony of nature. What these fools did, as Mach has said, was to save their successors the trouble of thinking.... It is necessary to think for those who do not like thinking, and as there are many, each one of our thoughts must be useful in as many circumstances as possible. For this reason, the more general a law is, the greater is its value.

> Henri Poincaré
> *Science and Method*
> translated by Francis Maitland

Contents

Preface	xiii

CHAPTER ONE
Physics of Propagating Waves — 3

Introduction	3
Discrete Wave-Propagating Systems	3
Approximation of Stress Wave Propagation in a Bar by a Finite System of Mass-Spring Models	4
Limiting Form of a Continuous Bar	5
Wave Equation for a Bar	5
Transverse Oscillations of a String	9
Speed of a Transverse Wave in a String	10
Traveling Waves in General	11
Sound Wave Propagation in a Tube	16
Superposition Principle	19
Sinusoidal Waves	19
Interference Phenomena	21
Reflection of Light Waves	25
Reflection of Waves in a String	27
Sound Waves	29
Doppler Effect	33
Dispersion and Group Velocity	36
Problems	37

CHAPTER TWO
Partial Differential Equations of Wave Propagation — 41

Introduction	41
Types of Partial Differential Equations	41
Geometric Nature of the PDEs of Wave Phenomena	42
Directional Derivatives	42
Cauchy Initial Value Problem	44
Parametric Representation	49
Wave Equation Equivalent to Two First-Order PDEs	51
Characteristic Equations for First-Order PDEs	55
General Treatment of Linear PDEs by Characteristic Theory	57

Another Method of Characteristics for Second-Order PDEs	61
Geometric Interpretation of Quasilinear PDEs	63
Integral Surfaces	65
Nonlinear Case	67
Canonical Form of a Second-Order PDE	70
Riemann's Method of Integration	73
Problems	82

CHAPTER THREE
The Wave Equation — 85

PART I
ONE-DIMENSIONAL WAVE EQUATION — 85

Factorization of the Wave Equation and Characteristic Curves	85
Vibrating String as a Combined IV and BV Problem	90
D'Alembert's Solution to the IV Problem	97
Domain of Dependence and Range of Influence	101
Cauchy IV Problem Revisited	102
Solution of Wave Propagation Problems by Laplace Transforms	105
Laplace Transforms	108
Applications to the Wave Equation	111
Nonhomogeneous Wave Equation	116
Wave Propagation through Media with Different Velocities	120
Electrical Transmission Line	122

PART II
THE WAVE EQUATION IN TWO AND THREE DIMENSIONS — 125

Two-Dimensional Wave Equation	125
Reduced Wave Equation in Two Dimensions	126
The Eigenvalues Must Be Negative	127
Rectangular Membrane	127
Circular Membrane	131
Three-Dimensional Wave Equation	135
Problems	140

CHAPTER FOUR
Wave Propagation in Fluids — 145

PART I
INVISCID FLUIDS — 145

Lagrangian Representation of One-Dimensional Compressible Gas Flow	146

Eulerian Representation of a One-Dimensional Gas	149
Solution by the Method of Characteristics: One-Dimensional Compressible Gas	151
Two-Dimensional Steady Flow	157
Bernoulli's Law	159
Method of Characteristics Applied to Two-Dimensional Steady Flow	161
Supersonic Velocity Potential	163
Hodograph Transformation	163
Shock Wave Phenomena	169

PART II
VISCOUS FLUIDS — 183

Elementary Discussion of Viscosity	183
Conservation Laws	185
Boundary Conditions and Boundary Layer	190
Energy Dissipation in a Viscous Fluid	191
Wave Propagation in a Viscous Fluid	193
Oscillating Body of Arbitrary Shape	196
Similarity Considerations and Dimensionless Parameters; Reynolds' Law	197
Poiseuille Flow	199
Stokes' Flow	201
Oseen Approximation	208
Problems	210

CHAPTER FIVE
Stress Waves in Elastic Solids — 213

Introduction	213
Fundamentals of Elasticity	214
Equations of Motion for the Stress	223
Navier Equations of Motion for the Displacement	224
Propagation of Plane Elastic Waves	227
General Decomposition of Elastic Waves	228
Characteristic Surfaces for Planar Waves	229
Time-Harmonic Solutions and Reduced Wave Equations	230
Spherically Symmetric Waves	232
Longitudinal Waves in a Bar	234
Curvilinear Orthogonal Coordinates	237
The Navier Equations in Cylindrical Coordinates	239

Radially Symmetric Waves	240
Waves Propagated Over the Surface of an Elastic Body	243
Problems	247

CHAPTER SIX
Stress Waves in Viscoelastic Solids — 250

Introduction	250
Internal Friction	251
Discrete Viscoelastic Models	252
Continuous Maxwell Model	260
Continuous Voigt Model	263
Three-Dimensional VE Constitutive Equations	264
Equations of Motion for a VE Material	265
One-Dimensional Wave Propagation in VE Media	266
Radially Symmetric Waves for a VE Bar	270
Electromechanical Analogy	271
Problems	280

CHAPTER SEVEN
Wave Propagation in Thermoelastic Media — 282

Introduction	282
Duhamel-Neumann Law	282
Equations of Motion	285
Plane Harmonic Waves	287
Three-Dimensional Thermal Waves; Generalized Navier Equation	293

CHAPTER EIGHT
Water Waves — 297

Introduction	297
Irrotational, Incompressible, Inviscid Flow; Velocity Potential and Equipotential Surfaces	297
Euler's Equations	299
Two-Dimensional Fluid Flow	300
Complex Variable Treatment	302
Vortex Motion	309
Small-Amplitude Gravity Waves	311
Water Waves in a Straight Canal	311
Kinematics of the Free Surface	316
Vertical Acceleration	317
Standing Waves	319

CONTENTS

Two-Dimensional Waves of Finite Depth	321
Boundary Conditions	322
Formulation of a Typical Surface Wave Problem	324
Example of Instability	325
Approximation Theories	327
Tidal Waves	337
Problems	342

CHAPTER NINE
Variational Methods in Wave Propagation — 344

Introduction; Fermat's Principle	344
Calculus of Variations; Euler's Equation	345
Configuration Space	349
Kinetic and Potential Energies	350
Hamilton's Variational Principle	350
Principle of Virtual Work	352
Transformation to Generalized Coordinates	354
Rayleigh's Dissipation Function	357
Hamilton's Equations of Motion	359
Cyclic Coordinates	362
Hamilton-Jacobi Theory	364
Extension of W to $2n$ Degrees of Freedom	370
H-J Theory and Wave Propagation	372
Quantum Mechanics	376
An Analogy between Geometric Optics and Classical Mechanics	377
Asymptotic Theory of Wave Propagation	380
Appendix: The Principle of Least Action	384
Problems	387

Bibliography	389
Index	391

Preface

THE PURPOSE of this volume is to present a clear and concise treatment of the various phenomena of wave propagation. By its very nature, wave propagation in various media involves the use of applied mathematics in solving the physics of traveling waves. The mathematics is wide in scope and has many aspects that seemingly have no relation to wave phenomena, as the contents list shows. But in a wider sense there is a deep relation between the mathematics and the physics of propagating waves. This is an interdisciplinary approach; it lies at the heart of any serious treatment of wave phenomena. In this sense we take a lesson from Lord Rayleigh, whose two classic works on the theory of sound present a rich variety of mathematical concepts that apparently have no direct relation to the phenomenon of sound. But these mathematical ideas are necessary to give the reader an insight into the theory of sound. We follow this approach and attempt to give the reader a tapestry of the mathematical principles that lie at the foundations of wave phenomena and explain its physics. It is true that Rayleigh's treatment is rather old-fashioned in the sense that it does not invoke characteristic theory, which is the basis of the partial differential equations (PDEs) of wave propagation. This is rather strange since Rayleigh must have been aware of the work of Riemann, who used characteristic theory in his great work in fluid dynamics. (There is a passing reference to Riemann in the treatment of waves in air, but no mention of characteristic theory.) Characteristic theory lies at the basis of our treatment of the mathematics of wave propagation. We present also a modern approach which makes use of vector notation, the modern terminology in treating PDEs, and variational methods.

We start with an exposition of the physics of wave propagation, which lays the physical basis for the deeper aspects of the mathematical treatment in later chapters. We then treat the PDEs of propagating waves with the emphasis on the theory of characteristics, which is the heart of the type of PDEs in wave propagation called "hyperbolic PDEs." The three types of PDEs are classified as hyperbolic, elliptic, and parabolic, according to the nature of the characteristic ordinary

PREFACE

differential equation. Characteristic theory is general enough to treat nonlinear phenomena. We start with linear PDEs and then go to the quasilinear and then the fully nonlinear PDEs. Next, we treat the wave equation (the Cauchy problem) by using characteristic theory, starting with one-dimensional waves and then going to two- and three-dimensional waves.

The phenomenon of wave propagation is based on the three fundamental laws of physics: conservation of mass (continuity equation), conservation of momentum (equations of motion), and conservation of energy (equations of state or constitutive equations). The first two conservation laws apply to wave phenomena in any medium; the third law defines the medium: gas, solid, or liquid. (We omit the conservation of charge since we are not dealing with wave propagation in an electromagnetic medium.) These conservation laws are expressed in either Eulerian or Lagrangian coordinates. They are fully described, as are their range of applications.

We then investigate wave phenomena in fluids. This chapter is divided into two parts: inviscid fluids and viscous fluids. Characteristic theory is used in treating one-dimensional compressible flow, two-dimensional flow, and shock wave phenomena. The topic of stress waves in solids offers a greater variety of waves than in fluids, since there are transverse and longitudinal stress waves, while inviscid fluids have only longitudinal waves. The next stage is the treatment of stress waves in viscoelastic solids, which combine the properties of elastic solids and viscous fluids. Then temperature effects are taken into account in the treatment of wave propagation in thermoelastic media.

There is also a chapter on water waves; we treat waves in a canal of infinite and finite depth and make use of perturbation theory in the investigation of shallow-water waves. The theory of tidal waves is treated first from the point of view of the equilibrium theory and then with the more realistic dynamical approach.

The final chapter discusses variational methods, which present a deeper treatment of the mathematics of wave propagation. The calculus of variations is developed. This leads to Fermat's principle, the principle of virtual work, D'Alembert's principle, Hamilton's principle, the equations of motion of Laplace, and Hamilton's canonical equations. All this leads to the Hamilton-Jacobi theory, which is the basis of the PDEs of wave propagation. The H-J theory is applied to wave propagation and to quantum mechanics, of which we present a brief survey. The method of asymptotic expansions is then applied to the asymptotic theory of wave propagation, where nonlinear waves can be treated.

The problems at the end of the chapters range from exercises to give the reader practice in mathematical techniques to more involved problems that tax the reader's ingenuity. This work is useful to the graduate student as well as the professional applied mathematician, physicist, and engineer. The engineer dealing with wave phenomena who is not too familiar with the method of characteristics and variational phenomena should profit from this work.

It is clear that this volume is too short to present all aspects of propagating waves. Our choice of material is subjective, in the sense that the material reflects the author's interests. In particular, wave propagation in electromagnetic media is not treated here. There are many good works on electromagnetic theory that discuss this subject. There is a book solely devoted to wave propagation in electromagnetic media, written by the author and entitled *Wave Propagation in Electromagnetic* (Springer-Verlag, 1990).

Mathematics of Wave Propagation

CHAPTER ONE

Physics of Propagating Waves

INTRODUCTION

In this chapter we shall discuss the physics of propagating waves, starting with simple physical models and then giving an elementary combined physical and mathematical treatment of waves traveling in continuous media. A mathematical treatment is reserved for subsequent chapters.

For our purposes a *continuous medium* is one in which there is a continuous distribution of matter in the sense that a differential volume of material (in the mathematical sense) has the same properties as the material in the large. This means that molecular and crystalline structures are neglected. It is known that electromagnetic (EM) waves travel in a vacuum with the speed of light. (A vacuum is a continuous medium with zero density of matter.)

A propagating medium involves oscillations of the material through which the wave travels, with a wave velocity characteristic of the material and the temperature. For example, sound waves travel with a wave velocity that depends on the temperature and the density of the medium (air or fluid). For EM waves traveling in a vacuum, we have an oscillating electric intensity vector and an oscillating magnetic intensity vector normal to the electric vector.

DISCRETE WAVE-PROPAGATING SYSTEMS

Although the main thrust of this book is a treatment of waves traveling in continuous media, it is useful to construct a physical model composed of a discrete set of oscillating masses coupled by springs. We shall neglect friction. The limit as the number of masses and springs becomes infinite in a finite region yields a continuous medium.

The simplest oscillating system consists of a spring fixed at one end and coupled to a mass. The small-amplitude oscillations of the mass

CHAPTER ONE

along the spring axis (due to initial spring compression or tension or velocity of the mass) are an example of longitudinal simple harmonic (sinusoidal) motion with a frequency equal to the square root of the ratio of the spring constant to the mass.

APPROXIMATION OF STRESS WAVE PROPAGATION IN A BAR BY A FINITE SYSTEM OF MASS-SPRING MODELS

We approximate the one-dimensional longitudinal wave propagation in a semi-infinite thin elastic bar by a discrete physical model composed of a linear array of coupled masses and springs displayed horizontally along the x axis, the initial mass being located at $x = 0$. Each mass and spring constant is the same. The array of mass-springs represents the elastic material of the bar. Suppose mass no. 1 at $x = 0$ is driven a small distance Δx to the right along the x axis by an external force. In engineering terminology this is called an *impact loading* problem; it has many interesting engineering applications. Mass no. 1 compresses spring no. 1 attached to mass no. 2. Mass no. 1 oscillates in simple harmonic motion while a signal is sent to mass no. 2, which moves to the right a small distance, compressing spring no. 2 (attached to mass no. 3), and so on. A *compressive stress wave* is set up that propagates to the right, leaving the oscillating masses in its wake. Note that the oscillating motion of the elastic bar (represented by the masses) is axial in the same direction as the propagating wave that progresses down the bar. That is why this type of wave is called a *longitudinal wave*. The *wave front* is the point on the x axis showing the boundary between the mass in motion and the masses to the right, which are at rest.

A *tensile stress wave* can be set up by moving mass no. 1 to the left, thus extending spring no. 1. A wave in tension is thus transmitted successively to the neighboring masses. Again, the wave front is the point where the furthermost right mass is activated.

The mathematical description of the coupled oscillations of a system of masses and springs with an impact boundary condition (BC) consists of a set of coupled ordinary differential equations (ODEs) with a boundary condition on mass no. 1. We shall not discuss the solution of this system except to say that there are standard methods in ODEs to solve such a system. One such method introduces a set of normal coordinates which decouples the system and makes it easier to solve.

Rather than pursue this approach we shall skip to the continuous bar, wherein the mathematics involves a single partial differential equation (PDE) for the displacement u of the bar that is a function of one-dimensional space x and time t, so that u is a function of x and t: $u = u(x,t)$. This PDE for $u = u(x,t)$ is called the *wave equation* and will be derived below. It will be shown that all solutions of the wave equation yield waves that travel with a wave speed characteristic of the material.

Limiting Form of a Continuous Bar

Suppose the total length of the mass-spring model is fixed and the number of masses and springs increases without limit. A continuous one-dimensional elastic medium is obtained in the limit. This is the model of an elastic bar, which may be subject to the same boundary conditions (impact loading, compressive or tensile) at the left end as in the above discrete model. To create a tensile stress the front end is pulled to the left.

Wave Equation for a Bar

We now derive the wave equation for longitudinal wave propagation in a thin bar. Let the axis of the bar be the x axis. We assume that each cross-section of the bar of area A remains constant during the passage of the wave, and the stress over it is uniform. When an element of the bar at x is disturbed by either a compressive or a tensile stress imposed on the boundary, that element is displaced by an amount u called the *displacement vector* at x. Since u is a function of space and time we have $u = u(x,t)$. We now assume that a compressive stress is created by pushing the front end ($x = 0$) to the right either by an impact load or by a constant applied velocity. In fig. 1.1a we see a longitudinal section of the bar and an element PQ of length dx in equilibrium. Let a compressive stress σ act on the face passing through P. A compressive stress equal to $\sigma + \sigma_x\,dx$ then acts on the face passing through Q. Figure 1.1b shows the unbalanced compressive stress on the element. (Note that from now on we shall use subscripts to denote differentiation with respect to the subscript. Thus $v_t = \partial/v\partial t$, etc.) Now let a tensile stress be applied to the element. Figure 1.1c shows the situation.

5

CHAPTER ONE

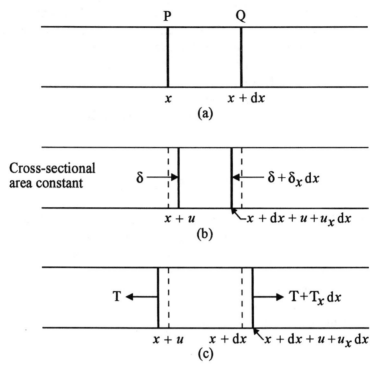

FIGURE 1.1. Compression and tension on element of a bar. a. Element of bar in equilibrium. b. Element in compression (due to compressive stress σ). c. Element in tension. Tensile stress $T = -\sigma$.

Coming back to a compressive stress, fig. 1.1b indicates that the net compressive stress on the element is $\sigma_x \, dx$. The net external force acting on the element is $\sigma_x A \, dx$. (Note that stress is force per cross-sectional area.) Let ρ be the density of the bar [assumed to remain constant (equilibrium value) during the motion of the wave]. The mass of an element or particle of the bar is $\rho A \, dx$. The velocity v of a particle of the bar is u_t, so that its acceleration a is $v_t = u_{tt}$. Note that to be rigorous, the acceleration of a particle is obtained by expanding dv/dt by the chain rule:

$$a = \frac{dv}{dt} = v_t + \frac{dx}{dt} v_x = v_t + v v_x.$$

6

The first term is the instantaneous time rate of change of velocity at a fixed x while the second term is the spatial rate of change of velocity at a fixed time—it is called the *convective term*. This term is nonlinear in the sense that it is the product of two small terms and is thus neglected in our linear analysis. Therefore we take the term u_{tt} as the acceleration of the element (particle). We equate the net external force acting on the element to the mass times the particle acceleration according to Newton's second law of motion, use the formula relating mass to density, and obtain

$$\rho u_{tt} = \sigma_x. \tag{1.1}$$

Equation (1.1) is a wave equation of sorts, but it contains two dependent variables u and σ. We can eliminate σ from this equation and obtain a PDE (the wave equation) for the displacement u by obtaining the stress as a function of strain, where the strain depends on u. We define the *strain* ε to be equal to the ratio of the difference between the undisturbed and the disturbed volume elements to the undisturbed volume element. We appeal again to fig. 1.1a, b, c. The volume of the undisturbed element (equilibrium configuration) is $A\,dx$, as shown in fig. 1.1a. For a compressive strain we see from fig. 1.1b that the volume of the element under compression is given by $A(x + dx + u + u_x\,dx) - A(x + u) - A\,dx = Au_x$. The strain ε is therefore given by

$$\varepsilon = \frac{Au_x\,dx}{A\,dx} = u_x. \tag{1.2}$$

A similar analysis holds for the tensile strain as can be seen from fig. 1.1c.

The ratio between the stress σ and the strain u_x is given by *Young's modulus E*, which is a positive constant characteristic of the material. Thus

$$E = \frac{\sigma}{\varepsilon} = \frac{\sigma}{u_x}. \tag{1.3}$$

This is called *Hooke's law*. Inserting eq. (1.3) into eq. (1.1) yields

$$\rho u_{tt} = E u_{xx}, \tag{1.4}$$

CHAPTER ONE

which can be rearranged to be

$$c^2 u_{xx} = u_{tt}, \qquad (1.5)$$

where

$$c^2 = \frac{E}{\rho}. \qquad (1.6)$$

Equation (1.5) is called the wave equation for the displacement u. c [given by eq. (1.6)] is called the *wave velocity*; it is the velocity with which the displacement wave travels down the bar.

We now give a heuristic proof that $c = \pm \sqrt{E/\rho}$ is the wave velocity or velocity of the moving wave front. To this end we displace the front end of the bar $x = 0$ to the right by an amount vt with a constant velocity for time t as shown in fig. 1.2. This is a BC of constant velocity rather than an impact. The particles of the bar from $x = 0$ to vt have a constant velocity. A compressive stress is initiated at $t = 0$ which travels through the bar. At time t the wave front has traveled a distance ct, where c is the velocity of the wave, which we shall determine. All particles of the bar behind the wave front (disturbed portion) travel with particle velocity v. The particles in front of the wave front have zero velocity. We now apply Newton's equation of motion to the disturbed particles behind the wave front. The change in particle momentum during time t is mv/t, where m is the mass of the material occupying the volume Act (the disturbed portion). It follows that $m = \rho Act$. The

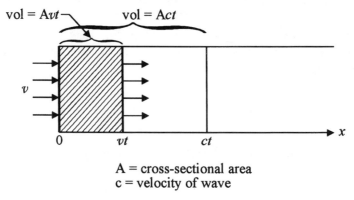

FIGURE 1.2. Constant velocity at front end of bar due to compressive stress.

decrease in volume is $A v t$. The original volume of the stressed material is Act.

Therefore the strain is given by

$$\varepsilon = \frac{v}{c}. \tag{1.7}$$

Using Hooke's law and Newton's equation of motion, $F = m(v/t)$, we get

$$F = A\sigma = AE\varepsilon = AE(v/c) = m(v/t) = \rho c t A(v/t),$$

yielding $c = \pm \sqrt{E/\rho}$. The positive square root is the speed of the traveling wave to the right (a *progressing wave*), while the negative square root is the speed of a wave traveling to the left (a *regressing wave*). It is seen that the wave velocity c depends on the material constants (ρ, E).

TRANSVERSE OSCILLATIONS OF A STRING

Consider a stretched string from A to B as shown in fig. 1.3. The string is assumed to be a one-dimensional elastic material and in its equilibrium configuration (the string or x axis). The boundaries A and B are fixed. Let a small portion of the string at A be distorted into a pulse (shown in fig. 1.3) which oscillates sinusoidally laterally to the string axis and propagates to the right without changing its shape (since we assume no friction to dampen the motion and thus change its shape). We have generated a *transverse wave* in the string. The waveform (in this case a pulse) is transmitted undistorted axially while the portion of the string influenced by the wave oscillates normal to the string axis. When the

FIGURE 1.3. Stretched string with pulse at the front end.

CHAPTER ONE

pulse reaches the boundary B it is reflected as a negative pulse which travels to the left. The boundary condition (BC) at B is fixed, meaning that the portion of the string in the neighborhood of B is motionless; therefore a negative pulse must be initiated at B (this cancels the positive pulse). When this pulse reaches the boundary at A it is reflected as a positive pulse, for the same reason. In general, a waveform of an arbitrary shape initiated at either boundary will propagate away from the boundary with unchanging shape and constant velocity (due to the string's uniform density). It is necessary that the amplitude of the waveform be small, otherwise nonlinear effects take over and the wave shape changes.

SPEED OF A TRANSVERSE WAVE IN A STRING

To derive the speed of a transverse wave in the string we consider a portion of the string acted on by a sinusoidal wave. To be specific, let u be the lateral (normal) displacement of the string due to the wave at a point x on the string axis at time t. Then $u = u(x,t)$. Consider the sinusoidal waveform in fig. 1.4a and focus on an element of the string of length Δs in fig. 1.4b which gives the configuration of the string at time t. The tensile stress T acts on the ends of the element as shown in fig. 1.4b, which shows the components of T at the ends. The normal component of the resultant tensile stress is

$$T \sin \theta_2 - T \sin \theta_1 = T(\sin \theta_2 - \sin \theta_1) \approx T(\tan \theta_2 - \tan \theta_1),$$

where we used the approximation $\sin \theta \approx \tan \theta$ for a small-amplitude waveform where θ is small. The magnitude of $\tan \theta_i$ is the slope of the curve at the points i (for $i = 1, 2$). The slopes at the end points are $(u_x)_1$ and $(u_x)_2$. The normal component of the resultant tensile stress becomes

$$T(u_x)_2 - T(u_x)_1 \approx Tu_{xx}.$$

In the limit $\Delta s \to ds$, u_x in the above expression tends to the second partial derivative of u with respect to x. Equating this normal component of the resultant stress to the string density times the normal component of particle acceleration (using Newton's equation of motion) yields

$$Tu_{xx} = \mu u_{tt},$$

PHYSICS OF PROPAGATING WAVES

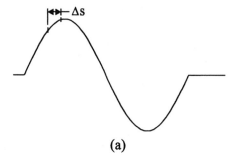

(a)

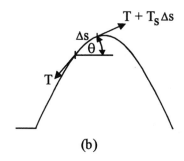

(b)

FIGURE 1.4. a. Sinusoidal wave form on string at time t. b. Element Δs of sine wave under tensile stress T.

where μ is the linear density of the string (mass per unit length). We rewrite this equation as eq. (1.5) where

$$c = \pm\sqrt{\frac{T}{\mu}}. \qquad (1.8)$$

The positive square root is the wave velocity of the propagating transverse wave (progressing wave). The proof of this statement is contained in the special form of the general solution of the wave equation, which we present below.

Traveling Waves in General

Having given an example of longitudinal wave propagation in a bar and transverse wave propagation in a string, we are now in a position to give a more general approach to the character of propagating waves. We

CHAPTER ONE

shall be concerned with one-dimensional wave propagation along the x axis. The wave front is said to be *planar*. This means that the wave front has a planar surface normal to the direction of the propagating wave (x axis). The wave travels with constant wave speed c. The mechanism that causes the wave to form (impact of a bar, plucking of a harp string, motion of a piston in a tube of gas, etc.) is assumed to have a small enough amplitude so that the resulting wave form is linear.

The mechanism that initiates a wave form is responsible for two types of motion:

(1) The oscillating particles of the medium, which are of two types: (a) in the direction of the propagating longitudinal wave (stress waves in a thin bar, sound waves in air, pressure waves in a gas tube, etc.), and (b) in the direction normal to that of the traveling wave, yielding transverse wave propagation (e.g., the vibrating string).

(2) The motion of a traveling wave, which moves with wave velocity c. If c is constant the medium in which the wave travels must be homogeneous and have constant material properties (in time and space) such as density and Young's modulus. For example, a sound wave traveling in air of nonuniform density in space has a speed c that depends on its position in space. If the density is nonuniform in time as well then c depends on both time and space. The wave equation is still a linear PDE but with a coefficient that depends on time and space. Problems in meteorology exhibit this phenomenon. The definition of a linear PDE will be reserved for chapter 2, where we investigate the PDEs of wave propagation in some detail. Note that we omit torsional waves because of certain complications.

To describe the behavior of a continuous medium after a wave passes through it we need to express certain dependent variables, such as particle displacement, strain, stress, pressure, etc., as functions of space and time. An appropriate choice of these dependent variables characterizes the state of the system under investigation in space and time. In general, for three-dimensional space each dependent variable is a function of four independent variables (space and time). In this section we consider one-dimensional space specified by the x coordinate.

Let f denote any dependent variable characterizing the deformed material (such as those mentioned above). Then $f = f(x, t)$. f is assumed to be a continuously differentiable function of (x, t), except for shock waves where we have discontinuities (to be treated in a later chapter). We shall now characterize the wave properties of $f(x, t)$. First

PHYSICS OF PROPAGATING WAVES

we study the wave properties of a sinusoidal waveform, and then we generalize to an arbitrary waveform.

For a sinusoidal waveform we consider a sine wave at time $t = 0$ as shown in fig. 1.5a, where f is plotted against x. We have

$$f(x,0) = f_0 \sin\left(\frac{2\pi x}{\lambda}\right), \qquad (1.9)$$

where λ is the wave length and f_0 is the amplitude of the sine wave (assume to be constant and small). (We could have chosen a cosine wave instead.) Suppose this sine wave moves to the right with a constant velocity without changing its waveform. Figure 1.5b shows the same waveform at time t. The wave has moved a distance ct to the right. This is a progressing wave. In the figure is a plot of $f(x,t)$ against x for a fixed t. It is easily seen that to get from the graph of $f(x,0)$ to that of

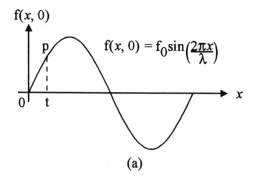

(a)

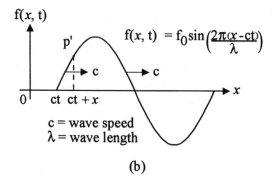

(b)

FIGURE 1.5. a. Sine wave at $t = 0$. b. Same sine wave at $t > 0$.

CHAPTER ONE

$f(x, t)$, we need only replace x by $x - ct$ in the above expression for $f(x, 0)$. It is clear that $f(x - ct)$ is the same function of the argument $x - ct$ that $f(x, 0)$ is of the argument x. Therefore the waveform does not change shape from its initial configuration to its configuration at time t. Indeed, since $f = f(x_1 - ct_1)$ at (x_1, t_1) and $f = f(x_2 - ct_2)$ at (x_2, t_2) for any two times, it follows that the waveform f does not change its form as it progresses into the medium, for any time—since t_1 and t_2 are arbitrary times. Setting $x - ct$ into the expression for $f(x, 0)$ yields

$$f(x, t) = f_0 \sin\left(\frac{2\pi(x - ct)}{\lambda}\right), \tag{1.10}$$

which is the expression for the progressing sinusoidal waveform at any time t. A similar analysis holds for a regressing wave. (We leave the proof of this statement to a problem.) The important point is that neither a progressing nor a regressing waveform changes its shape when its initial waveform is a sinusoidal one given by the above expression.

To emphasize the invariance of the waveform we perform the following analysis: Set

$$x - ct = \xi. \tag{1.11}$$

Then $f(x, t)$ becomes

$$f(x, t) = f_0 \sin\left(\frac{2\pi\xi}{\lambda}\right) = f(\xi). \tag{1.12}$$

[Note that this equation shows a mapping from the (x, t) variables to the single variable ξ.] If we let x and t vary such that $\xi = $ const, then $f(\xi)$ is a constant independent of x and t. By setting $t = 0$ we see that this constant value for $f(\xi)$ does not change as the wave travels a distance ct. Moreover, since the constant ξ is arbitrary it follows that each point on the graph for $f(x, 0)$ (characterized by a value of ξ) travels the same distance ct so that the wave is invariant in form.

If we increase x by Δx and decrease t by Δt and keep ξ constant, we get

$$x + \Delta x - c(t + \Delta t) = \xi = x - ct = \text{const}.$$

This gives

$$\frac{\Delta x}{\Delta t} = c \to \frac{dx}{dt},$$

which means the slope of the curve $\xi =$ const is $dx/dt = c$. We could obtain the same result, of course, by differentiating $x = ct = \xi$ with respect to t. We have thus shown that the above waveform is a progressing wave by substituting $x - ct$ for x in the initial waveform.

We can perform a similar analysis for a regressing wave. To this end we set

$$x + ct = \eta. \tag{1.13}$$

We then obtain

$$f(x,t) = f_0 \sin\left(\frac{2\pi(x+ct)}{\lambda}\right) = f_0 \sin\left(\frac{2\pi\eta}{\lambda}\right) = f(\eta). \tag{1.14}$$

By the above method we see that the initial waveform $f(x,0)$ is a regressing wave, since it travels to the left a distance $-ct$ in the time t. Also, the slope of the curve $\eta =$ const is $dx/dt = -c$.

The above analysis tells us that it is not necessary to restrict ourselves to a sinusoidal function as a waveform. Any arbitrary initial waveform $f(x,0)$ can be considered as long as its amplitude is small. Let $f(x - ct) = f(\xi)$ and let ξ be a variable (recall that $x - ct = \xi$). It is easily seen that $f(\xi)$ satisfies the wave equation, so that $f(x - ct)$ is a solution of the wave equation. We can verify this statement directly for $f(x - ct)$. Let $g(x,0)$ be another arbitrary initial waveform. It is easily seen that $g(\eta) = g(x + ct)$ also satisfies the wave equation. It follows that the general solution of the wave equation is

$$u(x,t) = f(x-ct) + g(x+ct) = f(\xi) + g(\eta) = u(\xi,\eta). \tag{1.15}$$

The mapping

$$x - ct = \xi, \qquad x + ct = \eta \tag{1.16}$$

maps the solution of the wave equation from x,t space to ξ,η space. The variables ξ,η are called *characteristic coordinates*. They play a central role in the theory of PDEs applied to wave phenomena and will be investigated in detail in a later chapter.

CHAPTER ONE

SOUND WAVE PROPAGATION IN A TUBE

We give a heuristic derivation of the velocity of a sound wave. The model is a semi-infinite one-dimensional tube containing air at a constant pressure p_0, density ρ_0, and temperature T_0. It is shown in fig. 1.6, where a diaphragm at the left boundary is oscillating axially, thus producing a longitudinal progressing sound wave in the tube. The diaphragm produces an axial pressure $p_0 + \Delta p$ where p_0 is the constant air pressure and Δp is the oscillating component of the pressure (alternately positive in compression and negative in tension). The particle of air in the neighborhood of the diaphragm is alternately compressed and expanded with a frequency equal to that of the oscillating diaphragm—sending signals that progress down the tube and produce a progressing sound wave with a wave velocity that we now proceed to calculate. Consider a small time t in which the diaphragm travels with a velocity Δv into the tube, compressing the adjacent particles with a pressure Δp.

The sound wave travels with a velocity c so that the volume occupied by the air disturbed by the wave during the time t is Act. The mass m of the disturbed air is ρAct. The relation between the pressure and volume change is $\Delta p = B \Delta v$, where the positive constant B is called the *bulk modulus*. We have

$$\Delta p = \frac{BAt \, \Delta V}{Act} = \frac{B \, \Delta V}{c}.$$

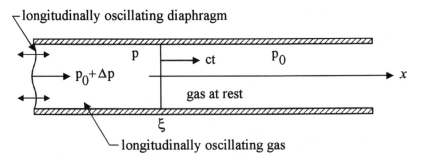

FIGURE 1.6. Longitudinal wave produced by an axially oscillating diaphragm. Wave front is at ξ and has wave speed c.

Newton's equation of motion becomes

$$\Delta F = A\,\Delta p = \frac{AB\,\Delta V}{c} = m\frac{(\Delta V)}{t} = \rho A c t\frac{(\Delta V)}{t}.$$

This yields the following formula for the sound velocity:

$$c = \pm\sqrt{\frac{B}{\rho}}. \qquad (1.17)$$

The positive square root gives a progressing wave while the negative square root yields a regressing wave. Note that replacing B by E gives the wave velocity for a longitudinal stress wave in a bar.

Newton first derived the above expression for c; but he got a result that did not agree with experiment since he used the isothermal instead of the adiabatic bulk modulus. *Isothermal* means at constant temperature, whereas *adiabatic* means there is no exchange of heat between the wave and the environment. In 1816 Laplace obtained the *t* correct expression for c in air when he showed that the alternate compressions and expansions of air at audio frequencies must be regarded as an adiabatic rather than an isothermal process, since the frequencies are too great to allow for an exchange of heat with the environment.

We now get the velocity of sound as a function of temperature T by appealing to some elementary thermodynamics. (The reader is referred to any text on thermodynamics or the chapter on this subject in any good intermediate physics text.) Another way of defining the bulk modulus B is that B is the ratio of stress σ to strain ε. (By convention $\sigma = -p$, where p is a compressive stress.) ε is defined as the differential change in volume dV per unit volume V. This gives

$$B = -\frac{V\,dp}{dV}. \qquad (1.18)$$

We consider an *ideal gas*. This is a gas where there is assumed to be no interaction between the molecules. This means the gas is fairly dilute. The relation between pressure p, temperature T, and volume V is

$$pV = nRT, \qquad (1.19)$$

CHAPTER ONE

where n is the number of moles of the gas and R is the *Boltzmann constant*, sometimes called the *gas constant*; $R = 1.987$ cal/g °C. The *equation of state* for a gas is the relationship between pressure and volume. For an ideal gas under an adiabatic condition the equation of state is

$$pV^\gamma = p_0 V_0^\gamma = \text{const}, \qquad (1.20)$$

where γ is the ratio of the specific heat at constant pressure to the specific heat at constant volume. The *specific heat* at constant pressure (volume) is defined as the amount of heat that must be added per unit mass per degree increase in temperature when the gas is heated and the pressure (volume) is kept constant.

Differentiating eq. (1.20) ultimately yields

$$\gamma p = -\frac{V\,dp}{dV}.$$

The negative sign occurs because p is a compressive stress and is negative, by convention. The bulk modulus becomes

$$B = \gamma p. \qquad (1.21)$$

Hence, the velocity of sound becomes

$$c = \pm\sqrt{\frac{\gamma p}{\rho}}. \qquad (1.22)$$

This is Laplace's expression for the velocity of sound. It is based on an adiabatic process (adiabatic equation of state). It agrees better with experiment than the isothermal one by Newton.

To obtain c as a function of T we invoke the ideal gas law given by eq. (1.19) and get

$$c = \sqrt{\gamma RT} = 331\sqrt{\frac{T}{273}} \text{ m/sec}, \qquad (1.23)$$

where T is in degrees Kelvin (the absolute temperature). Equation (1.23) tells us that the sound speed depends on the square root of the

TABLE 1.1
Measured Sound Speed in Air

Temperature (°C)	Speed (m/sec)	Speed (ft/sec)
0	331	1087
20	334	1129
100	366	1201
500	557	1814
1000	700	2297

absolute temperature. Table 1.1, taken from the *Smithsonian Physical Tables*, gives the measured speed of sound in air at various temperatures.

SUPERPOSITION PRINCIPLE

In discussing the wave equation it was mentioned that it is a linear PDE, since it is concerned with small-amplitude oscillations. A property of a linear PDE is the superposition principle: Suppose we have two solutions of a PDE represented by the functions v and w. Then a *linear combination* of these solutions is also a solution. This means that $u = c_1 v + c_2 w$ where c_1, c_2 are arbitrary constants and u is a general solution to the PDE. In general, if we have n independent solutions of a linear PDE then the general solution is a linear combination of these solutions. Clearly, the superposition principle holds for the wave equation as it is a linear PDE.

SINUSOIDAL WAVES

Before we discuss the interference of waves we present an elementary treatment of sinusoidal motion for those readers who are a bit rusty on this subject. (Those readers who may be bored with this rehash are free to skip this section and go to the subject of interference.) We start with simple harmonic motion. A model of this type of motion is a circle of radius r in the x, y plane whose center is the origin. From the point P on the circle in the first quadrant drop a vertical line to where it hits the horizontal diameter at point A, as shown in fig. 1.7. The line OA is the

CHAPTER ONE

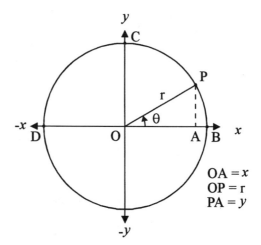

FIGURE 1.7. Simple harmonic motion.

distance x along the x axis. The line AP is the distance y parallel to the y axis. Let the angle POA be called θ. θ may be measured in degrees or in radians. As θ increases in the counterclockwise direction starting from B and going through a full circuit back to B, the angle traverses 360° or 2π radians. To understand what a radian is we consider a circle of radius $r = 1$. The circumference has length 2π. For the point P on the circle to make one round-trip on the circle it traverses an arc length equal to 2π. The angle θ consequently increases by the distance that P travels. θ is said to increase by 2π radians (corresponding to the distance $2\pi r$ traveled by the point P). Note that 2π radians is independent of the size of the circle. An angle of 180° corresponds to π radians, an angle of 90° to $\pi/2$ radians, etc. If the point P goes around the circle in a clockwise direction, the angle θ is negative. When P coincides with B, $\theta = 0$. When P coincides with C, $\theta = \pi/2$ radians, etc. As P makes its counterclockwise circuit starting from B it clearly performs circular motion. Corresponding to this motion of P, the motion of the point A is a back and forth one on the horizontal diameter, starting from B ($\theta = 0$), going to the left, crossing the origin, going to point D ($\theta = \pi$), and returning to its starting point B ($\theta = 2\pi$). Thus we have a mapping of the circular motion of all the points on the circle to the corresponding points on the x axis that exhibit "rectilinear" or back and forth motion. Along with this is the rectilinear motion of the projection of point P on the y axis. The coordinates r, θ are called *polar coordinates*; the corresponding x, y coordinates are called *Cartesian* or *rectangular coordinates*. The correspondence is $x = r \cos \theta, y = r \sin \theta$.

We now discuss the properties of a sine wave (a similar discussion holds for a cosine wave). We first treat the motion of a sine wave at a fixed point on the x axis. Let $u = A \sin(\omega t)$. A is the amplitude of the sine wave and ω is the *angular frequency* measured in radians/sec. Let T be the period (the time it takes for the wave to complete one cycle). Then $\omega T = 2\pi$ (starting from $u = 0$ and ending at $u = 0$). This means that $T = 2\pi/\omega = 1/f$ where f is the frequency of the sine wave in cycles/sec (cps) or hertz. We now consider a sine wave along the x axis for a fixed time and let $u = A \sin(2\pi x/\lambda)$. λ is the wavelength. It is the length of one cycle of the sine wave, as can be seen by setting $x = \lambda$. Now consider a sinusoidal progressing wave so that $u = A \sin(2\pi/\lambda)(x - ct)$. Clearly, when $x - ct = \lambda$ then $u = A \sin 2\pi = 0$. The relationship between wavelength λ, frequency f, and wave velocity c is $\lambda f = c$, where $f = \omega/2\pi$. We can then write u as $u = A \sin \omega[(x/c) - t]$. For a regressing wave $u = A \sin(2\pi/\lambda)(x + ct) = A \sin \omega[(x/c) + t]$. Finally, we bring in the *phase angle* ϕ. It represents the phase difference between two waves. This is seen as follows: Let $u = \sin(\theta - \phi)$ and $v = \sin \theta$. u and v are out of phase by the angle ϕ in the sense that when $u = 0$, $v = \sin \phi$, when $\phi = \pi$ $u = \sin(\theta - \pi)$ and $v = 0$, when $\phi = \pi/2$, u and v are 90° out of phase, etc. Clearly, account must be taken of the sign of ϕ. Figure 1.8a shows a progressing wave profile for u at time t; fig. 1.8b a profile at t for a regressing wave. A similar analysis holds for $\cos \theta$.

Interference Phenomena

The previous discussion of the superposition of solutions of the wave equation when applied to propagating waves can be stated as follows: When two or more waves travel simultaneously in a medium and their amplitudes are small enough so that no nonlinear effects occur, each wave proceeds independently in the sense that no other wave has an effect on it. The resultant displacement vector due to the wave at each point in the medium is the vector sum of the displacements of the other waves. This applies to stress waves in elastic media, small-amplitude water waves, sound waves, transverse waves in strings, etc. It applies

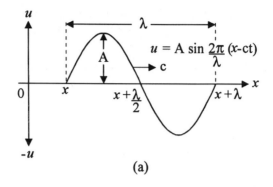

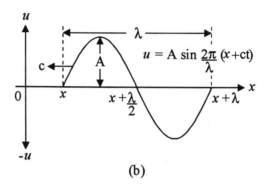

FIGURE 1.8. a. Progressing wave with wavelength λ and amplitude A traveling to the right with wave speed c. b. Regressing wave traveling to the left with wave speed $-c$.

rigorously to EM waves where we consider the displacement of the electric and magnetic vectors, except for large-amplitude lasers. The effect of the simultaneous propagation of two or more waves leads to the phenomenon of interference, which will now be discussed.

To fix our ideas we take traveling transverse waves along a string. Let the waves be represented by

$$u_1 = A_1 \sin \omega_1 \left[\left(\frac{x}{c}\right) - t\right], \qquad u_2 = A_2 \sin \omega_2 \left[\left(\frac{x}{c}\right) - t - \phi\right]. \quad (1.24)$$

These are two waves traveling in the same direction with different amplitudes, frequencies, and phase difference ϕ. First consider two waves in phase with the same frequency and different amplitudes

PHYSICS OF PROPAGATING WAVES

traveling in the same direction. Setting $\omega_1 = \omega_2 = \omega$, $\phi = 0$, the sum of these waves becomes

$$u = u_1 + u_2 = (A_1 + A_2)\sin \omega[(x/c) - t]. \tag{1.25}$$

The two waves are said to "interfere constructively." If we replace A_2 by $-A_2$ we get

$$u = (A_1 - A_2)\sin \omega\left[\left(\frac{x}{c}\right) - t\right]. \tag{1.26}$$

The two components are said to interfere destructively. A complete cancellation occurs if $A_1 = A_2$. Figure 1.9a, b, c shows cases of interference of two waves of the same amplitude traveling in the same direction with the same frequency.

We now consider two sine waves of the same amplitude, frequency, and phase traveling along the x axis in opposite directions. The progressing wave is

$$u_1 = A \sin \frac{2\pi}{\lambda}(x - ct).$$

The regressing wave is

$$u_2 = \sin \frac{2\pi}{\lambda}(x + ct).$$

The sum of these two components is

$$u = u_1 + u_2 = A \sin \frac{2\pi}{\lambda}(x - ct) + A \sin \frac{2\pi}{\lambda}(x + ct). \tag{1.27}$$

Recalling the trigonometric relations for the sum and difference of angles:

$$\sin(a + b) = \sin a \cos b + \cos a \sin b,$$

$$\sin(a - b) = \sin a \cos b - \cos a \sin b,$$

we can write eq. (1.27) in the forms

$$u = \left[2A \cos \omega\left(\frac{x}{c}\right)\right]\sin \omega t, \quad u = \left[2A \cos 2\pi\left(\frac{x}{\lambda}\right)\right]\sin \omega t. \tag{1.28}$$

23

CHAPTER ONE

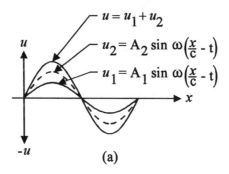

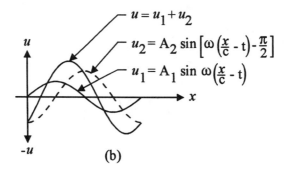

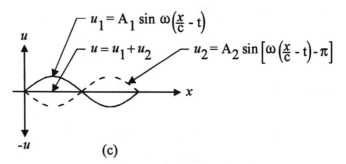

FIGURE 1.9. a. Constructive interference of two waves of same frequency and phase. b. Interference of two waves of same frequency and phase angle $\phi = \pi/2$. c. Destructive interference of two waves of same frequency and phase angle $\phi = \pi$.

The wave pattern given by eq. (1.28) is called a *standing wave*. The time-varying part of the wave is given by sin ωt, whereas the amplitude of this wave is a sinusoidal function of x with the parameters (ω, c) or (λ, c). In radio transmission this standing wave is called an *amplitude-modulated wave*. The motion associated with a standing wave is given in fig. 1.10 for four positions of the component traveling waves at times $t = 0$, $T/4$, $T/2$, and $(3/4)T$, where the period $T = 1/f$. The particles of the string influenced by the standing wave execute simple harmonic motion normal to the string axis with a frequency ω radians/sec or f cps (hertz).

Reflection of Light Waves

It is of interest to start this section by appealing to a geometric property of a light wave reflected from a mirror and relating it to a certain minimum principle. Figure 1.11 shows a horizontal line and the points A and B above the line. The problem is to find a point P on the line such that the lines AP and PB have a minimum length. So far this appears to have no relationship to reflected waves. But let us continue. To solve this problem we draw a vertical line from A crossing the horizontal line at C and continue down. Mark off a distance CA' equal to CA. The point A' is said to be the *image* of point A with respect to the horizontal line. By a similar construction obtain the point B' which is the image of the point B. Draw the straight line from A to B'. Also draw the straight line from B to A'. It turns out that they will cross the horizontal line at the point P. The line APB' is the shortest distance from A to B' since it is a straight line. Similarly, the line BPA' is the shortest distance from B to A'. Since A' is the image of A and B' is the image of B it follows that the path $AP + PB$ is the shortest distance compared to another point, say P', on the horizontal. This can easily be checked out by elementary geometry, using congruent triangles. Now, here is where light rays come in: Suppose a light ray starting at A goes along the path AP. Let the horizontal line be a mirror so that P is a point on the mirror. Then the light ray will be reflected at P and go along the line PB. AP is the *incident light ray* and PB is the *reflected light ray*. Erect a line normal to P going upward. Then the *angle of incidence* θ between the incident ray and the normal is equal to the *angle of reflection* ϕ between the reflected ray and the normal. This is a

CHAPTER ONE

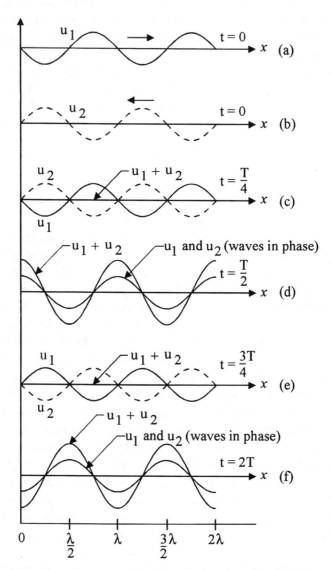

FIGURE 1.10. A standing wave is produced when two waves of the same frequency and amplitude travel in opposite directions. The traveling waves u_1 and u_2 are shown in (a) and (b) at $t = 0$. The resulting waveform $u_1 + u_2$ is shown during a complete cycle in (c), (d), (e), and (f).

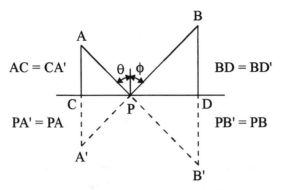

FIGURE 1.11. Construction to show that P is such that $AP + PB$ is a minimum length. AP is an incident light ray, PB is a reflected light ray. Angle of incidence θ = angle of reflection ϕ.

basic principle of optics. We may also shine an incident light ray from point B to P which is reflected to the point A. Thus we see how a purely geometric property of minimizing paths can be converted to a physical problem in optics. This construction illustrating the minimum principle was devised by the ancient Greek mathematician Hero. I do not know whether he related it to optics.

REFLECTION OF WAVES IN A STRING

Consider a stretched string of finite length along the x axis. The end points of the string are fixed. We saw previously what happens when a pulse travels to the end of the string. Let us look again at the situation. Let an incident pulse travel to the right. When it reaches the end it exerts a force on the support, which in turn exerts a reaction force on the string, thus setting up a reflected pulse that travels to the left with its displacement in the direction opposite to the original pulse. This is shown in fig. 1.12. Note that the shape of the pulse remains the same, since both pulses must be solutions of the wave equation. An incident downward pulse is reflected as an upward pulse. As mentioned, this change from a crest to a trough and vice versa must occur in order to satisfy the fact that the string does not move at the boundary. There is a 180° phase shift between the incident and reflected waves. If we replace a pulse by a continuous transverse wave train, the same relationship

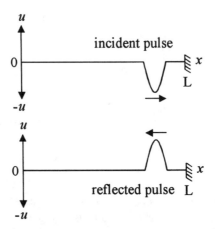

FIGURE 1.12. Incident pulse traveling to the right in a string reflected at the fixed end $x = L$. The reflected pulse has a 180° phase shift compared with the incident pulse. The waveform does not change.

between the incident and reflected waves must occur. When a continuous train of sinusoidal waves reaches a fixed boundary, a continuous train of reflected waves appears of the same wavelength and 180° phase shift and travels in the opposite direction. Since we have two wave trains traveling in opposite directions, a standing wave is produced. For our stretched string of finite length there are two fixed boundaries. This means that the fixed ends appear as nodes in any standing-wave pattern that may be formed. This statement implies that only for certain definite wavelengths will a standing wave be formed. The wavelength λ is equal to twice the distance between adjacent nodes. For a string of length L the standing-wave pattern is shown in fig. 1.13. The values of λ are (a) $\lambda = 2L$, (b) $\lambda = L$, (c) $\lambda = L/2$. Thus, the allowed values of λ that permit a standing wave to be produced in a string of length L are given by the general equation

$$\lambda = \frac{2L}{n}, \qquad n = 1, 2, 3, \ldots, \tag{1.29}$$

where n is any positive integer. Since frequency $f = c/\lambda$, the corresponding frequencies are $f = n(c/2L)$, $n = 1, 2, 3, \ldots$.

The standing wave patterns shown in the figure are called *normal modes of oscillation* of the string. Figure 1.13a shows the *fundamental mode*. Some higher modes are shown in the rest of the figure. For a sound wave the higher modes are called *harmonics*.

PHYSICS OF PROPAGATING WAVES

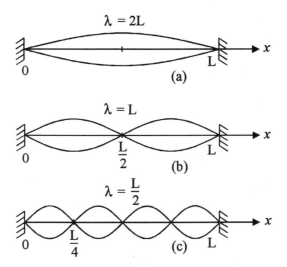

FIGURE 1.13. Standing-wave patterns in a string fixed at $x = 0$ and L. (a) Fundamental mode or first harmonic; (b) second harmonic; (c) fourth harmonic.

Example. A steel piano wire 1 meter long has a mass of 20 grams and is stretched with a force of 800 newtons. What are the frequencies of its fundamental mode and the next three higher modes?

$$c = \sqrt{\frac{T}{\mu}} = \sqrt{\frac{(800 \text{ N})}{(0.02 \text{ kg/m})}} = 200 \text{ m/sec}.$$

$\lambda = 1$ m, so that $f = c/\lambda = 100$ cps. The higher modes are 200, 300, 400, ..., cps.

SOUND WAVES

The vibrating string is a source of sound waves, as is shown by playing any stringed instrument, a piano, harpsichord, etc. We know that a vibrating string produces transverse waves. However, when we investigate the propagation of sound in air we see that sound waves are longitudinal waves which arise from the alternate compression and expansion of the air in a rapid enough manner to be an adiabatic process. A woodwind or brass instrument produces such waves by causing a column of air to oscillate. The frequencies produced by a musical instrument are clearly the same as those propagated through air. The following simple experiment shows that sound waves cannot be propagated in a vacuum: If an electric doorbell is suspended by fine wires inside a bell jar in such a way that the wires do not touch the walls of the jar, the sound of the bell ringing can be heard. However, if the jar is evacuated by a vacuum pump then the sound can no longer be heard. This shows that a physical medium such as air is needed to propagate sound waves. Now, if the jar is tilted so that the bell touches the wall of the jar, the sound can be heard again, thus indicating that a solid can transmit sound. Indeed, any elastic medium, be it solid, liquid, or gas, can transmit sound. A bell ringing under water clearly gives off sound waves.

Production of Sound by Oscillating Air Columns

We treated the propagation of sound waves above in the section "Sound Wave Propagation in a Tube." There we obtained the sound velocity c using the adiabatic equation of state for air, and also obtained c as a function of temperature. In this section we go a little deeper into the nature of sound waves in air columns.

We start by investigating the properties of sound produced in a column of air closed at both ends. The air in the column can be set into longitudinal oscillations of large amplitude by a small diaphragm vibrating with small amplitude at exactly the resonant frequency (thus causing large-amplitude oscillations of the air column) if the diaphragm is placed at an appropriate position inside the column. Figure 1.14 shows a closed air column and the fundamental mode of oscillation of the air column. Note that the BCs impose nodes on the oscillations at the

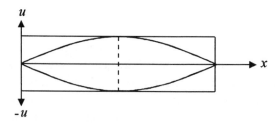

FIGURE 1.14. Fundamental mode for a closed air column, showing half a wavelength.

boundaries, since the column is closed at both ends. In addition, the fundamental mode shows an antinode in the center of the column (where the amplitude is a maximum). The displacement $u(x,t)$ for the fundamental mode is given by

$$u = A \sin\left(\frac{\pi x}{L}\right) \sin\left(\frac{2\pi t}{T}\right).$$

Note that the spatial part of this equation satisfies the BC that $u(0,t) = u(L,t) = 0$. The time-varying part of u is a sine wave of period T such that $u(x,T) = 0$ for $0 \le x \le L$. Also note that the wavelength $\lambda = 2L$. In fig. 1.15 the positions of various particles of air are plotted as a

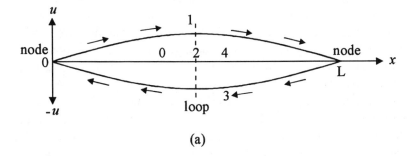

(a)

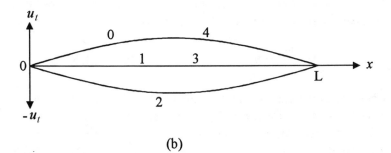

(b)

FIGURE 1.15. a. Displacement of particles in a closed pipe of length L for $L = \lambda/2$, showing u, the displacement to the right, maximum at the loop. Curve 0 is at $t = 0$. Curve 1 is at $t = T/4$. Curve 2 is at $t = T/2$ equilibrium. Curve 3 is at $t = 3T/4$, displacement to the left. Curve 4 is at $t = T$ equilibrium. b. Particle velocity in a closed pipe at various times. Curves 0, 1, 2, 3, 4 refer to times $t = 0, T/4, T/2, 3T/4, T$.

CHAPTER ONE

function of time. At $t = 0$ the particles are in the equilibrium configuration of no displacement. At $t = T/4$ they have maximum displacement to the right, at $t = T/2$, there is no displacement, and at $t = 3T/4$, there is maximum displacement to the left. At $t = T$, there is again no displacement. This is easily seen from the time-dependent part of the above equation. The profiles of u as a function of x at various times are indicated in fig. 1.15a by the curves $0, 1, 2, 3, 4$, corresponding to $t = 0, T/4, T/2, 3T/4, T$. The particle velocity

$$v = u_t = \left(\frac{2\pi}{T}\right) A \cos\left(\frac{2\pi t}{T}\right).$$

Note the fact that the time-dependent part of v is a cosine function. It follows that the profile of v as a function of x at the above times is given by fig. 1.15b. The particle displacement and velocity have maximum values at the center (antinodes) and nodes at the ends of the column. The oscillation in fig. 1.15a represents a standing wave in which the wavelength $\lambda = 2L$, the fundamental mode. Higher modes would have additional nodes equally displaced at and between the ends of the column, exactly as in the case of the stretched string.

Air columns employed in musical instruments are open at one or both ends in order to obtain more effective transmission of sound. Organ pipes have BCs of two types: (a) one end closed and one end open, called a *closed pipe*; and (b) both ends open, called an *open pipe*. The normal modes of oscillation of these two types are shown in fig. 1.16. The curves are maximum-displacement curves. The resonant frequencies of an open pipe can be written as

$$f = \frac{nc}{2L} = nf_0, \qquad f_0 = \frac{c}{2L}, \qquad n = 1, 2, 3, \ldots,$$

where f_0 is the fundamental frequency of the open organ pipe. $n = 2$ gives the first overtone, $n = 3$ gives the second overtone, etc.

For the case of the closed pipe, from the figure we see that the wavelength $\lambda = 4L/(2n + 1)$, $n = 0, 1, 2, 3, \ldots$. It follows that the resonant frequencies are

$$f = (2n + 1)c/4L, \qquad n = 0, 1, 2, 3, \ldots.$$

For $n = 0$ we get the fundamental wavelength $\lambda = 4L$, so that the fundamental frequency is $f = c/4L$. We see that, for the case of a

PHYSICS OF PROPAGATING WAVES

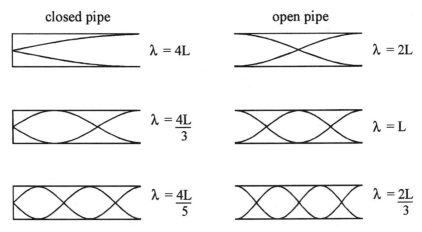

FIGURE 1.16. Normal modes of oscillation of organ pipes. The curves represent maximum displacements.

closed organ pipe only odd harmonics exist. $n = 1$ gives the second harmonic $f = 3c/4L$, etc. Let f_c be the fundamental frequency of the closed pipe. It follows that $f_0 = 2f_c$.

Example. Kundt's tube. This is a long glass tube filled with hydrogen gas, which allows us to calculate the speed of sound in hydrogen. One end of the tube is rigidly closed. At the other end is mounted a stiff metal diaphragm driven by a speaker. A small amount of light powder has been dusted into the tube. Suppose the temperature of the gas is 20°C. When a standing wave of audio frequency is set up in the tube, the powder collects sharply at the nodes where there is no motion. Thus we can measure the distance between nodes. One of the higher normal modes of oscillation gives nodes 10.5 cm apart when the frequency is 6200 hertz. It follows that $\lambda = 2 \times 10.5$ cm $= 0.210$ m. Then $c = \lambda f = 0.210 \times 6200 = 1300$ m/sec.

DOPPLER EFFECT

Everyone has observed that when a train whistle travels away from him/her the frequency of its sound or its pitch decreases, and when the train comes toward the observer, the frequency of the whistle increases. This change of frequency of the moving sound source (whistle) heard by the observer is called the *Doppler shift* or *Doppler effect*.

To understand the Doppler effect we start by considering a ringing bell as the sound source. Let the source be at point O in three-dimensional space of uniform air density, temperature, and pressure so that the velocity of the sound waves c is constant. If the bell moves to the right along the x axis with a speed v toward the observer, there are three cases:

(1) The source is stationary, $v = 0$
(2) The source moves subsonically, $v < c$
(3) The source moves supersonically, $v > c$

For case 1 the bell emits spherical sound waves whose wave fronts are spherical surfaces that expand uniformly with O as the fixed origin. Since the wave front expands with constant speed c, at times $t, 2t, 3t, \ldots$, the sound disturbances will have reached points that lie on the concentric spherical surfaces of center O and radius $ct, 2ct, 3ct, \ldots$, as shown in fig. 1.17a, which shows the projection of the spherical wave fronts in the plane of the paper.

For case 2, seen in fig. 1.17b the spherical surfaces will still have radii equal to $ct, 2ct, 3ct, \ldots$. But, since $v < c$, the wave fronts will never intersect or touch each other, so that as time increases these disturbances will eventually permeate all space. Since the bell moves to the right, the origins move to the right (the largest circle has traveled the most distance). For the case $v = c$ (the transition between case 2 and case 3) all the spherical surfaces will be tangent to the point O and eventually permeate the half space.

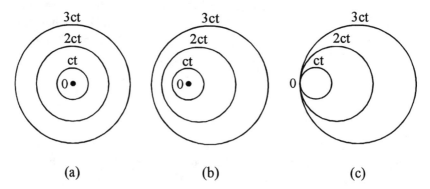

FIGURE 1.17. Planar projection of spherical wave fronts for (a) stationary, (b) subsonic, and (c) supersonic bell.

For case 3, as seen in fig. 1.17c, the sound waves will still lie on the surfaces of radii $ct, 2ct, 3ct, \ldots$, but the disturbances will never reach points outside the cone whose vertex is at O and whose surface is tangent to the spherical surfaces. This cone is called the *Mach cone*. As implied above, the Mach cone is the envelope of the expanding spherical sound waves. As v increases the cone angle gets smaller. This case occurs in supersonic aerodynamics. A plane or missile traveling faster than the speed of sound has the air rushing at the nose supersonically. The tip of the nose is the vertex of the Mach cone. The air aft of the nose is disturbed inside the Mach cone and not disturbed outside it.

To calculate the Doppler shift we need to get a relationship between the wavelength λ_s of the stationary source and the wavelength λ of the moving source (which is the wavelength seen by the observer). Suppose the source is moving toward the right toward the observer. The wavelength λ of the moving source is less than that of the stationary source λ_s by the distance the source moves in one period T, which is cT. Now $T = \lambda_s/c$. (Recall that $T = 1/f_s$, where f_s is the frequency of the stationary source.) This gives $\lambda = \lambda_s - vT = \lambda_s - v\lambda_s/c$. From this expression we obtain the ratio of the moving to the stationary wavelength as

$$\frac{\lambda}{\lambda_s} = \frac{(c-v)}{c}. \tag{1.30}$$

Note that this ratio equals unity if $v = 0$ (stationary source) and gets smaller as v gets larger. In the limiting or *transonic* case where $v = c$ the ratio is zero, and for $v > c$ the ratio is negative, which is impossible. This again shows that the bell must travel subsonically for the Doppler effect to be valid.

Suppose the bell moves toward the left away from the observer. It is clear that the above ratio becomes

$$\frac{\lambda}{\lambda_s} = \frac{(c+v)}{c}, \tag{1.31}$$

since the speed of the source moving away from the observer is the negative of the speed moving toward the observer.

Let f be the frequency heard by the observer corresponding to the wavelength λ, and let f_s be the frequency corresponding to λ_s. It is easily shown that

$$f = \frac{f_s c}{(c \pm v)}, \tag{1.32}$$

where the negative sign in the denominator corresponds to the source approaching the observer, and the positive sign corresponds to the source receding from the observer. Note that the velocity v of the source is positive.

Dispersion and Group Velocity

In dealing with the one-dimensional wave equations for stress waves in elastic media, it was shown that the waveform is invariant with respect to time. A *monochromatic waveform* was considered so that the transverse and longitudinal velocities were constant. In practice we have traveling waves with a continuous frequency distribution. Such a wave is called a *polychromatic waveform*. The wave speed, more accurately called the *phase speed*, varies with frequency so that the waveform changes its shape. This varying waveform is called *dispersion*, because each frequency travels with a different phase velocity, thus distorting or dispersing the wave form.

The phenomenon of dispersion is related to the *group velocity*, defined as the velocity with which a wave packet is propagated. To get an understanding of group velocity, we consider a wave packet that arises from the interaction of two monochromatic waves of the same amplitude and slightly different wave numbers and frequencies. Let $f(x,t)$ be the sum of these two waves. We may write

$$f(x,t) = \cos(\omega t - \zeta x) + \cos(\omega' t - \zeta' x),$$

where ω and ζ are the frequency and wave number of the first wave and $\zeta' = \zeta + \delta\zeta$, $\omega' = \omega + \delta\omega$. Using a trigonometric identity we get

$$f(x,t) = 2\cos\tfrac{1}{2}[(\delta\omega)t - (\delta\zeta)x]\cos\tfrac{1}{2}[(\omega + \omega')t - (\zeta + \zeta')x].$$

Since $\delta\omega$ and $\delta\zeta$ are small, we see that f is an amplitude-modulated wave with a slowly varying amplitude given by the first factor of the above equation (envelope). This is seen in fig. 1.18, where the ampli-

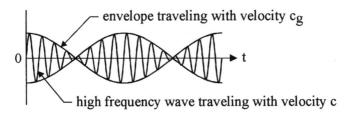

FIGURE 1.18. Amplitude-modulated wave as envelope of high-frequency waves.

tude-modulated wave is the envelope of the high-frequency wave with frequency $\delta\omega$ traveling with the wave speed c. The wave packet (envelope) has the frequency $\omega + \omega'$ travels with the group velocity c_g. To obtain c_g we consider one cycle of the first factor. We thus set

$$x\,\delta\zeta - t\,\delta\omega = 0, \qquad \text{or} \qquad \frac{x}{t} = \frac{\delta\omega}{\delta\zeta}.$$

We then define the group velocity c_g as

$$c_g = \lim \frac{\delta\omega}{\delta\zeta} = \frac{d\omega}{d\zeta}. \tag{1.33}$$

The group velocity may be given in terms of c and ζ or c and λ. Thus,

$$c_g = \frac{d\omega}{d\zeta} = c + \zeta\frac{dc}{d\zeta} = c - \lambda\frac{dc}{d\lambda}. \tag{1.34}$$

PROBLEMS

1. Verify that $u = f(x - ct) + g(x + ct)$ (where f and g are arbitrary functions of their arguments) is the general solution of the wave equation $c^2 u_{xx} = u_{tt}$.

2. Show that the characteristic coordinates $\xi = x - ct$, $\eta = x + ct$ transform the wave equation $c^2 u_{xx} = u_{tt}$ into the simpler PDE $u_{\xi\eta} = 0$. Then show that all solutions of this PDE must be of the form $u(\xi, \eta) = F(\xi) + G(n)$, where F and G are arbitrary functions of their respective coordinates.

CHAPTER ONE

3. A string 50 cm long with a mass of 5.0 g is under tension of 5 N. Find the speed of the transverse wave in this stretched string. (1 newton = 10^5 dynes.)

4. A rope is 10 m long. Experiment shows that a transverse wave passes from one end of the rope to another in 2 sec when the rope is under a tension of 20 N. Find the mass of the rope.

5. A cord 12 ft long weighs 0.1 lb. If tensile forces of magnitude 25 lbs are applied to the ends of the cord, find the speed of the transverse waves that are produced.

6. What is the magnitude of the tensile forces applied to the ends of the cord in problem 5 in order for transverse waves to move at a speed of 200 ft/sec?

7. A fisherman in an anchored boat observes that the crest of a wave passes him every 2 sec and it takes 4 sec for the crest to move the 40 ft length of the boat. Find the wavelength λ of the waves.

8. A fisherman sitting on a dock observes that his float makes 10 complete oscillations in 2 sec and that the distance between two consecutive wave crests is 2 ft. Find the speed of the wave.

9. A stretched string is forced to oscillate at one end by a mass-spring device that causes lateral oscillations with a frequency of 10 hertz. (Note: 1 hertz = 1 cps.) The speed of transverse waves in the string is 50 ft/sec. Find the wavelength of the traveling waves.

10. If the tension of the string in problem 9 were doubled, find the transverse wave speed, the wavelength of the waves produced when one end of the string is subjected to transverse oscillations at a frequency of 10 hertz (cps), at 1 hertz.

11. A sinusoidal wave train of frequency 15 hertz travels along a stretched string. The distance between a crest and the adjacent troughs is 5 ft. Find the wavelength and the wave speed.

12. Show that the displacement $u = A\cos(2\pi/\lambda)(x - ct) + A\cos(2\pi/\lambda)(x + ct)$ satisfies the wave equation $c^2 u_{xx} = u_{tt}$. Find the standing wave corresponding to these two wave components. Hint: Use the trigonometric relations for the sums and differences of two angles of the cosine waves.

PHYSICS OF PROPAGATING WAVES

13. Show that the longest wavelength of traveling transverse waves that can produce a standing wave in a string 10 ft long is 20 ft.

14. Sound of frequency 1000 hertz from a stationary source is reflected back from an object approaching the source at 30 m/sec to a stationary observer located at the source. Determine the frequency heard by the observer.

15. Locomotive A is moving southward at a speed of 100 ft/sec, while locomotive B is moving northward at a speed of 100 ft/sec. If the whistle on locomotive A has a fundamental frequency of 256 hertz, what will be the frequencies of the sounds heard by the engineer on locomotive B as the two locomotives pass each other? Take the speed of sound in air as 1100 ft/sec.

16. What is the length of a closed organ pipe if the fundamental frequency is 256 hertz when the speed of sound is 1080 ft/sec? What is the length of an open organ pipe under the same conditions?

17. A steel piano wire 1 meter long has a mass of 0.02 kg and is stretched with a force of 800 newtons. What are the frequencies of its fundamental mode of oscillation and the next three higher harmonics?

18. A long coiled spring lies along the x axis whose origin is the left end, which is attached to a steel ball supported vertically by a flexible metal blade clamped at the lower end. If the ball is given a small displacement δx to the left (or right) and then released, it will perform simple harmonic motion (compression and extension or tension) along the x axis with a frequency of f hertz. As a result a sinusoidal longitudinal wave train will be set up in the spring, which travels to the right. Each point x on the spring will perform simple harmonic motion of the same frequency. The wavelength λ is the distance between two adjacent maximum compressions (or extensions). Let ξ be the longitudinal displacement of a point of the spring from its equilibrium position x (before it is acted on by the wave). Then the equation for a traveling wave is

$$\xi = A \sin \frac{2\pi}{\lambda}(x - ct),$$

where A is the amplitude in ft, λ is the wavelength in ft, $\omega = 2\pi f$ is the radial frequency in radians/sec, and c is the wave speed. Show that this wave travels to the right (in the direction of increasing x). This is a

CHAPTER ONE

progressive wave. Hint: The waveform given by ξ is the same at a later time $t + \delta t$ and a later distance $x + \delta x$ and $c = \delta x/\delta t$. Find an analogous expression for ξ for a regressive wave (traveling in the direction of decreasing x). If $c = 6$ ft/sec and if the steel ball oscillates through a total distance of 0.1 ft at a frequency of 1.5 hertz, what is the wavelength of the longitudinal traveling wave and the radial frequency?

19. A source oscillating at a frequency of 12 hertz produces longitudinal waves of amplitude 0.04 m in a coiled spring. The speed of propagation of the wave is 20 m/sec. Write an equation describing the motion of a progressing wave and a regressing wave.

20. From the equation for a sinusoidal transverse wave in a string, derive an expression for the maximum velocity of a particle in the string. How does the maximum velocity depend on the amplitude of the wave? Upon the frequency?

21. Using the result of problem 19, calculate the maximum kinetic energy of a particle of mass $\mu \Delta x$ and show that the energy W per unit length of the string is given by $W = 2\pi^2 \mu f^2 A^2$.

22. Two transverse waves each of wavelength 2 m travel in opposite directions in a stretched string of length 8 m. How many loops are formed? Excluding the nodes at each fixed end, how many nodes appear in the string?

23. A standing transverse wave is set up in a stretched string 10 ft long. There are 7 loops set up in the standing wave. Find the wavelength producing this pattern.

24. A stretched string of length 10 m whose ends are rigid has a mass of 50 g and a tension of 10 N. If the string is plucked at the front end, thus setting up a pulse, find the time required for the pulse to travel to the back end, experience reflection, and travel back to the plucked end. Considering multiple reflections, find how many round-trips would be made each second.

25. Derive eq. (1.34) for the group velocity.

CHAPTER TWO

Partial Differential Equations of Wave Propagation

Introduction

In chapter 1 we treated the physics of wave propagation from a somewhat qualitative viewpoint, not using much mathematics except for deriving and discussing the one-dimensional wave equation for a bar, vibrating string, and sound waves in a tube. In this chapter we shall use a more quantitative approach to wave propagation phenomena. To this end we shall investigate in some detail the partial differential equations (PDEs) of wave propagation phenomena by introducing the method of characteristics that was alluded to in chapter 1. This method is the backbone of wave phenomena and therefore will be investigated in some detail starting with first-order PDEs and then going to second-order PDEs. We start by giving some basic properties of PDEs, introduce the concept of directional derivatives which is necessary to understand characteristic theory, and then investigate the Cauchy *initial value* (IV) problem (a key problem in wave propagation) by way of characteristic theory.

Types of Partial Differential Equations

Limiting our discussion to (x, y) space, a PDE is an equation for the dependent variable $u = u(x, y)$ of the form

$$F(x, y, u_x, u_y, u_{xx}, u_{xy}, u_{yy}, u_{xxx}, \ldots) = 0.$$

As mentioned in chapter 1, subscripts denote differentiation with respect to the subscript. The *order* of a PDE is defined by its highest derivative.

Example. $2u_{xxtt} + 3u_{xtt} + x^2 u_x + 7u = x \sin \omega t$ is a fourth-order PDE that is *nonhomogeneous* since the right-hand side is not zero. If it were zero then the equation would be *homogeneous*.

CHAPTER TWO

A PDE is *linear* if the coefficients of u and its partial derivatives are known constants or at most known functions of x, t. The above example is a linear PDE. Such an equation obeys the superposition principle described in chapter 1. A *nonlinear PDE* is one where the coefficients depend on u and the higher-order derivatives.

Example. $(u_{xx})^2 + uu_x + 5u = 0$ is nonlinear.

A *quasilinear PDE* is a nonlinear one where the highest derivatives are displayed linearly.

Example. $u(u_x)^2 u_{xx} + 3xu_{tt} + u_x = 0$ is a second-order quasilinear PDE.

The superposition principle does not hold for nonlinear or quasilinear PDEs.

Geometric Nature of the PDEs of Wave Phenomena

It was implied above that wave phenomena depend on an understanding of the properties of certain PDEs. For this discussion we consider a two-dimensional x, y plane. Recall that a PDE is an equation involving u and various derivatives of u with respect to x and y. The solution of a PDE is $u = u(x, y)$. This means we want to find a family of curves in the x, y plane (called *integral curves*) such that each point on a curve is a solution to the given PDE under certain boundary and initial conditions (BCs and ICs), and a particular value of a parameter defines a unique curve. Clearly, all the appropriate derivatives of u that are given in the PDE must also be satisfied at each point on the integral curves. It turns out that the PDEs of wave phenomena are usually second-order systems of PDEs of a special type. A single wave equation is an example of such a PDE. It can easily be shown that a second-order PDE such as the wave equation can be transformed into two first-order PDEs. Therefore an investigation of characteristic theory with regard to first-order systems will be undertaken.

Directional Derivatives

Since a solution of a PDE is given in terms of integral curves, it is of interest to examine these curves. We again consider the x, y plane. The

PDEs OF WAVE PROPAGATION

slope at any point on a curve is given by the ratio of a pair of numbers (a, b), which are called *direction numbers*. They are not unique; for, given any nonzero constant λ, it follows that $(\lambda a, \lambda b)$ are also direction numbers. Using λ we can *normalize* the direction numbers by setting $(\lambda a)^2 + (\lambda b)^2 = 1$. This yields

$$\lambda = \frac{1}{\sqrt{a^2 + b^2}}.$$

This means that we can find angles (α, β) such that

$$\lambda a = \cos \alpha, \qquad \lambda b = \sin \alpha = \cos \beta, \qquad \beta = \frac{\pi}{2} - \alpha.$$

$\cos \alpha, \cos \beta$ are called the *direction cosines*. It easily follows that $\cos^2 \alpha + \cos^2 \beta = 1$.

The slope b/a of a curve is given by the ODE $dy/dx = b/a$. If a and b are constants along the curve, then it is a straight line. If a or b varies with x, y, then the curve is not a straight line. Consider the variable t as a parameter. (We may think of t as time, although this is not necessary.) Then the above ODE is given in *parametric form* as

$$\frac{dx}{dt} = a(x, y), \qquad \frac{dy}{dt} = b(x, y). \tag{2.1}$$

The solution of this pair of first-order ODEs in parametric form in terms of the parameters t, τ is

$$x = x(t, \tau), \qquad y = y(t, \tau), \tag{2.2}$$

where τ is the constant of integration. It may be taken as the parameter that generates an initial value curve in that x, y plane (where $t = 0$). Fixing τ and varying t generates an integral curve given by eq. (2.2). Every point on such a curve is specified by a particular value of τ.

The *directional derivative* of u is du/dt. It is defined as the differentiation of u with respect to t along a curve $y = y(x)$. Expanding du/dt along the curve by the chain rule gives

$$\frac{du}{dt} = \left(\frac{dx}{dt}\right) u_x + \left(\frac{dy}{dt}\right) u_y, \tag{2.3}$$

where the direction numbers (a, b), representing the slope of the curve, are given by

$$\frac{dx}{dt} = a, \qquad \frac{dy}{dt} = b. \tag{2.4}$$

Suppose u is a solution to a first-order PDE such that $u = u_0$, which is a constant along an integral curve. Then differentiating u with respect to t according to the chain rule gives

$$\left(\frac{dx}{dt}\right)u_x + \left(\frac{dy}{dt}\right)u_y = \frac{du_0}{dt} = 0. \tag{2.5}$$

The PDE given by eq. (2.5) is constant along each integral curve given by a specific value of u_0. The direction cosines are given by eq. (2.4). Conversely, if we integrate the ODEs given by eq. (2.4), the solution represents a family of integral curves along which u [the solution to the PDE given by eq. (2.5)] is constant—a different constant for each curve. These remarks are quite important and will be further exploited when we investigate the Cauchy IV problem below.

CAUCHY INITIAL VALUE PROBLEM

The simplest Cauchy IV problem involves a single first-order homogeneous PDE, an initial value (IV) curve, and data in terms of a prescribed value of u along the IV curve called an initial condition (IC). It is defined as follows: Find the solution u that satisfies the PDE

$$a(x, y)u_x + b(x, y)u_y = 0, \tag{2.6}$$

where the coefficients (a, b) of the PDE are prescribed functions of x, y. The IV curve is given by

$$x = x(\tau), \qquad y = y(\tau) \qquad \text{for } t = 0. \tag{2.7}$$

Call this IV curve Γ. Along Γ we set $u = f(x)$, a prescribed function on the IV curve. This is the IC along Γ. Recall that varying the parameter τ generates Γ, along which $t = 0$. The solution of eq. (2.6) must satisfy the IC along the IV curve.

PDEs OF WAVE PROPAGATION

The coefficients (a, b) in the PDE given by eq. (2.6) are involved in the ODEs given by eq. (2.4); more specifically, they are the direction cosines of the integral curves. Since these ODEs are associated with the PDE given by eq. (2.6), they are called characteristic ODEs. Note that the PDE given by eq. (2.6) is homogeneous (the right-hand side is zero) and u is a constant along each integral curve (as shown above). This gives us an important point: The solution of a homogeneous PDE is reduced to solving the characteristic ODEs for the slopes of the characteristic curves and then, by integration, determining those characteristic curves which are integral curves. The solution is then determined by picking up the value of the initial data at the intersection of each characteristic curve with the IV curve.

Example. Let us solve the PDE

$$2u_x + 3u_y = 0$$

subject to the IC

$$u(x, 0) = f(x) \quad \text{at } \Gamma: \quad y = 0,$$

where $f(x)$ is a prescribed function of x on the IV curve Γ. The characteristic ODEs eq. (2.4) become

$$\frac{dx}{dt} = 2, \quad \frac{dy}{dt} = 3.$$

Integrating these characteristic equations gives $x = 2t + \tau$, $y = 3t$. Eliminating t from these equations gives

$$y = (\tfrac{3}{2})(x - x_i), \quad i = 1, 2, 3, \ldots,$$

where $\tau = x_i$, $i = 1, 2, 3, \ldots$. We see that the integration parameter $\tau = x_i$, which are points on the x axis (the IV curve) on which the data $u(x, 0) = f(x)$ are given. Figure 2.1 shows the first quadrant of the x, y plane, which is the solution domain, the data $f(x)$ on the x axis, and a few parallel characteristic curves, which are the lines $y = \tfrac{3}{2}(x - x_i)$. We pass a characteristic line from a *field point* whose coordinates are x, y (where we want to calculate the solution) back to where it intersects the x axis. At this point of intersection we pick up the value of the datum $f(x)$ and use it to calculate u at the field point

45

CHAPTER TWO

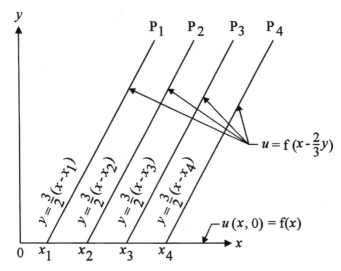

FIGURE 2.1. Solution of $2u_x + 3u_y = 0$, $u(x_1, y) = f(x)$ by the method of characteristics.

$u(x, y)$. u is the same function of x, y at the field point that it is at the point of intersection on the IV curve $[u(x_i, 0) = f(x_i) = f_i]$. Therefore the solution at the field point is

$$u(x, y) = f(x - \tfrac{2}{3}y).$$

(Note: We left off the subscript i since the field point is any point in the domain of the solution.) It is easily seen that this solution satisfies the IC by setting $y = 0$ for any point on the IV curve or x axis. The important point of this example is that *every characteristic line stemming from the IV curve and entering the domain of the solution picks up the initial datum $f(x)$ and carries it unchanged into the domain.* The reason for this is that the directional derivative $du/dt = 0$ so that u is a constant on a characteristic and therefore picks up the initial datum unchanged.

We now relate this example to a wave propagation problem by setting $y = t$ (the time) and letting the solution domain be $x > 0$, $t > 0$. To fix our ideas we let the IC be a sinusoidal function by setting $f(x) = A \sin(x/\lambda)$, where A is the amplitude of the waveform and λ is the wavelength. Let ω be the radial frequency in radians/sec. Then the solution is

$$u(x,t) = A \sin\left(\frac{x}{\lambda} - \frac{2}{3}\omega t\right).$$

The wave speed c is

$$c = \frac{dx}{dt} = \frac{2}{3}\omega\lambda = \left(\frac{4\pi}{3}\right)\omega\lambda,$$

where ω is the frequency in hertz (cps).

We now study the Cauchy problem for a more complicated linear first-order PDE, namely, one that is nonhomogeneous (the right-hand side is not zero). We consider the PDE

$$au_x + bu_y = cu, \qquad (2.8)$$

where the coefficients a, b, c are known functions of x, y. The ODEs for the characteristic curves are given by eq. (2.4). Using the PPE eq. (2.8), the directional derivative for u is $du/dt = cu$. We now have three characteristic ODEs, which are

$$\frac{dx}{dt} = a, \qquad \frac{dy}{dt} = b, \qquad \frac{du}{dt} = cu. \qquad (2.9)$$

The first two, as mentioned, give the slopes of the characteristic curves, while the third equation gives the directional derivative du/dt in terms of the right-hand side of the PDE. Note that the coefficients a, b involved in the ODEs for the characteristic slopes depend only on x, y since the PDE is linear. This means that the characteristic equations are independent of the solution. If the PDE were nonlinear, then one or more of a, b would be a function of u, so that the characteristic equations would depend on the solution, which makes nonlinear systems much more complicated than linear systems.

We now define the Cauchy problem for the case where a, b, c are constants:

$$au_x + bu_y = cu, \qquad \Gamma: \ y = 0, \qquad u(x,0) = f(x). \qquad (2.10)$$

The initial data for u is $f(x)$, a prescribed function of x on the IV curve Γ. To solve this problem we make use of the three characteristic ODEs

CHAPTER TWO

given by eq. (2.9). Integrating the first two ODEs yields the characteristic equations

$$x = at + \tau, \qquad y = bt, \qquad y = \left(\frac{b}{a}\right)x + \tau. \qquad (2.11)$$

The first two equations are the characteristic equations in parametric form. The last equation is a family of straight, parallel lines and is obtained by eliminating the parameter t from the first two equations. τ is the constant of integration. On the IV curve $x = \tau$. Integrating the third characteristic equation of (2.9) yields

$$u = \alpha e^{ct}, \qquad (2.12)$$

where α is a function of x, y to be determined from the IC. (Note that we are dealing with a PDE rather than an ODE so that α is not a constant.) The IC is obtained by setting $y = 0$. We get

$$u(x,0) = f(x) = f(\tau) = \alpha.$$

Using the first two characteristic equations we get $\tau = x - at = x - (a/b)y$. We can get off the x axis by substituting the above value of τ into the argument of f, yielding $f(x - (a/b)y)$. The solution then becomes

$$u(x, y) = f\left(x - \left(\frac{a}{b}\right)y\right)e\left(\frac{c}{a}\right)y. \qquad (2.13)$$

If we again let $y = t$, then u is stable only if the ratio c/b is zero or negative; for if this ratio is positive u will grow exponentially with time. If $c = 0$ then the PDE is homogeneous and the solution is given by $u(x, t) = f(x - (a/b)t)$. Again, if the IC is $f(x) = A\sin(x/\lambda)$, then the solution is

$$u(x,t) = f\left(x - \left(\frac{a}{b}\right)t\right)e\left(\frac{c}{b}\right)t.$$

The wave speed is $c = dx/dt = (a/b)\omega\lambda = (2\pi a/b)f\lambda$.

Example. Let us solve the Cauchy problem

$$xu_x + yu_y = cu, \qquad \text{IC:} \quad u(x,0) = f(x) \quad \text{on } \Gamma: \quad y = 0.$$

The characteristic equations (2.9) become

$$\frac{dx}{dt} = x, \qquad \frac{dy}{dt} = y, \qquad \frac{du}{dt} = cu.$$

Integrating these equations and using the IC gives

$$x = \tau e^t, \qquad y = 1 - e^t, \qquad u = f(\tau)e^{ct}.$$

We have thus obtained u in terms of the parameters (τ, t). But $\tau = x/(1-y)$, $e^{ct} = (1-y)^c$, so that we get $u(x, y)$ in the form

$$u(x, y) = f\left(\frac{x}{1-y}\right)(1-y)^c.$$

Parametric Representation

We now present a more systematic interpretation of the parametric representation of characteristic theory. We have been using the parametric form of the characteristic equations as shown by the ODEs given by eq. (2.9), where we introduced the parameter t. Also the parameter τ was introduced in eq. (2.11) as an integration constant. We thus have x, y as functions of the two parameters t, τ. This mapping to and from the x, y and t, τ planes will be explored below. The geometric representation of this approach is seen in fig. 2.2, where several characteristic curves are shown in the x, y plane crossing an IV curve given by Γ whose equation is $y = g(x)$. t plays the role of time so that on Γ $t = 0$, while τ generates the IV curve Γ. We may take τ as zero at the origin and increase it, thus generating Γ. On a given characteristic curve which we call C_τ, t varies from zero on the intersection with Γ and becomes more and more positive, thus generating the characteristic. Each characteristic curve is defined by a specific value of τ that is obtained from the intersection of the characteristic with the IV curve Γ.

Example. We now give a more complicated example, that of a Cauchy IV problem involving a quasilinear PDE:

$$uu_x + u_y = 1; \qquad u(\tau, 0) = U\tau \qquad \text{on } \Gamma: \quad x(\tau, 0) = \tau, \, y(\tau, 0) = 0,$$

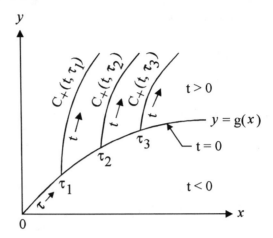

FIGURE 2.2. C_+ characteristics stemming from the IV curve Γ_1 $y = g(x)$ in parametric form.

where the IC U is constant, The characteristic ODEs are

$$\frac{dx}{dt} = u, \qquad \frac{dy}{dt} = 1, \qquad \frac{du}{dt} = 1.$$

Note that these equations are coupled, since the first one depends on the third. Moreover, the first two equations, which represent the characteristic slopes, involve the solution u, since the PDE is nonlinear (in this case, quasilinear). Each characteristic curve C_τ is given in parametric form by $x = x(\tau, t), y(\tau, t)$ by integrating the characteristic ODEs. The third equation gives

$$u(\tau, t) = t + U\tau.$$

Clearly this expression for u in parametric form satisfies the IC as can easily be seen by letting $t = 0$. To find the characteristic curves x, y as functions of the parameters τ, t, we need to integrate the first two characteristic ODEs and use the above solution for $u(x, t)$. We get

$$x = \frac{t^2}{2} + U\tau t + \tau, \qquad y = t.$$

Using these results, we obtain the solution for $u = u(x, y)$:

$$u = y + \left[\frac{U}{2(Uy + 1)}\right](2x - y^2).$$

Since the solution is simpler in the τ, t plane, namely, $u = t + U\tau$, we transform the Cauchy IV problem to τ, t coordinates. The PDE becomes

$$(u - t - U\tau)u_x + (Ut + 1)u_t = Ut + 1, \quad \text{IC:} \quad u(\tau, 0) = U\tau.$$

The details are left to a problem for the reader.

The above example has an interesting physical interpretation: If we let $y = t$ (time) and $u(x, t)$ be the particle velocity of a particle at x at time t, we have $dx/dt = u$. The particle acceleration is du/dt. Expanding this total derivative by the chain rule gives

$$\frac{du}{dt} = \left(\frac{dx}{dt}\right)u_x + u_t = uu_x + u_t,$$

where we used the fact that $dx/dt = u$. We see that the left-hand side of the PDE used in the above example is indeed the particle acceleration. We saw in chapter 1 in the section "Wave Equation for a Bar" that the particle acceleration consists of two terms. Recall that the first term, uu_x, is the convective term and represents the part of the acceleration at a fixed time where the particle velocity varies with x. The second term, u_t, represents the part of the acceleration at a fixed x where the velocity varies with time. The above problem tells us that the particle acceleration has an IV that varies linearly along the x axis.

WAVE EQUATION EQUIVALENT TO TWO FIRST-ORDER PDEs

It is possible, by introducing two new dependent variables, to convert the wave equation given by

$$c^2 u_{xx} - u_{tt} = 0 \tag{2.14}$$

to a pair of first-order PDEs. (c is the velocity of a traveling wave.) To give a physical motivation we revert to the section entitled "Wave

CHAPTER TWO

Equation for a Bar" in chapter 1 and recall that the strain ε of an element of the bar due to a stress was given by eq. (1.2), namely, $\varepsilon = u_x$. Since u is the particle displacement it follows that the particle velocity $v = u_t$. We thereby introduce the variables ε, v defined by

$$\varepsilon = u_x, \qquad v = u_t. \tag{2.15}$$

Inserting these expressions into the wave equation (2.14) and recognizing that $u_{xt} = u_{tx} = v_x$ and $u_{tx} = u_{xt} = \varepsilon_t$, we get the pair of first-order PDEs:

$$c^2 \varepsilon_x - v_t = 0, \qquad v_x - \varepsilon_t = 0. \tag{2.16}$$

A more symmetric formulation for this system is now given by letting $\phi = c\varepsilon$. Then eq. (2.16) becomes

$$c\phi_x = v_t = 0, \qquad cv_x - \phi_t = 0. \tag{2.17}$$

It is easily seen that both ϕ and v satisfy the wave equation, so that they are of the form $F(x - ct)$ and $G(x + ct)$, where F and G are arbitrary functions of their arguments.

We want to obtain the characteristic slopes (characteristic ODEs) for a pair of first-order PDEs involving the solutions ϕ, v as functions of x, y. This means that we must manipulate equations to satisfy the condition that the directional derivatives of ϕ and v must be in the same direction, that of the slope of the characteristics. In other words, we must set these directional derivatives equal to each other. Now we have the system (2.17) to deal with. We must try to obtain that linear combination of these PDEs which satisfies the condition that the directional derivatives of ϕ and v are equal. To this end we first expand these derivatives by the chain rule and get

$$\frac{d\phi}{dt} = \left(\frac{dx}{dt}\right)\phi_x + \phi_t = \frac{dv}{dt} = \left(\frac{dx}{dt}\right)v_x + v_t. \tag{2.18}$$

Multiplying the first equation of (2.17) by the unknown constant λ and adding the second equation gives

$$c(\lambda \phi_x + v_x) - (\lambda v_t + \phi_t) = 0. \tag{2.19}$$

Using eqs. (2.18) and (2.19) we have two possibilities:
(1) $\lambda = -1$, $dx/dt = c$. This gives

$$\left(\frac{dx}{dt}\right)(v - \phi)_x + (v - \phi)_t = 0. \tag{2.20}$$

This suggests introducing a new variable defined by $\Psi = v - \phi$. Then eq. (2.20) becomes

$$\left(\frac{dx}{dt}\right)\Psi_x + \Psi_t = \frac{d\Psi}{dt} = 0. \tag{2.21}$$

This means that on the family of C_+ characteristics whose slope is $dx/dt = c$, the function $\Psi = v - \phi = v - c\varepsilon = $ const. Recall that these characteristics are given by $x - ct = \xi$, so that $\Psi(\xi) = $ const on a characteristic defined by ξ.

(2) $\lambda = 1$, $dx/dt = -c$ Defining the variable $\Omega = v + \phi$, by a similar analysis we obtain

$$\left(\frac{dx}{dt}\right)\Omega_x + \Omega_t = \frac{d\Omega}{dt} = 0. \tag{2.22}$$

This means that on the family of C_- characteristics whose slope is $dx/dt = -c$ and whose equation is $x + ct = \eta$, the function $\Omega = \Omega(\eta)$ is constant on a characteristic defined by a particular value of η.

Summarizing, we have two distinct families of characteristics in the x, y plane, given by two distinct slopes. (1) On the C_+ family of characteristics defined by $x - ct = \xi$ we have $v - \phi = v - c\varepsilon = \Psi(\xi)$. (2) On the C_- family defined by $x + ct = \eta$ we have $\phi + v = c\varepsilon + v = \Omega(\eta)$. Since we are dealing with a pair of first-order PDEs, we have a certain linear combination of the two variables ε, v that is constant along a characteristic. Now both v and $\phi = c\varepsilon$ are solutions of the wave equation, which is a second-order PDE involving two distinct families of characteristics, instead of one as in a first-order PDE.

We have been dealing essentially with a first-order system of linear PDEs. We have demonstrated that, for a single first-order PDE, there exists a single family of characteristic curves in the x, y plane defined by the parameter τ, which gives the point of intersection of the characteristic with the IV curve Γ. Increasing the parameter t from 0 generates a characteristic for a given τ, and τ generates Γ on which $t = 0$. For a

CHAPTER TWO

pair of linear first-order PDEs we showed that there exist two distinct families of characteristics: the C_+ family generated by varying ξ and fixing η, and the C_- family generated by varying η and fixing ξ.

In treating a single first-order PDE we were able to obtain an ODE which gave the slope of the family of characteristics, as long as the coefficients of the PDE were real, given constants or functions of x, y. Now a question arises: Does every pair of first-order PDEs with real constant coefficients give rise to two families of characteristics? The answer is "no." To show this we give a counterexample in the form of a pair of PDEs for u, v:

$$u_x - v_y = 0, \qquad v_x + u_y = 0. \tag{2.23}$$

We shall attempt to get a linear combination of u and v that satisfies the condition that its directional derivative lies along the slope of a characteristic. To this end we introduce the dependent variable f such that $f = u + \lambda v$ and attempt to find the constant λ that satisfies the above condition. In order to determine the directional derivative we must make use of the given PDEs. We therefore form the same linear combination of these equations (2.23) by multiplying the second equation by λ and adding. We obtain

$$u_x + \lambda v_x + \lambda u_y - v_y = 0. \tag{2.24}$$

The directional derivative of f is

$$u_x + \lambda v_x + \left(\frac{dy}{dx}\right)(u_y + \lambda v_y) = \frac{df}{dx} = 0, \tag{2.25}$$

where we divided the equation by dx/dt to obtain the slope in the form dy/dx, yielding $df/dx = 0$. Now we cannot find a real value of the constant λ that satisfies the above conditions. Therefore no real characteristics exist in the x, y plane. So we try an imaginary value for λ by setting $\lambda = i$, where $i = \sqrt{-1}$. This is easily seen by making use of eq. (2.23). This gives $dy/dx = i$ for the slope, again showing no real characteristic. It follows that $f = u + iv$, so that f is a function of a complex variable whose real part is u and whose imaginary part is v. u and v are called *harmonic functions* since they both satisfy *Laplace's equation*:

$$u_{xx} + u_{yy} = \nabla^2 u = 0, \qquad v_{xx} + v_{yy} = \nabla^2 v = 0.$$

The operator $\nabla^2 = \partial^2/\partial x^2 + \partial^2/\partial y^2$ is called the *Laplacian*. If we set $z = x + iy$, z is called a *complex variable*, so that $f = u + iv = f(z)$; thus f is a function of the complex variable z. Moreover, f is an *analytic function* of z, which means that, at each point in the u, v plane where $f(z)$ is analytic, it has a unique derivative. This means that df/dz can be obtained by differentiating f with respect to x at constant y, and setting the result equal to the derivative obtained by differentiating f with respect to y at constant x (taking into account $z = x + iy$). We then equate real and imaginary parts of this equation. The calculations are as follows:

$$\frac{df}{dz} = f_x = u_x + iv_x = f_y = -i(u_y + iv_y) = v_y - iu_y.$$

We obtain eqs. (2.23), called the *Cauchy-Riemann equations*. These equations are the necessary and sufficient conditions that u and v be harmonic functions. All harmonic functions satisfy Laplace's equation and do not generate real characteristics. It is interesting to note that, for any generic dependent variable $u = u(x, y)$, the PDE $u_{xx} - u_{yy} = 0$ has two distinct families of characteristics and hence has wavelike properties, while merely changing the minus to a plus sign, giving the PDE $u_{xx} = u_{yy} = 0$, yields an equation with an entirely different property: no real characteristics, hence no wavelike properties. The essential difference between these two equations, the wave equation and Laplace's equation, brings up the subject of classifying second-order PDEs by the nature of their characteristics. This will be treated in the following section.

CHARACTERISTIC EQUATIONS FOR FIRST-ORDER PDES

We shall introduce this section by presenting a different analysis of the wave equation in terms of two first-order PDEs that does not rely on the stresses in a bar. It brings in the directional derivatives of u and v

CHAPTER TWO

in a direct way. Again considering the x, y plane, we set

$$u_x = p, \quad u_y = q. \tag{2.26}$$

Since $u_{xy} = u_{yx} = p_x = q_y$, we have the wave equation and this identity as the following pair of first-order PDEs:

$$L_1 = c^2 p_x - q_y = 0, \quad L_2 = p_y - q_x = 0, \tag{2.27}$$

where L_1 and L_2 stand for the left-hand sides of the appropriate PDEs. In order to obtain the characteristic ODEs (which give the slopes of the characteristic curves) for the system given by eq. (2.27), we must satisfy the following conditions: (1) The directional derivatives of u and v must be in the same direction. (2) We must find that linear combination of the PDEs (2.27), namely, $L_1 = 0$, and $L_2 = 0$, which is in the same direction as the directional derivatives. In order to satisfy the first condition we consider x as the independent variable and expand the directional derivatives in the form du/dx, dv/dx by the chain rule. We get

$$\frac{du}{dx} = u_x + \frac{dy}{dx} u_y,$$

$$\frac{dv}{dx} = v_x + \frac{dy}{dx} v_y.$$

This establishes the fact that these directional derivatives are in the same direction since their slopes are the same. To satisfy the second condition we introduce the constant λ (to be determined) and set $L = L_1 + \lambda L_2$ where L_1 and L_2 are defined by eq. (2.27). We get

$$L = c^2 p_x + \lambda p_y - \lambda q_x - q_y = 0. \tag{2.28}$$

Equation (2.28) is a linear combination of p_x, p_y, q_x, q_y. Specifically, the left-hand side is the sum of the expression $c^2 p_x + \lambda p_y = P$ and the expression $-(\lambda q_x + q_y) = Q$. From the nature of the first expression P, we must coordinate it with the expansion of the directional derivative dp/dx. To do this we must demand that the ratio of the coefficient of p_y to that of p_x in P must equal the same ratio in the expansion for dp/dx. This yields $\lambda/c^2 = dy/dx$. We do the same for Q, setting the ratio of the coefficients of q_y and q_x equal to the same ratio in the expansion for dq/dx. We obtain $1/\lambda = dy/dx$. These two results yields the following:

$$\lambda^2 = c^2, \qquad \left(\frac{dy}{dx}\right)^2 = c^{-2}.$$

We have determined the characteristic ODE, which is $(dy/dx)^2 = 1/c^2$. This yields two distinct characteristic slopes:

$$\frac{dy}{dx} = \pm\frac{1}{c}.$$

Again, if we think of y as the time then the characteristic slopes are $dx/dy = \pm c$, which is the wave velocity for a progressive or regressive wave, respectively.

Having determined the slopes and the corresponding linear combination of partial derivatives given by λ, we can now evaluate this combination L. First, we set $\lambda = c$. This gives

$$c(cp_x - q_x) + cp_y - q_y = 0.$$

Setting $df = c\,dp - dq$, we have $f = cp - q$. Since f is a function of x and y, we obtain $f_x = cp_x - q_x = (cp - q)_x$ and $f_y = cp_y - q_y = (cp - q)_y$. This give the following expression for L [eq. (2.28)]:

$$cf_x + f_y = 0. \tag{2.29}$$

This means that $f = cp - q = \text{const}$ on a C_+ characteristic whose slope is $dx/dt = c$.

Next we set $\lambda = -c$. A similar analysis yields the fact that the function $g = cp + q = \text{const}$ on a C_- characteristic.

If we set $p = \phi_x$ and $q = v_t$, we get the case of a bar discussed in the section entitled "Wave Equation Equivalent to Two First-Order PDEs."

General Treatment of Linear PDEs by Characteristic Theory

We are now in a position to treat general second-order linear PDEs by the method of characteristics in order to classify this PDE into three types as shown below. Note that since the characteristic ODEs depend

CHAPTER TWO

only on the highest order of a PDE, we concern ourselves only with the *principal part* of the second-order PDE defined by $ap_x + bp_y + bq_x + cq_y$. We set this principal part equal to zero, make use of the identity $p_y = q_x$, and obtain the first-order system

$$L_1 = ap_x + bp_y + bq_x + cq_y = 0,$$
$$L_2 = p_y - q_x = 0. \qquad (2.30)$$

We now form the linear combination $L = L_1 + \lambda L_2$, where, as usual, λ is a constant to be determined. L becomes

$$L = ap_x + (b + \lambda)p_y + (b - \lambda)q_x + cq_y = 0. \qquad (2.31)$$

We force the expression $ap_x + (b + \lambda)p_y$ to be in the same direction as the directional derivative dp/dx, and the expression $(b - \lambda)q_x + cq_y$ to be in the same direction as dp/dx and dq/dx, where

$$\frac{dp}{dx} = p_x + \left(\frac{dy}{dx}\right)p_y,$$

$$\frac{dq}{dx} = q_x + \left(\frac{dy}{dx}\right)q_y.$$

To this end we have the following rule concerning the ratio of the partial derivatives which must be coordinated with those in the directional derivatives:

$$\operatorname{coeff}(p_y) : \operatorname{coeff}(p_x) = \operatorname{coeff}(q_y) : \operatorname{coeff}(q_x) = \frac{dy}{dx}. \qquad (2.32)$$

(This is exactly what we did when investigating the wave equation as a pair of first-order PDEs, above.) Applying this rule gives

$$\frac{b + \lambda}{a} = \frac{c}{b - \lambda} = \frac{dy}{dx}. \qquad (2.33)$$

Setting the slope $dy/dx = z$, this expression gives the quadratic equation for z:

$$ax^2 - 2bz + c = 0, \qquad z = \frac{dy}{dx}. \qquad (2.34)$$

PDEs OF WAVE PROPAGATION

Using eq. (2.33), the linear combination of partial derivatives L becomes

$$L = \left(ap_x + \left(\frac{c}{z}\right)q_x\right) + z\left(aq_x + \left(\frac{c}{z}\right)q_y\right) = 0. \qquad (2.35)$$

We now let $ap + (c/z)q = F_\pm$, and insert this function into L in eq. (2.35). If the two roots for z in eq. (2.34) are real and unequal, then F_+ is constant on a C_+ characteristic whose slope is given by the positive root for z, while F_- is constant on a C_- characteristic whose slope is given by the negative root for z. It is easily seen that for the one-dimensional wave equation these functions reduce to $f = cp - q$ and $g = cp + q$.

Coming back to eq. (2.34), this quadratic equation for z is called the *characteristic equation* and is a pivotal result. It tells us that the roots for z depend on the coefficients a, b, c of the principal part of the PDE. It is clear that the roots for z are the characteristic ODEs, since they give the characteristic directions associated with the PDE. The roots are $z_\pm = b \pm \sqrt{D}$, where the discriminant $D = b^2 - ac$. There are three possibilities for D:

(1) $D > 0$, giving real and unequal roots
(2) $D = 0$, giving real and equal roots
(3) $D < 0$, giving complex conjugate roots.

This classification of the roots for z allows us to categorize the PDEs given by eq. (2.30), which is equivalent to the second-order PDE (upon using the definitions of p and q)

$$au_{xx} + 2bu_{xy} + cu_{yy} = 0. \qquad (2.36)$$

Table 2.1 summarizes the classification of this system.

For those readers who are not familiar with the heat equation, the unsteady heat equation is given by $\kappa u_{xx} = u_y$, where y is the time and the constant κ is called the thermal diffusivity. The heat equation expresses the phenomenon of diffusion, either of heat through a rod (since the equation is one dimensional in space) or a dye in a tube, etc.

This classification of second order PDEs into hyperbolic, elliptic, and parabolic illustrates the three basically different phenomena found in nature that involve dependent variables that vary both in space and time. The *hyperbolic PDE* represents the wavelike character of events

CHAPTER TWO

TABLE 2.1
Classification of Second-Order PDE or First-Order System

Discriminant $D = b^2 - ac$	Type of PDE	Nature of roots	Example
> 0	Hyperbolic	Real unequal	Wave equation $b = 0$, $ac < 0$
< 0	Elliptic	Complex conjugates	Laplace equation $b = 0$, $ac > 0$
$= 0$	Parabolic	Real equal	Unsteady heat equation $b = c = 0$, $a \neq 0$

that travel in space with a finite velocity, the wave velocity. A signal takes a finite time to travel a finite distance. Disturbances such as sound waves travel only along the real characteristics, progressing waves travel on C_+ characteristics, while regressing waves travel on C_- characteristics. One important aspect of the wave equation is that it runs backward as well as forward. This means that if we replace time by negative time the wave equation remains the same, and hence the solution merely involves the fact that disturbances propagate along the reversed families of characteristics. Therefore there is no thermal dissipation and disturbances are propagated without dissipation on both families of characteristics. It is possible to build in dissipation effects in the wave equation by adding a term that involves u_t, giving a frictional effect, which is a heat dissipation process. This does not affect the principal part of the PDE, so that a hyperbolic PDE can involve dissipation. The *elliptic PDE* has no real characteristics and hence shows no wavelike properties. This means that any event occurring in space is felt instantaneously at all points in space. This is what Newton's law of gravitational attraction tells us. For example, if a distant star were to suddenly explode, its effect would be felt throughout all space instantaneously, This, of course, is nonsense. This defect in Newton's law was rectified by Einstein in his general theory of relativity, which allows for the propagation of gravitational waves (which are so weak that they have not as yet been detected). The *parabolic PDE* is a bridge between the hyperbolic and elliptic, since it has but a single family of real characteristics. It represents the phenomenon of diffusion. A dye, for example, will diffuse throughout space with finite velocity. This dispersion process involves dissipation of energy, so that the further away from the source of the

PDEs OF WAVE PROPAGATION

dye, for example, the weaker the signal. The reason that there is only one family of characteristics is that if we reverse time we cannot reverse the process, as we can with the wave equation. In this sense, the parabolic equation is more realistic in that it allows for the thermodynamic process of dissipation or loss of thermal energy. The diffusion of heat throughout space is another example of this diffusion process. In fact, one of Einstein's important contributions before the relativity theory was to relate the dissipation of heat to a diffusion process by way of Brownian motion, which was discovered by Robert Brown, an English botanist. He discovered that tiny specks of material in pure water are buffeted in a random way. It was later shown by others that this buffeting process is due to the random impact of water molecules on the specks.

It is amazing that the purely mathematical concept of characteristic theory can lead to the classification of purely physical phenomena in nature. But isn't this the purpose of mathematics?

ANOTHER METHOD OF CHARACTERISTICS FOR SECOND-ORDER PDES

There is another way of developing the second-order PDE $au_{xx} + 2bu_{xy} + cu_{yy} = 0$ given by eq. (2.36). If the coefficients a, b, c are at most given functions of x, y, then the PDE is linear. If these coefficients are also functions of u, p, q, the PDE is said to be quasilinear. The above analysis of first-order PDES as well as the succeeding analysis hold for quasilinear systems as well as linear ones. The method we shall develop is based on Cauchy's IV problem, which we now define for the second-order PDE. Find $u = u(x, y)$ which is the solution to the Cauchy problem given by

$$au_{xx} + 2bu_{xy} + cu_{yy} = 0, \quad \Gamma: y = g(x), \quad u, p, q \text{ given.} \quad (2.37)$$

This means that we prescribe an IV curve Γ characterized by $y = g(x)$ on which we are given the ICs $u, p = u_x, q = u_y$. Since the PDE is second order we must prescribe the initial data on Γ as u and the first partial derivatives p, q. p and q are needed to calculate the derivative of u normal to Γ. The idea is to obtain a solution to eq. (2.36) in the neighborhood of Γ which is a strip surrounding Γ. This is seen in fig.

CHAPTER TWO

2.3, which shows Γ, the strip in the x, y plane, and the ICs. In order to obtain a solution in the strip off the Γ curve we must be able to calculate the higher-order derivatives on Γ using the ICs. Having done this, we use these higher-order derivatives to calculate the solution in the strip surrounding Γ by a Taylor series expansion in terms of these derivatives. Once a solution is obtained in the strip, by continuity we can extend the solution into the remainder of the solution domain. (Incidentally, we could set up the same type of analysis for the pair of first-order PDEs discussed in the previous section. But it is more convenient to proceed directly with a single second-order PDE.)

We now attempt to calculate these higher-order derivatives on Γ. To this end we extend the notation to second-order derivatives:

$$u_x = p, \quad u_y = q, \quad u_{xx} = r, \quad u_{xy} = s, \quad u_{yy} = t. \quad (2.38)$$

This means that, given the initial data u, p, q on Γ, our first task is to calculate $r, s, t, \ldots$ on Γ. For our purpose it is only necessary to calculate r, s, t. Inserting this notation into eq. (3.36) gives one equation and the three unknowns: r, s, t. We therefore need two additional equations. They are obtained by expanding dp/dx and dq/dx by the chain rule. Having done this, we obtain the system of three algebraic equations for r, s, t:

$$ar + 2bs + ct = 0,$$
$$r + zs = \frac{dp}{dx}, \quad (2.39)$$
$$s + zt = \frac{dq}{dx}, \quad z = \frac{dy}{dx}.$$

In the system given by eq. (2.39) the only unknowns are r, s, t, since dp/dx and dq/dx as well as the coefficients are known.

This is a system of three nonhomogeneous algebraic equations (the right-hand side is not zero). In order to solve for r, s, t, an elementary theorem in algebra tells us that the determinant of the coefficients of r, s, t in the left-hand side of eq. (2.39) shall not be zero. But what if it is zero? Then we cannot obtain a solution for r, s, t. Previously, we observed that the characteristics are curves for which the slopes are nowhere in the direction of Γ. In other words, if we were given the initial data u, p, q on a characteristic curve, then we could not solve the Cauchy IV problem. Therefore, we suspect that if we set the above-

PDEs OF WAVE PROPAGATION

mentioned determinant equal to zero and thus are not able to solve for r, s, t, the resulting equation will somehow or other generate the roots z (the slope of the characteristics). So let us set this determinant equal to zero and see what happens:

$$\begin{vmatrix} a & 2b & c \\ 1 & z & 0 \\ 0 & 1 & z \end{vmatrix} = 0. \qquad (2.40)$$

Expanding this determinant and setting it equal to zero gives

$$ax^2 - 2bz + c = 0, \qquad z = \frac{dy}{dx},$$

which is eq. (2.34), the characteristic ODEs for the slopes z. Our guess was correct. The condition that we cannot solve the Cauchy problem leads to the two families of characteristics whose slopes are the two real and unequal roots of the characteristic equation (2.34). From eqs. (2.34) and (2.39) we can obtain the relation between dp, dq, band z:

$$\frac{dq}{dp} = -\frac{a}{c} z. \qquad (2.41)$$

For the quasilinear case the characteristics depend on the solution u and probably on p and q. In general, the quasilinear system cannot be solved by analytical methods. A numerical analysis can be developed based on the above results to form a characteristic net in the x, y plane for the Cauchy problem for the quasilinear case. But this is beyond the scope of this book.

GEOMETRIC INTERPRETATION OF QUASILINEAR PDES

Geometric intuition is of great help in getting a better understanding of the role of PDEs in wave phenomena. To this end we present a geometric interpretation of first-order quasilinear PDEs for $u = u(x, y)$. It is convenient to use a more symmetric setting. We thereby introduce the (x, y, u) space. Our previous discussion of characteristic theory restricted us to the (x, y) plane. There we projected the characteristics on the (x, y) plane. In the (x, y, u) space we study *characteristic surfaces* and relate them to *integral surfaces*.

CHAPTER TWO

The most general first-order PDE in this space can be written in the implicit form

$$F(x, y, u, p, q) = 0, \quad p = u_x, \quad q = u_y. \quad (2.42)$$

For the quasilinear case F is linear in p and q and we may write the above equation in the following explicit form:

$$ap + bq = c, \quad (2.43)$$

where (a, b, c) are prescribed functions of (x, y, u). For the linear case (a, b, c) are at most functions of (x, y).

The geometric interpretation of eq. (2.43) is as follows: Each solution $u = u(x, y)$ yields an integral surface in the sense that, at each point P in (x, y, u) space on the integral surface, there exists a tangent plane given by the direction numbers (a, b, c), and whose normal has the direction numbers $(p, q, -1)$. In other words, the gradient at each point on the integral surface in (x, y, u) space has these direction numbers. Since the gradient is normal to the tangent plane at each point on the integral surface, we have the scalar product $(a, b, c) \cdot (p, q, -1) = 0$. This yields $ap + bc - c = 0$, which is the first-order quasilinear PDE given by eq. (2.43). This shows the connection between $(p, q, -1)$, (a, b, c), the PDE. Equation (2.43) tells us that the tangent planes of all the integral surfaces that pass through the point $P:(x, y, u)$ on the surface belong to a single *pencil of planes* whose axis is given by the characteristic directions according to the following ratios:

$$dx : dy : du = a : b : c.$$

These ratios hold at each point on the integral surface since the direction numbers of the tangent (a, b, c) satisfy the following ODEs:

$$\frac{dy}{dx} = \frac{b}{a}, \quad \frac{du}{dx} = \frac{c}{a}, \quad \frac{du}{dy} = \frac{c}{b}. \quad (2.44)$$

Equations (2.44) are the characteristic ODEs, which give the direction of the tangent to the integral surface. This characteristic direction is called the *Monge axis*. Clearly, each point $P:(x, y, u)$ on the integral surface may have a different direction for the Monge axis. The point P together with the direction of the Monge axis given by the above ratios is called a *characteristic line element*. The totality of directions of the

PDEs OF WAVE PROPAGATION

Monge axis forms a *direction field* in (x, y, u) space. The *characteristic curves* are obtained by integrating the characteristic ODEs (2.44). The family of these curves form a pencil that is the Monge axis. Introducing the parameter s measured along a characteristic curve, the above ratios become

$$\frac{dx}{ds} = a, \quad \frac{dy}{ds} = b, \quad \frac{du}{ds} = c, \qquad (2.44a)$$

which is the parametric form of the characteristic ODEs. The projections of the characteristic curves on the (x, y) plane are called the *characteristic base curves*. Integrating the PDE given by eq. (2.43) to yield a family of integral curves in (x, y, u) space is reduced to integrating the set of characteristic ODEs (2.44a) (which is clearly easier than integrating the PDE). In other words, integrating the PDE is reduced to finding the family of surfaces whose tangent planes are in the direction of the Monge axis, which is in the characteristic direction (tangent to the integral surface). It follows that (1) *Every surface generated by a one-parameter family of characteristic curves is an integral surface of the PDE*; and (2) *every integral surface is generated by a one-parameter family of characteristic curves* (*the parameter being s*). At the risk of being repetitious: Each integral surface is made up of characteristic curves, and no characteristic curve is not on an integral surface.

INTEGRAL SURFACES

In implicit form an integral surface in (x, y, u) space is represented by the equation

$$G(x, y, u) = 0. \qquad (2.45)$$

Analytical geometry tells us that the direction cosines (α, β, γ) of the normal to the surface at point (x, y, u) are given by the expressions

$$\cos \alpha = \frac{G_x}{\sqrt{\left(G_x^2 + G_y^2 + G_u^2\right)}}, \quad \cos \beta = \frac{G_y}{\sqrt{\left(G_x^2 + G_y^2 + G_z^2\right)}},$$

$$\cos \gamma = \frac{G_u}{\sqrt{\left(G_x^2 + G_y^2 + G_u^2\right)}}. \qquad (2.46)$$

CHAPTER TWO

To relate the normal and also the tangent to the surface $G = 0$ to the PDE, we expand $\dot{G}$ and $\dot{u}$ with respect to the parameter t. We obtain

$$\dot{G} = G_x\dot{x} + G_y\dot{y} + G_u\dot{u} = G_x\dot{x} + G_y\dot{y} + G_u(p\dot{x} + q\dot{y}).$$

From this equation we get the following relationship between the partial derivatives of G and the partial derivatives of u in the PDE:

$$p = -\frac{G_x}{G_u}, \qquad q = -\frac{G_y}{G_u}. \tag{2.47}$$

The equation for the tangent plane is

$$(\xi - x)G_x + (\eta - y)G_y + (\zeta - u)G_u = 0,$$

where (ξ, η, ζ) are running coordinates that generate an element of the tangent plane. In the limit we get $(\xi - x \to dx, \eta - y \to dy, \zeta - u \to du)$. With respect to the parameter t, $x = x(t)$, $y = y(t)$, $u = u(t)$, so that the equation for the tangent plane becomes

$$\dot{x}G_x + \dot{y}G_y + \dot{u}G_u = 0. \tag{2.48}$$

The expansion of the directional derivative is $\dot{u} = \dot{x}p + \dot{y}q$. Inserting this into the above equation for the tangent plane and equating coefficients of $(\dot{x}, \dot{y})$ yields the following expressions relating (p, q) to (G_x, G_y, G_u):

$$\frac{G_x}{G_u} = -p, \qquad \frac{G_y}{G_u} = -q. \tag{2.49}$$

The equation for the tangent plane (2.48) then becomes

$$\dot{x}p + \dot{y}q - \dot{u} = 0,$$

which is the expansion of the directional derivative of u. Again, we have the fact that $(\dot{x}, \dot{y})$ are the direction numbers of the tangent to the integral surface (in parametric form) and $(p, q, -1)$ are the direction numbers of its normal.

Nonlinear Case

The fully nonlinear case is more complicated than the quasilinear case because the coefficients (a, b, c) of the PDE are functions of (p, q) as well as (x, y, u). Recall that, for this case, a Monge axis can be constructed at each point P on the integral curve. The elements of the tangent planes of all the integral curves through P form a pencil through the Monge axis. As mentioned, for the fully nonlinear case the coefficients depend on (p, q), so that there is no longer a linear relation between p and q. The possible tangent planes through P on the integral surface form a one-parameter family enveloping a conical surface with P as the vertex. This cone is called the *Monge cone*. At each point on the integral surface there exists a Monge cone which is the envelope of all tangent planar elements through P.

We now obtain the *characteristic directions* which are the generators of the Monge cone. In contrast to the quasilinear case, where only one characteristic direction belongs to each point on the integral surface, in the fully nonlinear case there exists a one-parameter family of characteristic directions at each point. These characteristic directions are the five characteristic ODEs for dx/dt, dy/dt, du/dt, dp/dt, and dq/dt, which allow us to solve for (x, y, u, q, q). We now determine these ODEs. Recall that a one-parameter family of characteristic curves span (form) an integral surface. Let the parameter τ defines a characteristic curve. Let the parameter t be the distance along a characteristic curve, so that by fixing τ and varying t we generate a particular characteristic curve. An infinitely near characteristic curve is given by $\tau + d\tau$. The Monge cone at vertex $P:(x, y, u)$ is the envelope of all the characteristic curves from P with respect to τ. Since the point P on the integral surface is defined by (x, y, u), it follows that to get the envelope we consider p and q to be functions of τ. We now do the following procedures: (1) We differentiate u along a characteristic curve defined by the parameter τ. This gives the directional derivative du/dt. (2) We then differentiate this directional derivative with respect to t. (3) And finally we differentiate the PDE expressed by eq. (2.42) with respect to τ. The purpose of these calculations is to obtain the ODEs of the envelope of all the characteristics from $P:(x, y, u)$, which is the Monge

CHAPTER TWO

cone. The calculations are

$$\dot{u} = p(\tau)\dot{x} + q(\tau)\dot{x}, \qquad \dot{u} = \frac{du}{dt}, \qquad \text{etc.},$$

$$0 = p'\dot{x} + q'\dot{y}, \qquad p' = \frac{dp}{d\tau}, \qquad \text{etc.},$$

$$F_p p' + F_q q' = 0.$$

(Recall that only p and q are functions of τ.) Relating $\dot{x}, \dot{y}, \dot{u}$ to the corresponding coefficients, we obtain

$$\dot{x} = F_p, \qquad \dot{y} = F_q, \qquad \dot{u} = pF_p + qF_q. \qquad (2.50)$$

The system given by (2.50) is the three characteristic ODEs for $\dot{x}, \dot{y}, \dot{u}$. To get the other two ODEs (for p', q') we differentiate the PDE $F = 0$ first with respect to x and then with respect to y, and obtain

$$F_p p_x + F_q q_x + F_u p + F_x = 0,$$

$$F_p p_y + F_q q_y + F_u q + F_y = 0.$$

These equations hold identically on the integral surface. Using the first two equations of (2.50), they become

$$p' + F_u p + F_x = 0,$$

$$q' + F_u q + F_y = 0,$$

where we used the fact that $p_x = q_y$, $p_y = q_x$.

The five characteristic ODEs belonging to the fully nonlinear PDE $F(x, y, u, p, q) = 0$ are given here in parametric form $[x = x(t), \ldots]$

$$\dot{x} = F_p,$$

$$\dot{y} = F_q,$$

$$\dot{u} = pF_p + qF_q, \qquad (2.51)$$

$$p' = -(pF_u + F_x),$$

$$q' = -(qF_u + F_y).$$

We may rewrite these equations in terms of the following ratios:

$$\frac{dx}{F_p} = \frac{dy}{F_q} = \frac{du}{pF_p + qF_q} = \frac{dp}{-(pF_u + F_x)} = \frac{dq}{-(qF_u + F_y)}. \quad (2.52)$$

Example.

$$u^2(p^2 + q^2 + 1) - 1 = 0.$$

From the PDE and the characteristic ODEs in parametric form we get

$$\dot{x} = 2u^2 p, \quad \dot{y} = 2u^2 q, \quad \dot{u} = 2(1 - u^2),$$

$$\dot{p} = -2\frac{p}{u}, \quad \dot{q} = -2\frac{q}{u}.$$

From these expressions we obtain

$$\dot{u}p + \dot{p}u = (up)\dot{\,} = -2pu^2, \quad (uq)\dot{\,} = -2qu^2.$$

This yields

$$x - a = p, \quad y - b = q,$$

where (a, b) are constants. Squaring and adding these expressions gives

$$(x - a)^2 + (y - b)^2 = u^2(p^2 + Q^2).$$

Using the PDE in the above equations yields

$$(x - a)^2 + (y - b)^2 + u^2 - 1 = 0.$$

If we set $a = \lambda$ and $b = f(\lambda)$, where f is a prescribed function of the parameter λ which we vary continuously, we obtain a one-parameter family of integral surfaces that are unit spheres in x, y, u space whose centers lie on the curve $b = f(\lambda)$ in the x, y plane. For the case where $b = 0$ the family of unit spheres have their centers lying on the x axis. By differentiating the above equation with respect to λ and eliminating λ from these two equations we get the surface of the envelope that is a "tubular surface," a unit cylindrical surface.

CHAPTER TWO

Canonical Form of a Second-Order PDE

A canonical form of an equation is defined as its simplest or standard form. We learned previously that the second-order PDEs can be classified into three types according to the roots of the characteristic equation (2.34), which is a quadratic equation for the slopes of the characteristics. This quadratic guides us in transforming the PDE into its canonical form.

Consider the PDE

$$au_{xx} + 2bu_{xy} + cu_{yy} + d = 0, \qquad (2.53)$$

where the coefficients a, b, c are constants and d is a linear function of u, p, q, x, y. We let $L[u]$ be the principal part of this PDE, which consists of the terms involving the highest derivatives.

$$L[u] = au_{xx} + 2bu_{xy} + cu_{yy}, \qquad (2.54)$$

where L is the linear second-order differential operator operating on u. The principal part determines the classification of the PDE. The characteristic equation (2.34) can be written in the following factored form:

$$\left(\frac{dy}{dx} - z_+\right)\left(\frac{dy}{dx} - z_-\right) = 0.$$

The solution of this characteristic ODE consists of two families of straight lines which are real unequal roots of the characteristic equation for the hyperbolic case, complex conjugate roots for the elliptic case, and a single real characteristic for the parabolic case:

$$y - z_+ = \alpha, \qquad y - z_- = \beta. \qquad (2.55)$$

(α, β) are called the characteristic coordinates. They can be written in the implicit form

$$\phi(x, y) = \alpha, \qquad \Psi(x, y) = \beta. \qquad (2.55a)$$

The first equation is a one-parameter (α) family of solutions of the characteristic equation $dy/dx = z_+$ while the second equation is a one-parameter family (β) of solutions of $dy/dx = z_-$. This means that

PDEs OF WAVE PROPAGATION

if β is kept constant and α is varied a family of straight lines $y - z_+ = \alpha$ is generated. Conversely, if α is kept constant and β varied a family of straight lines $y - z_- = \beta$ is generated. The two-parameter families of straight parallel lines of slopes z_+, z_- in the (x, y) plane are mapped into lines parallel to the coordinate axes in the (α, β) or characteristic plane. (We are discussing the hyperbolic case where the slopes are real and unequal constants.)

In order to transform the PDE given by eq. (2.36), $au_{xx} + 2bu_{xy} + cu_{yy} = 0$, into the characteristic coordinates (α, β), we must transform the partial derivative in the (x, y) plane into the corresponding derivatives in the (α, β) plane, by using the chain rule, for example, $u_x = u_\alpha \alpha_x + u_\beta \beta_x$, etc. We obtain

$$Au_{\alpha\alpha} + Bu_{\alpha\beta} + Cu_{\beta\beta} = 0. \tag{2.56}$$

Omitting the calculations, the coefficients become

$$A = a(\phi_x)^2 + 2b(\phi_x \phi_y) + c(\phi_y)^2 = Q(\phi_x, \phi_y)$$
$$= a - 2bz_+ + z_+^2 c,$$
$$2B = a(\phi_x \psi_x) + 2b(\phi_x \psi_y + \psi_x \phi_y) + c(\phi_y \psi_y) \tag{2.57}$$
$$= a - 2b(z_+ + z_-) + z_+ z_-,$$
$$C = a(\psi_x)^2 = 2b(\psi_x \psi_y) + c(\psi_y)^2 = Q(\psi_x \psi_y)$$
$$= a - 2bz_- + z_-^2 c.$$

We observe that A has the same quadratic form Q in (ϕ_x, ϕ_y) that C has in (ψ_x, ψ_y), and that B is a mixed quadratic in $(\phi_x, \ldots, \psi_y)$.

By differentiating the equations (2.55a) we obtain the following system:

$$\phi_x + \left(\frac{dy}{dx}\right)\phi_y = 0; \quad \frac{dy}{dx} = z_+ = -\frac{\phi_x}{\phi_y} \quad \text{on } \alpha = \text{const},$$

$$\psi_x + \left(\frac{dy}{dx}\right)\psi_y = 0; \quad \frac{dy}{dx} = z_- = -\frac{\psi_x}{\psi_y} \quad \text{on } \beta = \text{const}.$$
$$\tag{2.58}$$

The system given by (2.58) is of the form

$$\frac{dy}{dx} = -\frac{f_x}{f_y}, \qquad (2.59)$$

where $f = \phi$ for $dy/dx = z_+$ and $f = \psi$ for $dy/dx = z_-$. Inserting eq. (2.59) into the quadratic for the characteristic ODE (2.34) yields

$$a(f_x)^2 + 2b(f_x f_y) + c(f_y)^2 = 0. \qquad (2.60)$$

Comparing eq. (2.60) with (2.56) we see that, for the hyperbolic case, we have

$$A = Q(\phi_x, \phi_y) = 0, \qquad C = Q(\psi_x, \psi_y) = 0. \qquad (2.61)$$

Now the characteristic ODE associated with the PDE given by eq. (2.56) is

$$AZ^2 - 2BZ + C = 0, \qquad Z = \frac{d\beta}{d\alpha}. \qquad (2.62)$$

We saw that, for the hyperbolic case, $A = C = 0$. Therefore eq. (2.62) reduces to $Z = 0$. This means that the characteristic curves in the (α, β) or characteristic plane are straight lines parallel to the axes, as mentioned above. The canonical form of the PDE (2.56) then becomes

$$u_{\alpha\beta} = 0. \qquad (2.63)$$

This is the canonical form of the one-dimensional wave equation.

It is of interest to consider the other two types of PDEs. For the elliptic case we have $b^2 - ac < 0$, so that the roots of the characteristic ODE are complex conjugates. We may define the characteristic coordinates by the relations

$$y - z_+ = \alpha + i\beta, \qquad y - z_- = \alpha - i\beta.$$

It is easily seen that the canonical form of the PDE (2.56) for the elliptic case is

$$u_{\alpha\alpha} + u_{\beta\beta} = 0. \qquad (2.64)$$

For the parabolic case the roots z are equal so that $b^2 - ac = 0$. The PDE (2.56) becomes

$$u_{\alpha\alpha} = 0. \tag{2.65}$$

The canonical forms of the second-order linear PDE with variable coefficients are obtained by the same treatment as above, but the characteristic slopes are no longer straight lines. For the hyperbolic case the characteristics are curvilinear both in the (x, y) plane and the $(\alpha\beta)$ plane. The quasilinear PDE is much more complicated since the characteristic slopes in both the Cartesian and characteristic planes depend on the solution to the PDE.

Riemann's Method of Integration

In this section we shall investigate an important method—formulated by G. Riemann (the great German mathematician)—of solving PDEs of the hyperbolic type. Riemann devised his approach as part of his classic studies of wave propagation in gases. In order to fully appreciate his theory of PDEs we first review some ideas in vector analysis and the theory of linear operators, which can be found in classic textbooks. But we present this review here for the benefit of the reader.

Divergence Theorem

We first review the divergence theorem enunciated by Gauss. There is no appreciable loss of generality in studying the two-dimensional case. This important theorem transforms the divergence of a vector in a given region to the normal component of the vector integrated over the surface bounding the region. Specifically, let $\mathbf{V}(x, y)$ be a vector, let $\mathbf{i}, \mathbf{j}$ be unit vectors in the x, y directions, and let P, Q be the x, y components of $\mathbf{V}$ so that $\mathbf{V} = \mathbf{i}P + \mathbf{j}Q$. Let $\mathfrak{R}$ be a simply connected region bounded by a closed curve C, and let $\mathbf{n}$ be the unit outward normal along C. Let dA be the element of area and ds be the element of arc length of C. Using the definition of $\mathbf{V}$ the divergence theorem becomes

$$\iint_{\mathfrak{R}} \nabla \cdot \mathbf{V} \, dA = \oint \mathbf{V} \cdot \mathbf{n} \, ds, \tag{2.66}$$

CHAPTER TWO

where

$$\mathbf{V} = \mathbf{i}\frac{\partial}{\partial x} + \mathbf{j}\frac{\partial}{\partial y}, \qquad \mathbf{n} = \left(\frac{dy}{ds}\right)\mathbf{i} - \left(\frac{dx}{ds}\right)\mathbf{j},$$

and

$$\iint_{\Re} (P_x + Q_y)\,dx\,dy = \oint_C (P\,dy - Q\,dx), \qquad (2.67)$$

where $dy/ds, -dx/ds$ are the direction cosines of $\mathbf{n}$.

Now consider the scalar functions $u(x, y), v(x, y)$ with continuous second derivatives. Suppose we choose $\mathbf{V}$ such that $\mathbf{V} = u\,\nabla v$. Using the vector identity

$$\nabla \cdot (u\,\nabla v) = u\,\nabla^2 v + (\nabla u) \cdot (\nabla v), \qquad (2.68)$$

eq. (2.66) becomes

$$\iint_{\Re} [u\,\nabla^2 v + (\nabla u) \cdot (\nabla v)]\,dA = \oint_C \mathbf{n} \cdot u\,\nabla v\,ds. \qquad (2.69)$$

Equation (2.69) is the first form of *Green's* identity. We want to obtain a more symmetric form of Green's identity. To this end we interchange u and v in eq. (2.69) and subtract the result from (2.69). We thus obtain the second form of Green's identity:

$$\iint_{\Re} [u\,\nabla^2 v\,dA - v\,\nabla^2 u\,dA] = \oint_C \left[u\frac{\partial v}{\partial n} - v\frac{\partial u}{\partial n}\right]ds. \qquad (2.70)$$

If we write the integrand in the left-hand side in extended form we obtain the following identity:

$$u(v_{xx} + v_{yy}) - v(u_{xx} + u_{yy}) = (uv_x - vu_x)_x + (uv_y - vu_y)_y = P_x + Q_y,$$
$$P = uv_x - vu_x, \qquad Q = uv_y - vu_y.$$

$$(2.71)$$

We have found a vector $\mathbf{V}$ such that $\mathbf{V} = \mathbf{i}P + \mathbf{j}Q$. It follows that the integrand in the left-hand side of eq. (2.70) can be written as

$$u\,\nabla^2 v - v\,\nabla^2 u = P_x + Q_y = \nabla \cdot \mathbf{V}. \qquad (2.72)$$

74

PDEs OF WAVE PROPAGATION

The Laplacian operator $\nabla^2 = \partial^2/\partial x^2 + \partial^2/\partial y^2$ in the left-hand side is an example of a second-order linear differential operator in two dimensions so that $L[u] = u_{xx} + u_{yy}$, $L[v] = v_{xx} + v_{yy}$. This means that we can write eq. (2.70) in the form

$$\iint_{\mathfrak{R}} \{uL[v] - vL[u]\}\, dA = \oint \left[u\frac{\partial v}{\partial n} - v\frac{\partial u}{\partial n} \right] ds. \qquad (2.73)$$

The Laplacian operator $L = \nabla^2$ has the property that $L[u] = L[v]$ in order for the integrand $uL[v] - vL[u]$ to be equal to $\nabla \cdot \mathbf{V} = P_x + Q_y$.

In general, this property of the Laplacian operator does not have to be valid for all second-order linear differential operators L. Then $L[v]$ is replaced by a more general linear operator L^*, such that $uL^*[v] - vL[u] = \nabla \cdot \mathbf{V}$. L^* is called the *adjoint operator*. It operates on v and relates to the operator operating on L in the sense that $uL^*[v] - vL[u]$ equals the divergence of the vector $\mathbf{V}$. For the Laplacian we have $L[u] = L^*[v]$. The Laplacian operator is said to be *self-adjoint*, since the adjoint operator equals L.

An example of an operator that is not self-adjoint operating on u is given by

$$L[u] = u_{xx} - u_y. \qquad (2.74)$$

L is the one-dimensional unsteady Fourier heat transfer operator where x is space and y is time. Let us assume that $L^*[v]$ takes the form

$$L^*[v] = v_{xx} + (\alpha v)_x + (\beta v)_y, \qquad (2.75)$$

where α, β are functions of x, y to be determined. We now want to find the components of $P(x, y)$ and $Q(x, y)$ such that $uL^*[v] - vL[u] = P_x + Q_y$. We have

$$u[v_{xx} + (\alpha v)_x + (\beta v)_y] - v[u_{xx} - u_y] = P_x + Q_y.$$

This gives

$$\begin{array}{cc} P_x = uv_{xx} - vu_{xx}, & P = uv_x - vu_x, \\ Q_y = uv_y + vu_y, & Q_y = uv, \quad \alpha = 0, \quad \beta = 1. \end{array} \qquad (2.76)$$

CHAPTER TWO

This yields

$$L^*[v] = v_{xx} + v_y. \qquad (2.77)$$

The above adjoint expression $L^*[v]$ for the heat transfer operator has a rather curious physical interpretation: If u stands for temperature then the adjoint problem is the solution of the PDE $L^*[v] = 0$ where the "temperature" $v(x, y)$ is the solution of the unsteady heat transfer equation with time y "running backward." [The author has investigated the more general study of thermal stresses where $v(x, y)$ plays the role of a Green's function, which is not discussed here.]

We write the canonical form of a general second-order linear PDE as

$$u_{xy} + au_x + bu_y + cu = f, \qquad (2.78)$$

where a, b, c, f are known functions of (x, y). It is clear that (x, y) are the characteristic coordinates. For the special case where $a = b = c = f = 0$, eq. (2.78) reduces to the wave equation in one dimension. It is easily seen that the L operator operating on u for this case is self-adjoint.

Returning to the general case given by eq. (2.78), we now construct the adjoint operator L^*. To this end we assume the following expression for $L^*[v]$:

$$L^*[v] = v_{xy} + \alpha v_x + \beta v_y, \qquad (2.79)$$

where α and β are constants to be determined. The terms c, u, and f do not enter the following calculations. Recall that $uL^*[v] - vL[u] = P_x + Q_y$. Our task is now to determine P and Q. We have

$$u(v_{xy} + \alpha v_x + \beta v_y) - v(u_{xy} + au_x + bu_y)$$
$$= (-vu_y + auv)_x + (uv_x + buv)_y = P_x + Q_y, \qquad (2.80)$$

where

$$P = -vu_y + auv, \quad Q = uv_x + buv, \qquad (2.81)$$

PDEs OF WAVE PROPAGATION

$\alpha = -a$, and $\beta = -b$. $L^*[v]$ becomes

$$L^*[v] = v_{xy} - av_x - bv_y. \tag{2.82}$$

L need not be a second-order operator. As an example, take L as a first-order operator operating on u, giving

$$L[u] = au_x + bu_y. \tag{2.83}$$

It is easily seen that the adjoint operator operating on v is

$$L^*[v] = -(av_x + bv_y), \tag{2.84}$$

where a and b are constants. It follows that

$$P_x = -a(uv)_x, \quad Q_y = -b(uv)_y; \quad P = -auv, \quad Q = -buv. \tag{2.85}$$

We now have all the ammunition for Riemann's method of attack on the second-order PDE. Specifically, we consider Riemann's method of solving the Cauchy IV problem for the canonical form of the PDE given by eq. (2.78). The principal part of this equation is u_{xy}. The Cauchy problem may now be formulated as follows: Determine the solution $u = u(x, y)$ of eq. (2.78) that satisfies the IC u, p, q on the IV curve Γ which nowhere has the characteristic direction.

Figure 2.3 shows the portion of the IV curve in the (x, y) or rectangular plane on which are prescribed u, p, q. We consider points A and B near each other on Γ. Let the domain of the solution be the region above the IV curve. From A draw the characteristic line segment $\eta = $ const into the solution domain, and from B draw the line segment $\xi = $ const. These line segments will intersect at the field point P whose coordinates are (ξ, η). The domain of dependence of the field point P is the triangular region D bounded by the characteristics AP and BP and the portion of the IV curve intercepted by these characteristics. The range of influence is this portion of AB of Γ. Figure 2.3 thus describes the Cauchy IV problem, which is to find $u = u(x, y)$ at the field point P that satisfies the canonical PDE and the IC on the range of influence AB on Γ.

We now go to Riemann's method of solving this problem. Riemann called the function $v = v(x, y)$ a *test function* (recall that the adjoint

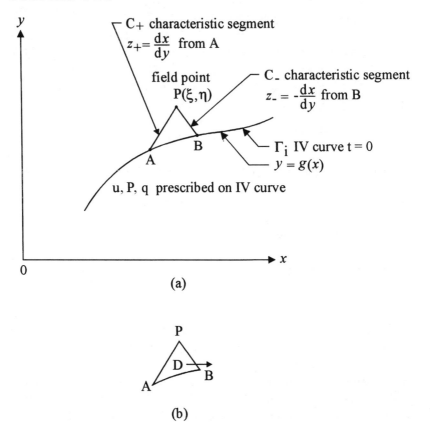

FIGURE 2.3. a. Field point P obtained from the intersection of C_+ from A and C_- from B. b. Domain of dependence (D) and range of influence (arc AB).

operator operates on v). Riemann's method is to calculate u_P at the field point given the IV in the range of influence. To this end he multiplied the PDE (2.78) by v, integrated over the domain of dependence D, transformed the integral by Green's identity so that u appears as a factor in the integrand, and tried to determine the test function to satisfy the conditions of the problem. The resulting algorithm involves integrating over the domain D making use of the test function v.

To carry out this procedure we first simplify the test function by setting $L^*[v] = 0$. We also have $L[u] = u_{xy} + au_x + bu_y + cu = f$. According to fig. 2.3, D contains the closed curve $ABPA$. $d\xi = 0$ on BP,

PDEs OF WAVE PROPAGATION

$d\eta = 0$ on PA, and $L[u] = f$. The divergence theorem eq. (2.73) then becomes

$$\iint_D vf\,dx\,dy = \int_{AB}\left[(-uv_y + auv)\,dy - (vu_x - buv)\,dx\right]$$
$$+ \int_{BP}(uv_y + auv)\,dy + \int_{PA}(vu_x + buv)\,dx. \quad (2.86)$$

We now integrate the line integral along the path PA by parts, in order to separate v from u in the integrand and isolate u_P. We obtain

$$\int_{PA} v(u_x + bu)\,dx = v_A u_A - v_P u_P - \int_{PA} u(v_x - bv)\,dx.$$

Inserting this expression into eq. (2.86) and isolating u_P yields

$$v_P u_P = v_A u_A + \int_{BA}\left[v(u_x + bu)\,dx + u(v_y - av)\,dy\right]$$
$$+ \int_{BP} u(v_y - av)\,dy - \int_{PA} u(v_x - bv)\,dx + \iint_D vf\,dx\,dy.$$
$$(2.87)$$

The plan is to get u_P in terms of the integral of the area D and the IC on Γ. This means that we would like the line integrals along the characteristics BP and PA to vanish. We set

$$v_y - av = 0 \text{ on } BP \text{ where } x = \xi,$$
$$v_x - bv = 0 \text{ on } PA \text{ where } y = \eta, \quad (2.88)$$
$$v_P = 1.$$

These equations define v. Equation (2.87) becomes

$$u_P = v_A u_A + \int_{AP}\left[v(u_x + bu)\,dx + u(v_y - av)\,dy\right] + \iint_D vf\,dx\,dy.$$
$$(2.89)$$

CHAPTER TWO

This is not the form we want since it involves u at only one end point A of the range of influence. We want a symmetric form that also involves u at the other end point, B. To this end we manipulate the integrand in the line integral in eq. (2.89), recognizing that $vu_x\, dx + uv_y\, dy = d(uv) - uv_x\, dx + vu_y\, dy$. Calling this line integral I and integrating by parts we obtain

$$I = u_B v_B - u_A v_A - \int_{AB} \left[u(v_x - bv)\, dx + v(u_y + au)\, dy \right].$$

Inserting this expression into eq. (2.89) yields

$$u_P = v_B u_B - \int_{AB} \left[u(v_x - bv)\, dx + v(u_y + au)\, dy \right] + \iint_D vf\, dx\, dy. \tag{2.90}$$

Finally, adding (2.89) and (2.90) we obtain

$$u_P = u(\xi, \eta) = \tfrac{1}{2}(v_A u_A + v_B u_B) + \tfrac{1}{2}\int_{AB} \left[(vu_x - uv_x + 2buv)\, dx \right.$$

$$\left. + (uv_y - vu_y - 2auv)\, dy \right] + \iint_D vf\, dx\, dy. \tag{2.91}$$

Equation (2.91) represents the solution u_P at the field point P in terms of the Riemann or test function v on Γ and in D. If the forcing function $f = 0$, then the solution can be obtained by integrating the data and v along the range of influence AB. Otherwise, we have the additional integral over D of vf.

The defining equations (2.88) for the Riemann function show that v is a continuous differentiable function of two sets of variables: (x, y), the coordinates of the IV curve Γ, and (ξ, η), the coordinates of the field point P. We may rewrite the defining equations of v as

$$L_{(x, y)}[v(x, y; \xi, \eta)] = 0,$$
$$v_y(\xi, y; \xi, \eta) - a(\xi, y)v(\xi, y; \xi, \eta) = 0 \quad \text{on char. } BP,$$
$$v_x(x, \eta; \xi, \eta) - b(x, \eta)v(x, \eta; \xi, \eta) = 0 \quad \text{on char. } PA, \tag{2.92}$$
$$v(\xi, \eta; \xi, \eta) = 1 \quad \text{at field point } P,$$

where we have used the obvious notation $L_{(x,y)}$ to mean differentiation with respect to x, y. The second and third equations of (2.92) are ODEs along the respective characteristics. Their solutions are

$$v(\xi, y; \xi, \eta) = e^{\int_\eta^y a(\lambda, \xi) d\lambda} \quad \text{along } BP,$$
$$v(x, \eta; \xi, \eta) = e^{\int_\xi^x b(\lambda, \eta) d\lambda} \quad \text{along } PA.$$
(2.93)

In summary, the system (2.91) and (2.93) allows us to solve the Cauchy IV problem in characteristic coordinates (x, y) in terms of the Riemann function $v(x, y; \xi, \eta)$.

As our first example we take the one-dimensional homogeneous wave equation in characteristic coordinates in the canonical form $L[u] = u_{xy} = 0$. It is clear that the operator L is self-adjoint so that $L^*[v] = v_{xy} = 0$ (from the definition of v, $L^*[v] = 0$). Then v has the form

$$v(x, y; \xi, \eta) = F(x; \xi, \eta) + G(y; \xi, n).$$

It is easily seen from the defining equations for the Riemann function v that $v(x, y; \xi, \eta) = 1$. Then the solution for u in characteristic coordinates for the one-dimensional wave equation becomes

$$u_P = u(\xi, \eta) = \tfrac{1}{2}[u_A + u_B] + \tfrac{1}{2}\int_\Gamma (u_x \, dx - v_y \, dy).$$

As another example, we consider the PDE

$$L[u] = u_{xy} + cu = 0,$$

where c is a given constant. It is easily seen that L is self-adjoint. Therefore $L^*[v] = v_{xy} + cv = 0$. The above PDE for u is a special case of eq. (2.78), where $a = b = 0$. Therefore the characteristic equations (2.93) tell us that along the characteristics BP and PA $v(0,0) = 1$. Moreover, v depends on the relative position of P with respect to A and B. Therefore v depends on the coordinates $\bar{x} = x - \xi$, $\bar{y} = y - \eta$.

CHAPTER TWO

v is symmetric in x and y. The simplest assumption is that

$$v = w(z), \quad \text{where } z = \overline{xy}.$$

This yields the ODE

$$z \frac{d^2 w}{dz^2} + \frac{dw}{dz} + cw = 0.$$

The solution for w that has no singularities at the origin is

$$w = J_0\left(\sqrt{4cxy}\right),$$

where $J_0(\ )$ is Bessel's function of the first kind of order zero.

PROBLEMS

1. Solve the following IV problem:

$$xu_x + yu_y = 0, \quad \text{IC: } u(x, 1) = f(x).$$

Let the IV curve be represented by $x = \tau$, $y = 1$. Solve the characteristic ODEs in parametric form (parameters are t, τ). Show and prove that the solution is $u(x, y) = f(x/y)$.

2. Solve the Cauchy problem

$$xu_x + yu_y = \alpha u, \quad \text{IC: } u = f(x) \text{ for } y = 1.$$

Use the parameters t, τ.

3. Solve the Cauchy problem

$$au_x + bu_y = 1, \quad \text{IC: } U = f(X) \text{ on the IV curve } y = mx.$$

4. Find the general solution of $yu_x - xu_y = 0$.

5. Work out the details of the quasilinear PDE given in the example in the section "Parametric Representation."

6. Show that the complete integral of the PDE

$$u^2(p^2 + q^2 + 1) = 1, \quad p = u_x, \quad q = u_y$$

is

$$(x - a)^2 + (y - b)^2 + u^2 = 1,$$

which is a two-parameter family of spheres of radius 1 in x, y, z space whose centers have the parameters a, b and lie in the x, y plane. As these parameters vary continuously a tubular surface is generated. Use characteristic theory to show that $d(x + up) = d(y + uq) = d(p/q) = 0$, obtaining $x - a = -up$, $y - b = -uq$, $p = cq, \ldots$ (a, b, c are constants of integration).

7. The PDE $p^2 + q^2 = (\nabla u)^2 = 1$ is the *eikonal equation* for u in two dimensions. Its importance in optics is shown in chapter 8. Show that the characteristic ODEs are

$$\frac{dx}{dy} = \frac{dp}{dq} = \frac{du}{1}, \quad \frac{dp}{dx} = 0, \quad \frac{dq}{dx} = 0.$$

From these ODEs show that the characteristic curves are straight lines, which turn out to be light rays for a medium of refractive index equal to unity. Show directly, by separating u into the sum of a function of x and a function of y, that $u = ax + \sqrt{1 - a^2}\, y + w(a)$. Differentiating this equation with respect to the parameter a, setting $a = p$, $q = \sqrt{1 - a^2}$, we get a family of characteristic strips (consider x as the independent variable). This leads to the characteristic curves.

8. (Problem 7 continued.) Eliminate p and q from the PDE and the characteristic ODEs and obtain for the functions $u(x)$ and $y(x)$ the Monge equation which belongs to our PDE:

$$\left(\frac{du}{dx}\right)^2 - \left(\frac{du}{dy}\right)^2 = 1.$$

Its solutions are those curves whose tangents form an angle of 45° with the x, y plane at all points. These are called *focal* or *caustic* curves and are involutes of the curves $u = $ const.

CHAPTER TWO

9. Find the general solution of $p^2 + yq - u = 0$. Show that the characteristic ODEs lead to $q = \text{const} = \alpha$, $du = \sqrt{u - \alpha y}\, dx + \alpha\, dy$, and that the solution is $u = \alpha y + \frac{1}{4}(x + \beta)^2$, where β is a constant of integration.

10. Classify the following PDEs according to type:

(a) $\sqrt{x^2 + y^2}\, u_{xx} + 2(y - x)u_{xy} + \sqrt{x^2 + y^2}\, u_{yy}$,

(b) $u_{rr} + \dfrac{1}{r} u_r - u_{tt} = 0$,

(c) $x^2 y u_{xx} + xy^2 u_{yy} = 0$,

(d) $u_{xy} - xy u_{yy} = 0$,

(e) $y u_{xx} - x u_{yy} + xy u_x = 0$.

11. Reduce the following PDEs to canonical form:

$$u_{xx} + 2u_{xy} + u_{yy} + (\sin x)u_x + y^2 u_y = 0,$$

$$u_{xx} + a u_{xy} + b u_{yy} + c u_x + d u_y = 0,$$

$$u_{xy} - 2xy^2 u_{yy} = 0,$$

$y u_{xx} + x u_{yy} = 0$, for $x > 0, y > 0$; $x > 0, y < 0$; $x < 0, y > 0$.

12. Work out the details leading to eq. (2.57).

CHAPTER THREE

The Wave Equation

Part I ONE-DIMENSIONAL WAVE EQUATION

In chapter 1 we derived the one-dimensional wave equation for longitudinal stress waves in a bar and transverse waves in a vibrating string. We also derived expressions for the wave speed in a bar, in a vibrating string, and for sound waves. We showed that all solutions of the wave equation $c^2 u_{xx} - u_{tt} = 0$ must be of the form $u(x,t) = f(x - ct) + g(x + ct)$ for arbitrary functions f and g which depend on the ICs. In chapter 2 we investigated the method of characteristics for first- and second-order PDEs and used this method to classify PDEs into hyperbolic, elliptic, and parabolic types. In part I of this chapter, we shall explore other methods of solving the one-dimensional wave equation, in addition to the method of characteristics, which will be more easily applicable to IV (Initial Value) and BV (Boundary Value) problems. In part II we shall investigate the wave equation in two and three dimensions.

FACTORIZATION OF THE WAVE EQUATION AND CHARACTERISTIC CURVES

In this section we cast the one-dimensional wave equation in the setting of chapter 2 where the concepts of the directional derivative and characteristic theory were discussed. The wave equation can be factored as follows:

$$(cD_x - D_t)(cD_x + D_t)u = 0, \tag{3.1}$$

where the partial derivative operators are given by

$$D_x \equiv \frac{\partial}{\partial x}, \quad D_t \equiv \frac{\partial}{\partial t}.$$

CHAPTER THREE

Set

$$(cD_x + D_t)u \equiv cu_x + u_t = v. \tag{3.2}$$

The directional derivative of u is du/dt. This is the total derivative of u with respect to t in the direction along the tangent to a curve in (x,t) space. This curve is called a characteristic curve. It plays an important role in the PDEs of wave propagation and will be described below. (See chapter 2 for more details.) We expand this directional derivative and obtain

$$\dot{u} = u_x \dot{x} + u_t, \tag{3.3}$$

where

$$\dot{u} \equiv \frac{du}{dt}, \qquad \dot{x} \equiv \frac{dx}{dt}.$$

If eq. (3.3) for the directional derivative is correlated with the PDE given by eq. (3.2), then we must have

$$\dot{x} = c. \tag{3.4}$$

Moreover, if the second factor of the left-hand side of eq. (3.1) is equal to zero then $v = 0$ in eq. (3.2), which gives

$$cu_x + u_t = 0. \tag{3.5}$$

Inserting eq. (3.3) into eq. (3.5) tells us that the directional derivative $du/dt = 0$ or $u = \text{const}$ when eq. (3.5) holds. Integrating (3.4) yields

$$x - ct = \xi, \tag{3.6}$$

where ξ is the constant of integration. The ordinary differential equation (ODE) given by eq. (3.4) is called the characteristic ODE corresponding to the PDE given by eq. (3.5). The integral of eq. (3.4) is given by eq. (3.6) and is the characteristic curve (straight line in this case) in the (x,t) plane. u has a constant amplitude everywhere on this characteristic curve defined by a given value of ξ. Therefore the amplitude of u is $u(\xi)$. On another characteristic the value of the amplitude of u is another constant. Suppose we are given the initial value (IV) of u, for example, $u(x,0) = f(x)$, a prescribed function of x. $f(\xi)$ is the initial amplitude of u at $x = \xi$. [It is clear from eq. (3.6) that $x = \xi$ when

THE WAVE EQUATION

$t = 0$.) Everywhere on this characteristic stemming from the point $(\xi, 0)$ as t increases, $u = u(\xi)$. Using eq. (3.6) the solution of the PDE (3.5) along the characteristic is

$$u(x,t) = f(x - ct).$$

This tells us that in time t the wave has traveled a distance ct along the x axis so that at time t, u is the same function of $f(x - ct)$ that it was at $t = 0$, namely, $f(x)$, which means the wave shape is invariant along the characteristic. By varying the parameter ξ, we generate a one-parameter family of straight-line characteristics in the (x, t) plane emanating from the x axis. This is seen in fig. 3.1a, where t is plotted as the ordinate against the abscissa x [the standard method of expressing characteristics in the (x, t) plane]. Three characteristic lines are plotted for the parameters (ξ_1, ξ_2, ξ_3), whose slope is dt/dx. This formulation is an example of the Cauchy initial value (IV) problem, which plays a pivotal role in the theory of wave propagation, and which we discussed more completely in chapter 2.

Now set the first factor of the left-hand side of eq. (3.1) equal to zero. Using eq. (3.2) we obtain

$$cv_x - v_t = 0. \tag{3.7}$$

This is a first-order linear PDE for $v(x,t)$. It is easily seen by expanding the directional derivative dv/dt that, if we impose the restriction that $dv/dt = 0$, we must obtain

$$\dot{x} = -ct. \tag{3.8}$$

Equation (3.8) is the characteristic ODE corresponding to the PDE given by eq. (3.7), where $-\dot{x}$ is substituted for c. This type of wave travels with unchanging form along the negative x axis with a wave velocity $c = -dx/dt$ and is called a regressing wave. The integral of eq. (3.8) is given by

$$x + ct = \eta, \tag{3.9}$$

where η is the constant of integration. By varying η we generate a one-parameter family of characteristic lines. Figure 3.1b gives a plot of three characteristic lines for this case stemming from the x axis, whose slope is $-dt/dx$.

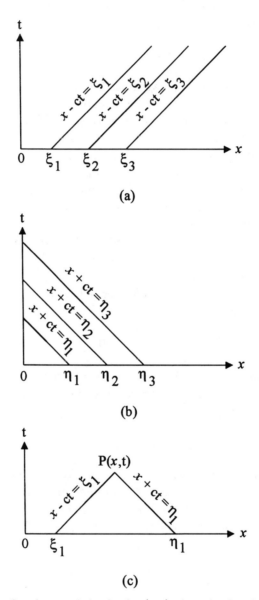

FIGURE 3.1. a. C_+ characteristics in the (x, t) plane. b. C_- characteristics in the (x, t) plane. C_+ and C_- characteristics intersecting at field point P.

By combining figs. 3.1a and 3.1b we obtain fig. 3.1c, which involves a progressing wave traveling with positive velocity and a regressing wave traveling with negative velocity as a result of setting both factors in the left-hand side of eq. (3.1) equal to zero.

This geometric approach to solving the wave equation tells us that we have two sets of parameters (ξ, η). Fixing η and varying ξ generates a family of characteristics which we call C_+ since they all have a positive slope. Varying η for a fixed ξ generates a family of C_- characteristics which has a negative slope. (ξ, η) are called characteristic coordinates. The Cauchy IV problem for $v(x,t)$ tells us that the IV of v, for example, $v(x,0) = g(x)$, is given by prescribing $g(x)$ on the x axis. This is seen in fig. 3.1b, where the characteristics stem from three values of $g(x)$ on the x axis.

The above analysis suggests a transformation of coordinates from (x,t) to (ξ, η) by

$$x - ct = \xi,$$
$$x + ct = \eta. \tag{3.10}$$

It is easily seen that the wave equation $c^2 u_{xx} - u_{tt} = 0$ in the (x,t) plane is transformed into the following wave equation in the (ξ, η) plane:

$$xu_{\xi\eta} = 0. \tag{3.11}$$

From eq. (3.11) we get the general solution for u in terms of the characteristic coordinates:

$$u(\xi, \eta) = f(\xi) + g(\eta), \tag{3.12}$$

where f and g are arbitrary functions of their arguments. Note that when $t = 0$ the IV of the progressing wave is $f(x)$ and that of the regressing wave is $g(x)$. The solution of the wave equation in the (x,t) plane is obtained by inserting eq. (3.10) into the right-hand side of eq. (3.12). We get

$$u(x,t) = f(x - ct) + g(x + ct). \tag{3.13}$$

It is easily shown that any functions of the form $f(x - ct)$ and $g(x + ct)$ satisfy the one-dimensional wave equation $c^2 u_{xx} - u_{tt} = 0$. It follows that *only* functions having the arguments $(x - ct, x + ct)$ satisfy this

CHAPTER THREE

wave equation. This means that the general solution of the wave equation is given by eq. (3.13) where $f(x - ct)$ and $g(x + ct)$ depend on the IC and BC. Note that the wave equation is second order in x and in t and thus requires knowing two ICs and two BCs. For the case of an IV problem the range of x is $-\infty < x < \infty$. (The BCs at infinity require boundedness of the solution.)

We illustrate the above analysis by examining the behavior of a progressing wave for an initial pulse. Let the initial waveform be

$$f(x) = \begin{cases} \sin 2\pi x & \text{for } 0 \leq x \leq \pi, \\ 0 & \text{for } x > \pi. \end{cases}$$

This gives us a positive sinusoidal pulse of unit wavelength. The solution for u in the appropriate range is

$$u(x,t) = \sin 2\pi(x - ct).$$

To study the behavior of the wave for this special IC we plot u versus x for the pulse for several values of t. Figure 3.2a is a plot of $u(x, 0)$ versus x. Figure 3.2b is a plot of $u(x, t)$ versus x, where each point on the pulse in fig. 3.2a moves a distance equal to ct to the right. Figure 3.2c is a plot of $u(x, 2t)$, where each point of the pulse at $t = 0$ moves a distance $2ct$. These simple diagrams further illustrate the invariance of the wave form as t increases. A similar example can be constructed for a regressive wave using any IC you might make up.

VIBRATING STRING AS A COMBINED IV AND BV PROBLEM

We now discuss another method of solving the wave equation, namely, the method of separation of variables. We shall use the vibrating string as an example. Lord Rayleigh mentioned the importance of studying the vibrating string. Indeed, he said, "Among vibrating bodies there are none that occupy a more prominent position than the Stretched String. From earliest times they have been employed for musical purposes...." Consider a stretched string of length L fixed at both ends and given an arbitrary initial displacement and initial velocity. The mathematical

THE WAVE EQUATION

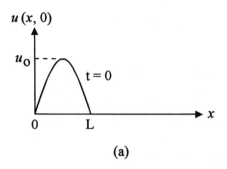

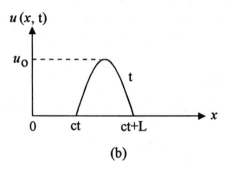

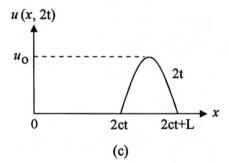

FIGURE 3.2. Progressing pulse of length L and amplitude u_0 moving with velocity c. a. Initial position of pulse. b. Pulse at time t. c. Pulse time $2t$.

CHAPTER THREE

formulation is

$$c^2 u_{xx} - u_{tt} = 0, \quad 0 \leq x \leq L, \quad t \geq 0,$$
$$t = 0: \quad u(x,0) = f(x), \quad u_t(x,0) = g(x), \quad (3.14)$$
$$x = 0: \quad u(0,t) = 0,$$
$$x = L: \quad u(L,t) = 0.$$

The initial displacement is a prescribed function $f(x)$ and the initial velocity is a prescribed function $g(x)$. The BCs are homogeneous in the sense that the string is fixed at $x = 0$ and L.

We wish to find the lateral displacement of the string $u(x, t)$ that satisfies the problem given by eq. (3.14). We shall first investigate the method of separation of variables, which is quite a useful approach to a linear system where the space and time variables can be separated, and then compare this technique to another method, D'Alembert's approach, which can be compared to the method of characteristics.

Separation of Variables

We look for solutions of the wave equation (3.14) which are of the form

$$u(x,t) = v(x)w(t). \quad (3.15)$$

This means that we attempt to find solutions of the wave equation that are the product of a function v of x and a function w of t. The only reason for this assumption is that it works. It will turn out that these functions will satisfy certain ODEs which are easier to solve than a PDE. Inserting eq. (3.2) into the wave equation gives

$$c^2 w\ddot{v} - v\ddot{w} = 0,$$

where $\ddot{v} \equiv d^2v/dx^2$, $\ddot{w} \equiv d^2w/dt^2$. We rewrite this equation as

$$c^2 \frac{\ddot{v}}{v} = \frac{\ddot{w}}{w}.$$

The left side of this equation varies only with x and is independent of t, while the right side varies with t and is independent of x. The only way this can be is that both sides must be equal to the same constant, which we call k. We have

$$c^2 \frac{\ddot{v}}{v} = \frac{\ddot{w}}{w} = k,$$

which yields the following two ODEs:

$$\ddot{v} - \frac{k}{c^2}v = 0, \qquad 0 \le x \le L \qquad (3.16)$$

and

$$\ddot{w} - kw = 0, \qquad 0 \le t < \infty. \qquad (3.17)$$

Since eq. (3.16) is a second-order ODE for $v(x)$, it requires two BCs. These are supplied by the BCs of the problem, which are $u(0,t) = u(L,t) = 0$. Therefore, setting $v(0) = v(L) = 0$ satisfies these BCs, and the solution of eq. (3.16) must be of the form

$$v(x) = A\cos(\lambda x) + B\sin(\lambda x), \qquad \lambda = \frac{1}{c}\sqrt{-k},$$

where $k < 0$. A, B, and λ are constants to be determined. It is clear that if $k > 0$ the solution for v is given in terms of exponentials, so that both BCs cannot be satisfied. In order to satisfy the BC $v(0) = 0$ we must have $A = 0$. To satisfy the BC $v(L) = 0$ we must have $\sin(\lambda L) = 0$. This condition implies that $\lambda L = n\pi$ for $n = 0, 1, 2, 3, \ldots$. This gives a discrete set of values of λ which depend on n, namely,

$$\lambda = \lambda_n = \frac{n\pi}{L}, \qquad k = -c^2\lambda_n. \qquad (3.18)$$

For each value of n there is a solution of the form

$$v_n(x) = A_n \sin\left(\frac{n\pi x}{L}\right). \qquad (3.19)$$

The term λ_n is called the *n*th *eigenvalue* corresponding to the *n*th *eigenfunction* v_n given by eq. (3.19). Each of the n eigenfunctions corresponding to its eigenvalue separately satisfies the two BCs. This means that the spatial part of the solution depends only on the particu-

CHAPTER THREE

lar values of λ, the eigenvalues, and the corresponding eigenfunctions. No other values of λ yield solutions. We also see that the constant A_n depends on n.

A similar analysis of the ODE (3.17) tells us that the nth eigenfunction $w_n(t)$ corresponding to the eigenvalue λ_n is given by

$$w_n(t) = C_n \cos\left(\frac{n\pi ct}{L}\right) + D_n \sin\left(\frac{n\pi ct}{L}\right). \tag{3.20}$$

Having determined the nth eigenfunction for the spatial part of the solution $v_n(x)$ and the time part $w_n(t)$, we form their product, which is the nth eigenfunction of the solution $u_n(x,t) = v_n(x)w_n(t)$. The solution of our mixed BV, IV problem is the linear combination

$$\sum_{n=0}^{\infty} u_n(x,t). \tag{3.21}$$

Inserting the above results into eq. (3.21) yields

$$u(x,t) = \sum_{n=1}^{\infty} A_n \sin\frac{n\pi x}{L}\left(C_n \cos\frac{n\pi ct}{L} + D_n \sin\frac{n\pi ct}{L}\right).$$

By combining coefficients the solution becomes

$$u(x,t) = \sum_{n=1}^{\infty} \sin\frac{n\pi x}{L}\left(a_n \cos\frac{n\pi ct}{L} + b_n \sin\frac{n\pi ct}{L}\right). \tag{3.22}$$

Equation (3.22) is the solution that satisfies the BCs of our problem. But we have yet to satisfy the two ICs. The discrete sets of constants a_n, b_n will be determined from these ICs. Note that the solution (3.22) is expressed as an infinite series whose time part is a complete sinusoidal function of t (linear combination of sines and cosines) which is modulated by a sine function of x (because of the two BCs). This series is called a *Fourier series* in x and t. It represents a standing wave pattern (see chapter 1, "Interference Phenomena"). The theory of Fourier series tells us that this series is convergent. In order to satisfy the two ICs we need to calculate the particle velocity u_t. This is given by

$$u_t = \sum_{n}\left(\frac{n\pi c}{L}\right)\sin\frac{n\pi x}{L}\left(-a_n \sin\frac{n\pi ct}{L} + b_n \cos\frac{n\pi ct}{L}\right). \tag{3.23}$$

THE WAVE EQUATION

We are now in a position to apply the ICs $u(x,0) = f(x)$ and $u_t(x,0) = g(x)$. Setting $t = 0$ in eq. (3.22) gives

$$f(x) = \sum_n a_n \sin \frac{n\pi x}{L}. \qquad (3.24)$$

To calculate g we set $t = 0$ in eq. (3.23) and obtain

$$g(x) = \sum_n \left(\frac{n\pi c}{L}\right) b_n \sin \frac{n\pi x}{L}. \qquad (3.25)$$

We now determine the a_ns that satisfy eq. (3.24) and the b_ns that satisfy eq. (3.25). This can easily be done since they are the *Fourier coefficients* of $f(x)$ and $g(x)$, respectively. We obtain

$$a_n = \frac{2}{L} \int_0^L f(x) \sin \frac{n\pi x}{L} dx, \qquad (3.26)$$

$$b_n = \frac{2}{L} \int_0^L g(x) \sin \frac{n\pi x}{L} dx. \qquad (3.27)$$

To show that the solution (3.22) represents progressing and regressing waves, we make use of the trigonometric identities

$$\sin \alpha \cos \beta = \tfrac{1}{2}[\sin(\alpha + \beta) + \sin(\alpha - \beta)],$$

$$\sin \alpha \sin \beta = \tfrac{1}{2}[\cos(\alpha - \beta) - \cos(\alpha + \beta)].$$

Equation (3.22) becomes

$$u(x,t) = \frac{1}{2} \sum_n a_n \left[\sin \frac{n\pi}{L}(x - ct) + \sin \frac{n\pi}{L}(x + ct)\right]$$

$$+ \frac{1}{2} \sum_n b_n \left[\cos \frac{n\pi}{L}(x - ct) - \cos \frac{n\pi}{L}(x + ct)\right]. \qquad (3.28)$$

It is easily seen that this form of the solution satisfies the BCs and ICs of the problem.

CHAPTER THREE

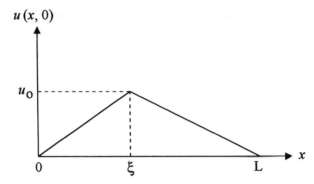

FIGURE 3.3. Initial configuration of harp string of length L and amplitude u_0.

Example. A plucked harp string. As an example of the method of separation of variables we investigate a taut harp string of length L initially plucked a lateral distance from the x axis at the point $x = \xi$ by an amount u_0. The string is initially at rest. Thus, at $t = 0$ the string forms two sides of a triangle as shown in fig. 3.3. We point out that $0 < u_0/L \ll 1$ in order for the linear wave equation to be valid. $u(x, 0) = f(x)$ becomes

$$f(x) \begin{cases} \dfrac{u_0}{\xi} x, & 0 \le x \le \xi, \quad 0 < \xi < L, \\ \dfrac{u_0}{L - \xi}(L - x), & \xi \le x \le L. \end{cases} \tag{3.29}$$

Since the initial velocity of the string $g(x) = 0$, this means that the Fourier coefficients $b_n = 0$. The Fourier coefficients a_n become

$$a_n = \frac{2}{L} \int_0^L f(x) \sin \frac{n\pi x}{L} dx$$

$$= \frac{2}{L} \int_0^\xi \frac{u_0}{\xi} x \sin \frac{n\pi x}{L} dx + \frac{2}{L} \int_\xi^L \left(\frac{u_0}{L - \xi} L - x \sin \frac{n\pi x}{L} \right) dx$$

$$= \frac{2 u_0}{\xi(L - \xi) n^2} \sin \frac{n\pi \xi}{L}, \quad n = 1, 2, \ldots. \tag{3.30}$$

THE WAVE EQUATION

The solution for the lateral displacement of the plucked harp string them becomes

$$u(x,t) = \frac{2u_0}{\xi(L-\xi)} \sum_n \frac{1}{n^2} \sin\frac{n\pi\xi}{L} \sin\frac{n\pi x}{L} \cos\frac{n\pi ct}{L}. \quad (3.31)$$

This solution is the sum of progressing and regressing waves, and can be expressed as such by using the appropriate trigonometric identities. We get

$$u(x,t) = \frac{u_0}{\xi(L-\xi)} \sum_n \frac{1}{n^2} \left[\sin\frac{n\pi}{L}(x-ct) + \sin\frac{n\pi}{L}(x+ct) \right]. \quad (3.32)$$

D'ALEMBERT'S SOLUTION TO THE IV PROBLEM

Jean Le Rond D'Alembert, a French mathematical physicist (1717–83), devised a method of solving the Cauchy IV Problem. For convenience we reformulate this problem as a pure IV problem by using the whole range of x and imposing the mild condition that the solution be bounded at infinity.

$$c^2 u_{xx} - u_{tt} = 0, \quad -\infty < x < \infty, \quad 0 < t < \infty,$$
$$\text{ICs:} \quad u(x,0) = f(x), \quad u_t(x,0) = g(x), \quad -\infty < x < \infty. \quad (3.33)$$

To solve this problem we start with the general solution $u(x,t) = F(x - ct) + G(x + ct)$. Setting $t = 0$ and using the IC $u(x,0) = f(x)$ gives

$$F(x) + G(x) = f(x), \quad (3.34)$$

while we satisfy the second IC by differentiating the general solution with respect to t and then setting $t = 0$. We get

$$-cF'(x) + cG'(x) = g(x). \quad (3.35)$$

Equations (3.34) and (3.35) are two equations for the unknown functions F and G in terms of the given ICs $f(x)$ and $g(x)$. To determine

CHAPTER THREE

them we first differentiate eq. (3.34) and obtain

$$F' + G' = f'(x). \qquad (3.36)$$

Next, we rewrite eq. (3.35) as

$$-F' + G' = \frac{1}{c}g(x). \qquad (3.37)$$

Subtracting (3.37) from (3.36) gives

$$F' = \frac{1}{2}f'(x) - \frac{1}{2c}g(x), \qquad (3.38)$$

while adding (3.36) and (3.37) yields

$$G' = \frac{1}{2}f'(x) + \frac{1}{2c}g(x). \qquad (3.39)$$

To obtain $F(x)$ and $G(x)$ we integrate eqs. (3.38) and (3.39) from 0 to x and account for the constants of integration, which we call α and β. We get

$$F(x) = \frac{1}{2}f(x) - \frac{1}{2c}\int_0^x g(z)\,dz + \alpha,$$

$$G(x) = \frac{1}{2}f(x) + \frac{1}{2c}\int_0^x g(z)\,dz + \beta.$$

Now $F(x) + G(x) = f(x)$, so that $\alpha + \beta = 0$ or $-\beta = \alpha$.

If we now consider the x, t plane where t is the ordinate, we have made use the of two ICs on the x axis. The above results yield $F(x)$ and $G(x)$, still on the x axis. To obtain the solution at any field point off the x axis (for $t > 0$), we merely observe that the function $F = F(x - ct)$ and $G = G(x + ct)$ (recall how we introduced these arbitrary functions). This gives us

$$F(x - ct) = \frac{1}{2}f(x - ct) - \frac{1}{2c}\int_0^{x-ct} g(z)\,dz + \alpha$$

and

$$G(x + ct) = \frac{1}{2}f(x + ct) + \frac{1}{2c}\int_0^{x+ct} g(z)\,dz - \alpha.$$

Adding these two equations yields

$$F(x - ct) + G(x + ct)$$
$$= \frac{1}{2}[f(x - ct) + f(x + ct)] + \frac{1}{2c}\int_{x-ct}^{x+ct} g(z)\,dz. \quad (3.40)$$

We know that the general solution is $u(x,t) = F(x - ct) + G(x + ct)$. Therefore, upon using eq. (3.40), D'Alembert's solution becomes

$$u(x, y) = \frac{1}{2}[f(x - c) + f(x + ct)] + \frac{1}{2c}\int_{x-ct}^{x+ct} g(z)\,dz. \quad (3.41)$$

Equation (3.41), D'Alembert' solution, has a very interesting geometric interpretation. This is seen in fig. 3.4. At the field point $P:(x,t)$ we draw a C_+ characteristic, whose equation is $x - ct = \xi$, to where it intersects the x axis at $x = \xi$. Also from P we draw a C_- characteristic, $x + ct = \eta$, to where it intersects the x axis at $x = \eta$. The expression $(1/2)[f(x - ct) + f(x + ct)] = (1/2)[f(\xi) + f(\eta)]$ is clearly the average value of $u(x,t)$ of the IC $f(x)$ at $x = \xi$ and $x = \eta$ if the IC $g(x) = 0$. The expression $(1/2c)\int_\xi^\eta g(z)\,dz$, where $x - ct = \xi$ in the lower limit of integration and $x + ct = \eta$ in the upper limit, is the solution for $u(x,t)$ if the IC $f(x) = 0$. Notice that this part of the solution involves integrating the IC $g(x)$ along the interval cut out by the intersection of the two characteristics from P (the interval $[\xi, \eta]$). Adding both these results gives D'Alembert's solution of the Cauchy IV problem for both the given ICs.

The concept of a *characteristic net* will now be discussed. Consider again the quarter plane $x \geq 0$, $t \geq 0$. Let $u = f(x - ct)$. Then $f(x - ct)$ is constant along the line $x - ct = \xi =$ const, which is a C_+ characteristic. For each assigned value of ξ there is a C_+ curve. Along this characteristic $u = f(x - ct)$ (a progressing wave) travels undistorted from its IC where $u = f(\xi)$ n the x axis. A similar analysis holds for $u = f(x + ct)$ (a regressing wave) along the family of C_- characteris-

CHAPTER THREE

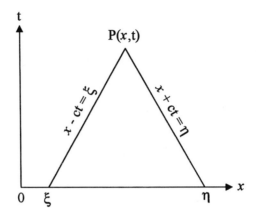

FIGURE 3.4. Field point (x, t) resulting from the intersection of a C_+ and a C_- characteristic stemming from the IC at $x = \xi, \eta$.

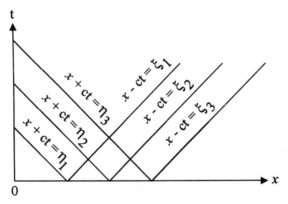

FIGURE 3.5. Three C_+ and C_- characteristic curves illustrating a characteristic net in the (x, t) plane.

tics, stemming from $t = 0$ where $u = g(\eta)$. These two distinct families of characteristic curves form a characteristic net, shown in fig. 3.5, where a set of three C_+ and C_- characteristics are shown. For the IC $u(x, 0) = 0$, $u_t(x, 0) = g(x)$ we have $(1/2C)\int_\xi^\eta g(z)\,dz$, where the initial data (ID) involve integration of g along the segment (ξ, η), as explained above. Note that it is a property of second-order hyperbolic PDEs that two families of distinct characteristics exist (the slopes being different).

THE WAVE EQUATION

DOMAIN OF DEPENDENCE AND RANGE OF INFLUENCE

We stress again that disturbances on the x axis (the IV curve) caused by the ID (initial data) are propagated only along the characteristics. These disturbances may be a sound wave source, initial waveforms for a vibrating string, etc. This property of hyperbolic PDEs that disturbances can only propagate along characteristics sets them off from elliptic and parabolic PDEs. It has a very important physical consequence, namely, that disturbances along the IV curve in the x, t plane divide the region $t > 0$ into two subregions, the region affected by the disturbances due to the ID, the *range of influence* of segment (ξ, η), and the region not affected by the ID. Specifically, we see that in fig. 3.6 the ID on the segment $[\xi, \eta]$ of the x axis affects the field point $P:(x,t)$. Moreover, all field points inside the triangle bounded by the characteristics C_+, C_- and the segment $[\xi, \eta]$ (for $t > 0$) are affected by the ID on the segment $[\xi, \eta]$ (fig. 3.6). In addition, no field points in the region $t > 0$ outside this triangle are affected by the ID on the segment $[\xi, \eta]$.

From this analysis we deduce the interesting fact that the solution for u at the field point P depends only on the initial data on the segment of the initial line obtained by drawing the characteristics backward from P to the initial line. The triangle $(\xi P \eta)$ formed by the two characteristics from the field point P and the segment (ξ, η) is called the *domain of dependence* of P as seen in fig. 3.6. This has an important consequence: (1) All initial data on the initial line segment (ξ, η) affect the solution in the domain of dependence. (2) No initial data outside this initial segment can affect the solution in the range of influence.

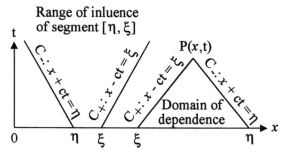

FIGURE 3.6. Range of influence of segment $[\eta, \xi]$ and domain of dependence.

CHAPTER THREE

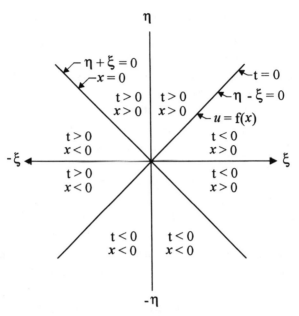

FIGURE 3.7. Various regions in the (ξ, η) plane.

CAUCHY IV PROBLEM REVISITED

It is instructive to solve the Cauchy IV problem given by eq. (3.33) in the (ξ, η) or characteristic plane. To this end we make use of the mapping given by eq. (3.10). We first observe that the regions of the (x, t) plane map into the corresponding regions of the (ξ, η) plane as shown in fig. 3.7, which shows the corresponding regions in the (x, t) plane. Note that the inverse mapping of eq. (3.10) is

$$x = \frac{1}{2}(\xi + \eta), \qquad t = \frac{1}{2c}(\eta - \xi).$$

The formulation of the Cauchy IV problem in characteristic coordinates becomes

$$u_{\xi\eta} = 0 \qquad \text{for } \eta > \xi, \quad t > 0,$$

$$u = \frac{1}{2}f\left(\frac{1}{2}(\xi + \eta)\right), \qquad -u_\xi + u_\eta = \frac{1}{c}g\left(\frac{1}{2}(\xi + \eta)\right) \quad (3.42)$$

$$\text{for } \eta - \xi = 0, \quad t = 0,$$

since $u_t = c(-u_\xi + u_\eta)$. The region $t \geq 0$ maps into the region $\eta \geq \xi$. The line $t = 0$ (x axis) maps into the line $\eta - \xi = 0$ (as seen in fig. 3.7). The solution of eq. (3.42) is

$$u(\xi, \eta) = F(\xi) + G(\eta), \tag{3.43}$$

where F and G depend on the ICs. To satisfy these ICs we must have

$$F(\xi) + G(\eta) = f(\xi) \qquad \text{for } \eta - \xi = 0,$$

$$-F'(\xi) + G'(\eta) = \frac{1}{c}g(\xi) \qquad \text{for } \eta - \xi = 0.$$

Integrating the second equation gives

$$-F(\xi) + G(\eta) = \frac{1}{c}\int_0^\xi g(z)\,dz - F(0) + G(0) \qquad \text{for } \eta = \xi.$$

We shall see that the constants $F(0), G(0)$ do not appear in the solution. Solving for F and G in terms of the IC yields

$$F(\xi) = \frac{1}{2}f(\xi) - \frac{1}{2c}\int_0^\xi g(z)\,dz + F(0) - G(0),$$

$$G(\eta) = \frac{1}{2}f(\eta) + \frac{1}{2c}\int_0^\eta g(z)\,dz - F(0) + G(0).$$

Adding these two equations gives the solution

$$u(\xi, \eta) = \frac{1}{2}[f(\xi) + f(\eta)] + \frac{1}{2c}\int_\xi^\eta g(z)\,dz. \tag{3.44}$$

This is D'Alembert's solution of Cauchy's IV problem in characteristic coordinates.

We now give some examples of the superposition principle which, in general, states that for a linear PDE a linear combination of solutions is also a solution. This is not true for a nonlinear system.

First we use the superposition principle to prove the *uniqueness* of the Cauchy IV problem. Suppose there are two different solutions of the Cauchy problem given by $u_1(x, t)$ and $u_2(x, t)$. Then $v(x, t) = u_1(x, t) - u_2(x, t)$ is also a solution. It is easily seen that the ICs on v are $v(x, 0) = 0$, $v_t(x, 0) = 0$. Clearly, this means that $v(x, t) = 0$ or

CHAPTER THREE

$u_1(x, t) = u_2(x, t)$, which contradicts the assumption that u_1 and u_2 are different, which proves uniqueness.

We now use the superposition principle to prove that the solution of the Cauchy IV problem with two nonhomogeneous ICs (not zero) is the sum of the solutions for the following two problems: (a) The Cauchy IV problem for $g(x) = 0$. (b) The Cauchy IV problem for $f(x) = 0$. Suppose (a) that $v(x, t)$ is the solution of the Cauchy IV problem for the ICs $v(x, 0) = f(x)$, $v_t(x, 0) = 0$ (one homogeneous and one nonhomogeneous IC), and (b) that $w(x, t)$ is the solution for the ICs $w(x, 0) = 0$, $w_t(x, 0) = g(x)$. Using the superposition principle we state that, if $u = v + w$, then u is the solution of the Cauchy IV problem for the more general nonhomogeneous ICs given by $f(x)$ and $g(x)$. Clearly, this result tells us that the Cauchy problem defined for two nonhomogeneous ICs is the sum of two simpler problems, each of which involves only one nonhomogeneous IC. This illustrates the principle of superposition of solutions. We can do this because the wave equation is linear.

We now give a *geometric interpretation* of the Cauchy IV problem in terms of traveling waves in the (x, t) plane. The general solution is $u(x, t) = F(x - ct) + G(x + ct)$. The case $u(x, t) = F(x - ct)$, as mentioned above, is a progressing wave whose waveform does not change with time. This is seen in fig. 3.8a, where the waveform expressed by $F(x - ct)$ travels a distance $c(t_2 - t_1)$ from its position at t_1 to its position at t_2. The case $u(x, t) = G(x + ct)$ is a regressing wave, as shown in fig. 3.8b.

Suppose we have a solution of the Cauchy problem for the ICs $u(x, 0) = f(x)$, $u_t(x, 0) = 0$. As mentioned above, D'Alembert's solution is $u(x, t) = (1/2)[f(x - ct) + f(x + ct)]$. Suppose the initial waveform is given by fig. 3.9a. This could be the case of an oscillating string initially at rest with an initial waveform given by $f(x)$ as shown in fig. 3.9a. Figure 3.9b shows the waveform (solution) for t small enough so that the two waves $f(x - ct)$ and $f(x + ct)$ do not separate. A small portion of the waveform maintains the same shape and amplitude while the rest of the wave splits into half the amplitude the progressing and regressing waves travel in opposite directions with the same wave speed c and same waveform. At a later time the progressing and regressing waveforms separate; they have the same waveforms but half the amplitude of the original wave and travel in opposite directions with the same wave speed. A three-dimensional plot of $u(x, t) = (1/2)[f(x - ct) + f(x + ct)]$ is given in fig. 3.10 in the form of an isometric drawing.

THE WAVE EQUATION

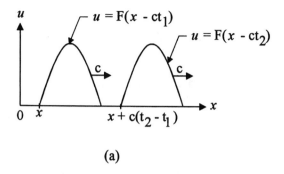

(a)

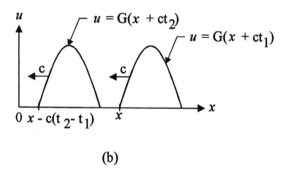

(b)

FIGURE 3.8. a. Progressing wave at times t_1 and $t_2 > t_1$. b. Regressing wave.

Next we consider the case where the ICs are $u(x,0) = 0$, $u_t(x,0) = g(x)$. This could be the case of an oscillating string initially in equilibrium, but receiving an initial velocity whose waveform is $g(x)$. Figure 3.11 shows an example of $g(x)$ versus x for $t = 0$ as a step function (a positive constant between $-a$ and a) and the domain of dependence and range of influence. Recall that the solution for this case is given by $u(x,t) = (1/2c)\int_{x-ct}^{x+ct} g(z)\,dz$. Figure 3.12 shows a three-dimensional isometric plot of $u(x,t)$.

SOLUTION OF WAVE PROPAGATION PROBLEMS BY LAPLACE TRANSFORMS

Another method of solving wave propagation problems is by Laplace transforms. The Laplace transform is a linear integral operator that operates on a function of t, for example, and converts it to a corre-

CHAPTER THREE

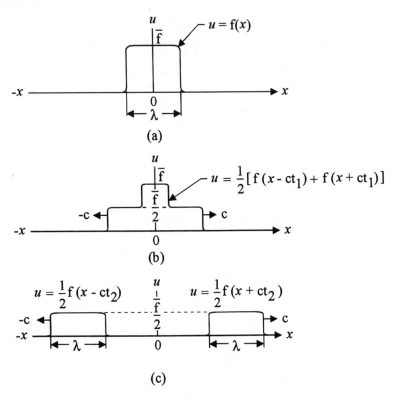

FIGURE 3.9. Waveform $u = f(x)$ at (a) $t = 0$, (b) $t = t_1$ showing partial separation, (c) $t_2 > t$ showing two separate traveling waves.

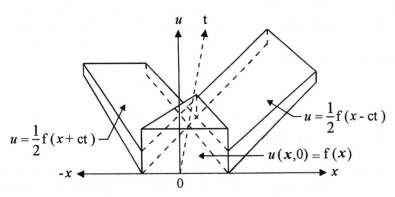

FIGURE 3.10. Isometric drawing of surfaces $u = (1/2)[f(x - ct) + f(x + ct)]$.

THE WAVE EQUATION

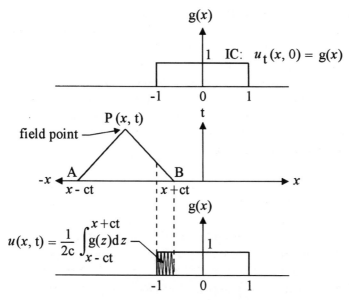

FIGURE 3.11. Domain of dependence (A, B) and range of influence (triangle APB) and area intercepted by the step function $g(x)$ that represents the solution at (x, t).

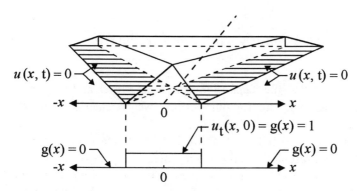

FIGURE 3.12. Isometric drawing of the surface

$$u(x,t) = \frac{1}{2c} \int_{x-ct}^{x+ct} g(z)\, dz, \qquad g(x) = \begin{cases} 1, & |x| \le 1, \\ 0, & |x| > 0. \end{cases}$$

107

CHAPTER THREE

sponding function of a parameter s called the Laplace transform of the function. For example, the Laplace transform of an ODE is an algebraic equation in the Laplace transform which reduces the complexity of the problem. Taking the inverse Laplace transform yields the solution of the ODE. The Laplace transform method is useful in linear PDEs since it reduces the number of independent variables. For example, the Laplace transform of the one-dimensional wave equation with respect to t is reduced to an ODE, the three-dimensional wave equation is reduced to a two-dimensional PDE, etc.

Laplace Transforms

In order to get an understanding of the Laplace transform method we start by considering a function of a single variable $u = u(t)$ defined on the interval $0 \leq t < \infty$ such that the integral $\int_0^T |u(t)|\, dt$ exists for every positive T, meaning that $u(t)$ is absolutely integrable. Also let $u(t) = O(e^{s_0 t})$ as $t \to \infty$, where s_0 is a real constant (positive, negative, or zero). This means that $u(t)$ is of exponential order as t goes to infinity; that is, a constant s_0 exists such that $e^{-s_0 t}|u(t)|$ is bounded for all t greater than some number T, however large. Thus $u(t)$ does not grow more rapidly than $Ae^{s_0 t}$, where A is a positive constant. With these mild conditions on $u(t)$ we are ready to define the Laplace transform of $u(t)$.

The Laplace transform of $u(t)$ is

$$\tilde{u}(s) = \mathscr{L}\{u(t)\} = \int_0^\infty e^{-st} u(t)\, dt, \qquad (3.45)$$

where s is called the *Laplace transform parameter* and is, in general, a complex number. But for our purpose we consider s to be real and positive.

Equation (3.45) tells us that $\tilde{u}(s)$ is the Laplace transform of $u(t)$, which is represented by the linear integral operator $\mathscr{L} = \int_0^\infty e^{-st}\{\ \}\, dt$ operating on $u(t)$ to produce $\tilde{u}(s)$. Clearly, this Laplace transform operator exists since we postulated that $u(t)$ is absolutely integrable. We may write

$$\tilde{u}(s) = \mathscr{L}\{u(t)\}, \qquad u(t) = \mathscr{L}^{-1}\{\tilde{u}(s)\}, \qquad (3.46)$$

where the operator $\mathscr{L}^{-1}$ is the *inverse transform operator*, which transforms $\tilde{u}(s)$ to $u(t)$.

We state the following property of the Laplace transform (hereinafter called the "transform"): Let $u^{(n)}(t)$ be the nth derivative of $u(t)$. Then

$$\mathscr{L}\{u^{(n)}(t)\} = s^n \tilde{u}(s) - \sum_{k=0}^{n-1} s^k u^{(n-k-1)}(0), \qquad (3.47)$$

where $u^{(k)}(0)$ is the kth derivative of u evaluated at $t = 0$. This property of the transform is most important in solving linear ODEs. It tells us that the nth derivative of $u(t)$ is transformed into $s^n \tilde{u}(s)$ so that an nth-order ODE is transformed into an nth-degree polynomial in s times the transform plus terms involving the ICs.

To prove that $\mathscr{L}\{\dot{u}(t)\} = s\tilde{u}(s) - u(0)$, we integrate by parts, obtaining

$$\mathscr{L}\{\dot{u}(t)\} = \int_0^\infty e^{-st} \dot{u}(t) \, dt = \lim_{T \to \infty} \int_0^T e^{-st} \dot{u}(t)$$

$$= \lim_{T \to \infty} \left\{ e^{-st} u(t) \Big|_0^T + s \int_0^T e^{-st} u(t) \, dt \right\}$$

$$= \lim_{T \to \infty} \left\{ e^{-sT} u(T) - u(0) + s \int_0^T e^{-st} u(t) \, dt \right\}$$

$$= s\tilde{u}(s) - u(0).$$

To prove that $\mathscr{L}\{\ddot{u}(t)\} = s^2 \tilde{u}(s) - su(0) - \dot{u}(0)$, we let $v(t) = \dot{u}(t)$. Then

$$\mathscr{L}\{v(t)\} = s\tilde{v}(s) - v(0)$$

$$= s[s\tilde{u}(s) - u(0)] - \dot{u}(0)$$

$$= s^2 \tilde{u}(s) - su(0) - \dot{u}(0).$$

The transform of higher-order derivatives may be proved by mathematical induction.

To get the transform of $u = \sin \omega t$, we recognize that u satisfies the following ODE with the appropriate ICs:

$$\ddot{u} + \omega^2 u = 0, \qquad u(0) = 0, \qquad \dot{u}(0) = \omega.$$

CHAPTER THREE

The transform of this system is

$$s^2\tilde{u} - \omega + \omega^2\tilde{u} = 0, \qquad \tilde{u} = \frac{\omega}{s^2 + \omega^2}.$$

We leave it to a problem to show that the transform of $\cos \omega t$ is $s/(s^2 + \omega^2)$.

We now state the translation property of the transform:

$$\mathscr{L}\{e^{at}u(t)\} = \tilde{u}(s - a).$$

This means that if $u(t)$ is multiplied by e^{at}, then the transform is translated or shifted by an amount $-a$ (the constant a is independent of s; it may be a function of another variable such as a space variable for a PDE). We leave the proof to a problem.

Let $H(t - a)$ be the Heaviside unit step function defined by

$$H(t - a) = \begin{cases} 0, & t < a, \\ 1, & t > a. \end{cases}$$

It operates on $u(t)$, transforming it to $u(t - a)$ for $t > a$, 0 for $t < a$. The transform of $H(t - a)$ is

$$\mathscr{L}\{H(t - a)\} = \frac{e^{-st}}{s}.$$

The proof is left to a problem.

As an example, take $\tilde{u}(s) = 1/s^2$; then $u(t) = t$. We have

$$\mathscr{L}^{-1}\left\{\frac{e^{-as}}{s^2}\right\} = H(t - a)t = \begin{cases} t - a, & t > a, \\ 0, & t < a. \end{cases}$$

With this brief introduction to Laplace transforms, we turn to using this method to solve the linear one-dimensional wave equation. To this

THE WAVE EQUATION

end we apply the Laplace transform to the function of t, leaving the function of x alone. The result is an ODE in x with the parameter s. We the take the inverse transform to obtain the solution.

APPLICATIONS TO THE WAVE EQUATION

Semi-Infinite Bar

We consider a semi-infinite bar with a BC of a constant stress applied to the front end ($x = 0$) and zero ICs. The formulation of this BV problem is

$$c^2 u_{xx} - u_{tt} = 0, \quad 0 \leq x < \infty, \quad 0 \leq t < \infty,$$
$$\text{ICs:} \quad u(x,0) = 0, \quad u_t(x,0) = 0, \quad (3.48)$$
$$\text{BCs:} \quad E u_x(0,t) = \sigma_0, \quad u(\infty,t) = 0,$$

where σ_0 is the stress at the front end $x = 0$ (taken here to be a compressive stress (although the same analysis holds for a tensile stress) and E is the Young's modulus of the bar. Hooke's law applied to a one-dimensional medium tells us that the stress is equal to the Young's modulus times the strain (see chapter 1). The physical significance of this problem is to find the particle displacement $u(x,t)$ of a semi-infinite one-dimensional bar initially with zero displacement and at rest where a compressive stress of magnitude σ_0 is applied to the front end. The wave or phase velocity is given by $c = \sqrt{E/\rho}$.

The transformed problem is obtained by taking the Laplace transform of the PDE and the IC and BC. The transformed problem becomes

$$\frac{d^2 \tilde{u}(x,s)}{dx^2} - \lambda^2 \tilde{u}(x,s) = 0, \quad \lambda = \frac{s}{c},$$
$$E \frac{d\tilde{u}(0,s)}{dx} = \frac{\sigma_0}{s}, \quad \tilde{u}(\infty,s) = 0. \quad (3.49)$$

The solution of the transformed problem is

$$\tilde{u}(x,s) = \frac{c\sigma_0}{s^2 E} e^{-(s/c)x}. \quad (3.50)$$

CHAPTER THREE

The transform of the stress $\sigma(x, t)$ in the bar is

$$E \frac{d\tilde{u}(x, s)}{dx} = \frac{\sigma_0}{s} e^{-(s/c)x}. \tag{3.51}$$

The inverse transform of $\tilde{u}(x, s)$ is the solution

$$u(x, t) = \frac{\sigma_0 c}{E} H\left(t - \frac{x}{c}\right) t$$

$$= \begin{cases} \dfrac{\sigma_0 c}{E}\left(t - \dfrac{x}{c}\right), & t > \dfrac{x}{c}, \\ 0, & t < \dfrac{x}{c}, \end{cases} \tag{3.52}$$

where $H(t - a)$ is the Heaviside step function. The stress $\sigma(x, t)$ produced in the bar due to the applied force at the front end is obtained by taking the inverse transform of eq. (3.51), yielding

$$\sigma(x, t) = E u_x(x, t) = H\left(t - \frac{x}{c}\right)\sigma_0. \tag{3.53}$$

Setting $x = 0$ in eq. (3.53) yields the BC at the front end.

We now give a physical interpretation of the solution $u(x, t)$ for the particle displacement. During the time when the constant force is applied to the front end, the particle displacement is zero until a time $t_0 = x_0/c$, after which it increases linearly with time (as shown by the solution). We see that t_0 is the time for the signal at the front end (produced by the applied stress) to propagate a stress wave a distance x_0. x_0 is to be considered as the position at time t_0 of the stress wave which travels through the bar with a velocity c. At t_0 the region of the bar to the left of x_0 under stress, while the region to the right is in equilibrium (unstressed), where $u = 0$. x_0 is to be considered the position of the wave front at time t_0. This is consistent with the method of characteristics previously described. In this method we plot the C_+ characteristic in the part of the (x, t) plane where $t \geq 0$, $x \geq 0$. The leading characteristic is given by $x - ct = 0$. In the region where $x - ct > 0$, $u(x, t) = 0$, the *dead space*. In the region $x - ct < 0$, the values of $u(x, t)$ are picked up by the C_+ characteristics emanating from the t axis (the front end) and are proportional to t. This is shown in fig. 3.13. In the region $x - ct < 0$, typical characteristic curve called

THE WAVE EQUATION

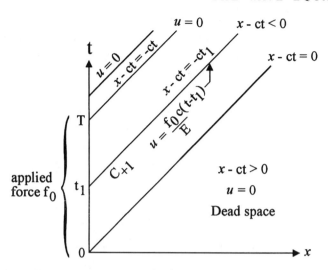

FIGURE 3.13. Characteristics in the (x, t) plane for a force f_0 applied from $t = 0$ to $t = T$ on a semi-infinite bar.

C_{+1} is shown. Along this characteristic line the solution is $u = f_0(c/E)(t - t_1)$. Note that at $x = 0$, $t = t_1$ and $u = 0$.

We note that due to the linearity of the wave equation there will come a time (if the applied stress remains) when the values of u will not be of small amplitude and hence the solution will be invalid.

Suppose the applied stress at the front end is zero at some time $t = T$ and remains zero for all $t > T$. Then, since the characteristics emanating from $x = 0$ for $t > T$ carry zero information, the region above the characteristic $ct - x = T$ is a dead space, in which $u(x, t) = 0$.

Finite One-Dimensional Bar

We consider the BV problem for a finite one-dimensional bar of length L with a constant velocity applied to the front end while the back end remains rigid. The formulation is

$$c^2 u_{xx} - u_{tt} = 0, \quad 0 \le x \le L, \quad 0 \le t < \infty,$$

$$\text{ICs:} \quad u(x, 0) = 0, \quad u_t(x, 0) = 0, \quad (3.54)$$

$$\text{BCs:} \quad x = 0, \quad u_t(0, t) = v_0, \quad x = L, \quad u(L, t) = 0.$$

CHAPTER THREE

An appropriate scaling parameter must multiply $u(x,t)$ to make it small enough to satisfy the linear theory. But this will be neglected here. This problem is important in the field of continuum mechanics and also has military applications. The transformed problem is

$$\frac{d^2\tilde{u}(x,s)}{dx^2} - \lambda^2 \tilde{u}(x,s) = 0, \qquad \lambda = \frac{s}{c}, \qquad (3.55)$$

$$\text{BCs: } s\tilde{u}(0,s) = \frac{v_0}{s}, \qquad \tilde{u}(L,0) = 0.$$

The solution of this ODE is

$$\tilde{u}(x,s) = A \cosh \lambda x + B \sinh \lambda x,$$

where A and B are functions of s to be determined from the BCs. They are $A = v_0/s^2$, $B = -(1/s^2)v_0 \coth \lambda L$. The solution of the transformed problem is

$$\tilde{u}(x,s) = \frac{v_0 \sinh \lambda L \cosh \lambda x - \cosh \lambda L \sinh \lambda x}{\sinh \lambda L}$$

$$= \frac{v_0}{s^2} \frac{\sinh \lambda(L-x)}{\sinh \lambda L}. \qquad (3.56)$$

The solution for $u(x,t)$ is obtained by taking the inverse transform of $\tilde{u}(x,s)$. In this approach we expand the right-hand side of eq. (3.56) as follows:

$$\frac{v_0}{s^2} \frac{\sinh \lambda(L-x)}{\sinh \lambda L} = \frac{v_0}{s^2} \left[\frac{e^{-\lambda x} - e^{-\lambda(2L-x)}}{1 - e^{-2\lambda L}} \right]$$

$$= \frac{v_0}{s^2} [e^{-\lambda x} - e^{-\lambda(2L-x)}] \sum_{n=0}^{\infty} e^{-2n\lambda L}$$

$$= \frac{v_0}{s^2} \sum_{n=0}^{\infty} [e^{-\lambda(2nL+x)} - e^{-\lambda(2(n+1)L-x)}] = u(x,t).$$

$$(3.57)$$

Using the fact that $\mathscr{L}^{-1}\{e^{-as}\} = H(t-a)$, eq. (3.57) becomes

$$u(x,t) = [ct-x] - [ct-(2L-x)] + [ct-(2L+x)]$$
$$- [ct-(4L-x)] + [ct-(4L+x)]$$
$$- \cdots - [ct-(2nL-x)] + [ct-(2n+x)] + \cdots, \quad (3.58)$$

where the function $[ct - y]$ is defined by

$$[ct - y] = \begin{cases} 0 & \text{for } ct \le y, \\ ct - y & \text{for } ct \ge y. \end{cases} \quad (3.59)$$

Note that the solution given by eq. (3.58) is not really an infinite series since, according to the definition of the function $[ct - y]$, for any finite value of t there are only a finite number of terms in the series. The regions for the first few terms are

0 terms: region (0), $ct \le x$
1 term: region (1), $x \le ct \le 2L - x$
2 terms: region (2), $2L - x \le ct \le 2L + x$
3 terms: region (3), $2L + x \le ct \le 4L - x$
4 terms: region (4), $4L - x \le ct \le 4L + x$.

Using eq. (3.58), and taking into account the definition of the function $[ct - y]$, the solution and the range of time in the various regions are shown in table 3.1.

In region 1 the solution is obtained from the C_+ characteristics emanating from $x = 0$ and there is no reflection off $x = L$. In region 2 the solution is obtained by adding the data from the C_+ characteristic in region 1 to that obtained from the C_- characteristic emanating from the reflection at $x = L$ in region 2. In region 3 the solution is obtained

TABLE 3.1
Various Regions in the (x, t) Plane for the Finite Bar of Length L

Region	$u(x, t)$	Range of t at $x = 0$	Range of t at $x = L$
0	0	0	0 to L/c
1	$ct - x$	0 to $2L/c$	L/c
2	$2(L - x)$	$2L/c$	L/c to $3L/c$
3	$ct - 3x$	$2L/c$ to $4L/c$	$3L/c$
4	$4(L - x)$	$4L/c$	$3L/c$ to $5L/c$

CHAPTER THREE

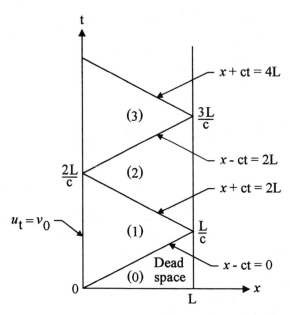

FIGURE 3.14. Stress waves in a finite bar. BCs: $x = 0$, $u(0, t) = v_0$; $x = L$, $u(L, t) = 0$.

by adding the data from region 1, the reflection from $x = L$ in region 2 and the data from $x = 0$ in region 3. In region 4 the solution is obtained by adding the data from region 1, the reflection from $x = L$ in region 2, the reflection at $x = 0$ in region 3, and the reflection at $x = L$ in region 4. This information is shown in fig. 3.14 by the appropriate characteristics.

NONHOMOGENEOUS WAVE EQUATION

We now consider the Cauchy IV problem with a prescribed external force per unit mass $F(x, t)$. As an example, take the semi-infinite string where the front end (at $x = 0$) oscillates laterally due to an external spring. In this example the external force is an oscillatory one applied to the front end. The IV problem of $u(x, t)$ is formulated as follows:

$$c^2 u_{xx} - u_{tt} = F(x, t), \quad -\infty < x < \infty, \quad t \geq 0,$$
$$t = 0, \quad u(x, 0) = f(x), \quad u_t(x, 0) = g(x), \quad (3.60)$$
$$x = 0, \quad u(0, t) = 0; \quad x = \infty, \quad u(\infty, t) = 0.$$

THE WAVE EQUATION

Notice that we have nonhomogeneous ICs. We simplify and solve the problem for homogeneous ICs. The reason is the superposition principle: For homogeneous BCs the general solution of a linear nonhomogeneous PDE with nonhomogeneous ICs is equal to the solution of the homogeneous PDE with nonhomogeneous ICs plus the solution of the nonhomogeneous wave equation with homogeneous ICs. We can also formulate this principle for a nonhomogeneous PDE with nonhomogeneous BCs and then combine it with the corresponding IV problem.

Applying this principle to our IV problem, we set $u(x,t) = v(x,t) + w(x,t)$, where $v(x,t)$ is the solution of the homogeneous wave equation with nonhomogeneous IC, and $w(xt)$ is the solution to the nonhomogeneous wave equation with homogeneous ICs. The solution of $v(x,t)$ is given in two places in this chapter: eq. (3.28), as the sum of progressive and regressive waves given as Fourier series, and eq. (3.41), D'Alembert's solution, which was interpreted geometrically in terms of characteristic theory. This formulation of the solution is more pertinent to our investigation of the solution of $w(x,t)$.

The formulation of the problem for $w(x,t)$ is

$$c^2 w_{xx} - w_{tt} = F(x,t), \quad t > 0, \quad 0 \le x < \infty,$$
$$t = 0, \quad w(x,0) = 0, \quad w_t(x,0) = 0, \quad (3.61)$$
$$x = 0, \quad w(0,t) = 0; \quad x = \infty, \quad w(\infty,t) = 0.$$

To solve this problem we transform it to characteristic coordinates. Recall that

$$\xi = x - ct, \quad \eta = x + ct; \quad 2x = \xi + \eta, \quad 2t = -\xi + \eta. \quad (3.62)$$

The wave equation in characteristic coordinates for this problem is

$$4 w_{\xi \eta} = F\left(\frac{\xi + \eta}{2}, \frac{-\xi + \eta}{2} \right). \quad (3.63)$$

We may either integrate the wave equation with respect to ξ and then η or vice versa.

Recall that the characteristic region of the solution corresponding to the region $0 \le x < \infty$, $0 \le t < \infty$ is shown in fig. 3.7, where various regions in the (ξ, η) plane are shown in the section "Cauchy IV Problem Revisited." $t = 0$ maps into the line $\xi = \eta$ and the region of

CHAPTER THREE

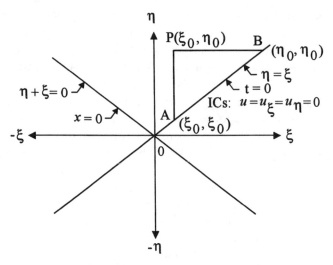

FIGURE 3.15. Field point P in the characteristic plane.

solution in the (x, t) plane maps into the part of the (ξ, η) plane above the line $\xi = \eta$. The ICs in the (x, t) plane are $u(x, 0) = 0$, $u_t(x, 0) = 0$. It is clear that we can add the condition $u_x(x, 0) = 0$ for all x, from $u(x, 0) = 0$ since t is held fixed. Since the mapping from the (x, t) to the (ξ, n) plane is linear, it follows that u_ξ and u_η are linear combinations of u_t and u_x. Therefore the ICs in the characteristic plane become

$$u = 0, \quad u_\xi = 0, \quad u_\eta = 0 \quad \text{for } \xi = \eta. \tag{3.64}$$

The integration with respect to the characteristic coordinates is made clear by studying the diagram in fig. 3.15. From the field point $P(\xi_0, \eta_0)$, drop a vertical line downward until it hits the initial line ($\xi = \eta$) at point $A(\xi_0, \xi_0)$. Then draw a horizontal line to the right until it hits the initial line at point $B(\eta_0, \eta_0)$. Integrating the wave equation in eq. (3.63) first with respect to η means integrating along the line AP. We get

$$4u_\xi(\xi_0, \eta_0) = \int_{\xi_0}^{\eta_0} F(\xi_0, \bar{\eta}) \, d\bar{\eta}$$

since one IC is $u_\xi(\xi_0, \xi_0) = 0$.

Now we integrate with respect to ξ from point B in the figure along the horizontal line to point P_0 (the field point), and use the IC

THE WAVE EQUATION

$u(\eta_0, \eta_0) = 0$. We then obtain the solution in characteristic coordinates:

$$u(\xi, \eta) = -\tfrac{1}{4} \iint_\Delta F(\xi, \eta)\, d\xi\, d\eta, \tag{3.65}$$

where the integration is taken over the triangle APB formed by the vertical characteristic from A to P and the horizontal characteristic from B to P. To map the solution to the (x, t) plane, we use the transformation of the element of area in the characteristic plane to the corresponding element in the (x, t) plane. We get

$$d\xi\, d\eta = J\, dx\, dt,$$

where J is the Jacobian of the transformation and is given by

$$J = \frac{\partial(\xi, \eta)}{\partial(x, t)} = \begin{vmatrix} \xi_x & \xi_t \\ \eta_x & \eta_t \end{vmatrix} = \begin{vmatrix} 1 & -c \\ 1 & c \end{vmatrix} = 2c.$$

(A detailed explanation of the Jacobian as a mapping function is given in chapter 4 in the section "Characteristics in the Hodograph Plane.") Then the solution in the (x, t) plane becomes

$$u(x, t) = -\frac{c}{2} \int_0^t dt \int_{x-ct}^{x+ct} F(\bar{x}, \bar{t})\, d\bar{x}. \tag{3.66}$$

The transformation from characteristic coordinates to rectangular coordinates maps the domain of dependence, which is the triangle formed by the characteristics $\xi = \text{const}$, $\eta = \text{const}$, and the portion of the line $\xi = \eta$ intercepted by these two characteristic, into the corresponding triangle formed by the C_+ and C_- characteristics and the portion of the x axis intercepted by these two characteristics. This is shown in fig. 3.16, where ct is plotted versus x. (We use ct instead of t as the ordinate so that distance is plotted against distance.) The triangle $P'A'B'$ is equilateral since in this plane the characteristics have slopes ± 1. The triangle PAB in the characteristic plane maps into the triangle $P'A'B'$ in the (x, ct) plane. The coordinates of A' are $(x - ct, 0)$ and the

119

CHAPTER THREE

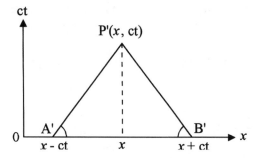

FIGURE 3.16. Mapping of fig. 3.15 into the (x, ct) plane.

coordinates of B' are $(x + ct, 0)$, since the C_+, C_- characteristics have the equations $x - ct = \xi$, $x + ct = \eta$, respectively. The zone of influence on the x axis is $A'B'$ and has a length equal to x.

WAVE PROPAGATION THROUGH MEDIA WITH DIFFERENT VELOCITIES

Thus far we have been dealing with a homogeneous medium. We now consider one-dimensional waves propagating in a medium with two different wave or phase velocities. Important engineering problems arise from this situation. Let $x = 0$ be the boundary between the two velocities. Let $x > 0$ be region (1) and let $x < 0$ be region (2). We formulate the problem:

$$c^2 u_{xx} - u_{tt} = 0, \quad c = c_1 \quad \text{for } x > 0, \quad c = c_2 \quad \text{for } x < 0,$$

$$\text{IC:} \quad u = u_t = 0 \quad \text{for } x > 0, \quad t = 0,$$

$$u = F_2(-x) \quad \text{for } x < 0, \quad t = 0, \tag{3.67}$$

$$F_2(\xi) = 0 \quad \text{for } \xi < 0,$$

$$\text{Jump Conds:} \quad u_1 - u_2 = [u] = 0,$$

$$u_{1x} - u_{2x} = [u_x] = 0 \quad \text{for } x = 0.$$

THE WAVE EQUATION

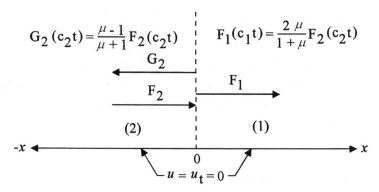

FIGURE 3.17. One-dimensional wave propagation in regions of two different wave speeds: c_1 in region (1), c_2 in region (2). F_2 is the prescribed incident wave in (2); F_1 is the transmitted wave in (1); G_2 is the reflected wave in (2); F_1 and G_2 are unknowns. $\mu = c_1/c_2$.

Clearly, $u = 0$ for $x > c_1 t$; hence

$$u_1(x,t) = F_1(c_1 t - x) = F_1(-\xi) \quad \text{for } 0 < x < c_1 t,$$
$$u_2(x,t) = F_2(c_2 t - x) + G_2(x + c_2 t) \quad (3.68)$$
$$= F_2(-\xi) + G_2(\eta) \quad \text{for } x < 0.$$

The situation is shown in fig. 3.17. The incident wave F_2 traveling to the right in (2) is prescribed. The transmitted wave F_1 traveling to the right in (1) and the reflected wave G_2 traveling to the left in (2) are to be found.

The jump conditions for $x = 0$ yield

$$F_2(c_2 t) = G_2(c_2 t) = F_1(c_1 t),$$
$$c_2[F_2'(c_2 t) - G_2'(c_2 t)] = c_1 F_1'(c_1 t). \quad (3.69)$$

This gives

$$F_1(c_1 t) = \frac{2\mu}{1 + \mu} F_2(c_2 t),$$
$$G_2(c_2 t) = \frac{\mu - 1}{\mu + 1} F_2(c_2 t), \quad \mu = \frac{c_1}{c_2}. \quad (3.70)$$

CHAPTER THREE

The "transmission" coefficients $2\mu/(1 + \mu)$ and $(\mu - 1)/(\mu + 1)$ indicate how much of the incoming wave is transmitted and reflected. If c_1/c_2 is large, the incoming wave is transmitted with its amplitude nearly doubled, and the amplitude of the reflected wave is nearly that of the incident wave. If c_1/c_2 is small, the amplitude of the transmitted wave is small and nearly all of the transmitted wave is reflected with opposite sign.

ELECTRICAL TRANSMISSION LINE

In this section we shall investigate an electrical transmission line. We define a transmission line to be a double electrical parallel cable extending along the infinite x axis. Let $u(x, t)$ be the voltage between opposite points of the lines and let $i(x, t)$ be the current in the cables (flowing in opposite directions in each cable). Let the parameters (assumed to be constants) C, R, L, and G be the capacitance, resistance, inductance, and leakage conductance per unit length, respectively, of the transmission line.

The differential equations for u and i are obtained by considering a closed rectangular element of the circuit between points x and $x + dx$ and applying Ohm's and Kirchhoff's laws and the law of conservation of charge to this circuit. The voltage difference per unit length from x to $x + dx$ is u_x. According to Ohm's law, the voltage due to the resistance of this element of length is Ri. According to Faraday the voltage due to induction L is Li_t. Ohm's and Kirchhoff's laws lead to a first-order PDE for u and i,

$$Li_t + Ri + u_x = 0,$$

while the conservation of charge leads to another first-order PDE,

$$Cu_t + Gu + i_x = 0.$$

Differentiating the first equation with respect to x and the second with respect to t, subtracting, and then eliminating i from the first equation yields the second-order PDE for v:

$$u_{xx} - LCu_{tt} - (RC + LG)u_t - RGu = 0.$$

Using the notation

$$c^2 = \frac{1}{LC}, \quad a = RC + LG, \quad b = RG,$$

the PDE becomes

$$u_{xx} - c^{-2}u_{tt} - au_t - bu = 0. \tag{3.71}$$

Equation (3.71) is called the *telegraph equation*. If $a = b = 0$, then there is no resistance in the circuit so that $R = 0$ and $1/G = 0$, and the equation reduces to the wave equation with a wave or phase velocity $c = 1/\sqrt{LC}$. If there is resistance then both a and b are not zero. It will be shown that the term au_t represents dissipation of energy or damping, while the term bu represents dispersion. Setting

$$u(x, t) = f(t)v(x, t), \tag{3.72}$$

we shall show that this change of variables yields a new equation with no first-derivative terms. Performing the appropriate differentiations on eq. (3.72) and using eq. (3.71) yields

$$fv_{xx} - c^{-2}fv_{tt} - (2c^{-2}\dot{f} + af)v_t - (c^{-2}\ddot{f} + a\dot{f} + bf)v = 0.$$

Thus far $f(t)$ is arbitrary, so that we may choose it such that the coefficient of the v_t term vanishes. This gives $2c^{-2}\dot{f} + af = 0$, which yields

$$f = e^{-ac^2 t/2}. \tag{3.73}$$

Then the PDE for v has no first-derivative term and reduces to

$$v_{xx} - c^{-2}v_{tt} + kv = 0, \quad k = b - \frac{a^2c^2}{4}. \tag{3.74}$$

CHAPTER THREE

Therefore the solution $u(x,t)$ of the telegraph equation can be expressed in terms of the solution $v(x,t)$ of eq. (3.74) by means of the formula

$$u(x,t) = e^{-ac^2 t/4} v(x,t). \tag{3.75}$$

If we set $k = 0$ then eq. (3.74) becomes the wave equation for v, so that

$$u(x,t) = e^{-ac^2 t/2}[F(x - ct) + G(x + ct)]. \tag{3.76}$$

As we recall, F represents a progressing wave while G represents a regressing wave. There is a damping factor represented by the exponential term as long as a is positive (a is the damping coefficient). The waves travel undistorted but damped. These waves are therefore called *relatively undistorted*.

Physically, the assumption that $k = 0$ means that $RC = LG$. This tells us that we can distribute inductances in each cable at appropriate intervals (1/4 or 1/2 mile). This idea is due to the physicist Pupin.

Suppose $k \neq 0$. Then if we look for solutions of the form $v = F(x - ct)$, we find that

$$\ddot{F} - \ddot{F} + kF = 0 \quad \text{so that } F = 0 \text{ if } k \neq 0.$$

Similarly, $G(x + ct)$ is also zero. Therefore the waves are not relatively undistorted. However, if we set $v = F(x - \gamma t)$, where γ is different from c, we find

$$\ddot{F} + \frac{kc^2}{c^2 - \gamma^2} F = 0.$$

This tells us that we have only sines, cosines, or hyperbolic functions as long as $c \neq \gamma$. The waves do not all propagate with the same velocity (since the parameter γ may vary). In fact, if we multiply the solution for v by some function $A(\gamma)$ and integrate over γ this is equivalent to using

124

THE WAVE EQUATION

the Fourier transform (which we shall not go into here). Waves of this nature (for $k \neq 0$) are called *dispersive waves* since they travel with a variable phase velocity. The term kv in eq. (3.74) produces dispersion.

PART II THE WAVE EQUATION IN TWO AND THREE DIMENSIONS

TWO-DIMENSIONAL WAVE EQUATION

In part I, we investigated the one-dimensional wave equation. In this section we shall discuss the two-dimensional wave equation, which is

$$u_{xx} + u_{yy} - c^{-2}u_{tt} = 0. \tag{3.76}$$

We interpret the solution u to be the vertical displacement of any point on a *membrane* from its equilibrium position in the (x, y) plane. A membrane is a thin piece of stretched elastic material whose edges are clamped on a closed curve. It does not resist shear stresses. If it resists shear it is called a plate and is governed by a higher-order PDE. Rather than look for a general solution to eq. (3.76), we focus our attention on solving a combined IV and BV problem for special cases.

Consider the following problem:

$$u_{xx} + u_{yy} - c^{-2}u_{tt} = 0,$$

ICs: $u(x, y, 0) = f(x, y), \quad u_t(x, y, 0) = g(x, y),$ (3.77)

BCs: $u(x, y, t) = 0 \quad$ on boundary C.

This is the mathematical formulation of the vibrating membrane in Cartesian coordinates in the region R bounded by the closed curve C on which the lateral displacement u vanishes, since the membrane is fixed on the boundary.

CHAPTER THREE

REDUCED WAVE EQUATION IN TWO DIMENSIONS

The two-dimensional wave equation can be written as

$$c^2 \nabla^2 u - u_{tt} = 0, \quad \text{where } \nabla^2 = \frac{\partial^2}{\partial x^2} + \frac{\partial^2}{\partial y^2}. \quad (3.78)$$

∇^2 is the Laplacian in two dimensions. We separate the time-dependent from the space-dependent part of the two-dimensional wave equation by setting

$$u(x, y, t) = v(x, y)w(t).$$

Inserting this expression for u into eq. (3.78) yields

$$c^2 w \nabla^2 v - v\ddot{w} = 0,$$

which can be written as

$$\frac{\nabla^2 v}{v} = \frac{\ddot{w}}{c^2 w} = \lambda.$$

Since the left-hand side of the above equation is independent of t and the right-hand side is independent of x and y, we set the expression equal to a constant, which we call λ (to be determined). This gives us the following ODE for w and PDE for v:

$$\ddot{w} - \lambda c^2 w = 0 \quad (3.79)$$

and

$$\nabla^2 v - \lambda v = 0. \quad (3.80)$$

Equation (3.80) is the reduced wave equation or Helmholtz equation in two dimensions. As in the case of the vibrating string (discussed in part

126

THE WAVE EQUATION

I), the set of values of λ are the eigenvalues corresponding to the eigenfunctions w and v.

THE EIGENVALUES MUST BE NEGATIVE

To show that all the eigenvalues λ must be negative, we use the Green's identity in two dimensions, which tells us that

$$\iint_R v\nabla^2 v \, dA = \int_C v \frac{\partial v}{\partial n} \, ds - \iint_R (\nabla v)^2 \, dA.$$

Since $v = 0$ on the boundary C and $\nabla^2 v = \lambda v$ in the region R, the above equation reduces to

$$\lambda = -\frac{\iint_R (\nabla v)^2 \, dA}{\iint_R v^2 \, dA}.$$

It follows that the eigenvalues λ are negative since both integrals are positive.

If we set $\lambda^2 = -k$, then the solution of eq. (3.79) becomes

$$w(t) = A \cos kct + B \sin kct, \tag{3.81}$$

where A and B are constants to be determined from the ICs. k is the wave number.

RECTANGULAR MEMBRANE

The rectangular membrane is not used on a drumhead, but we study this type of region for its instructive value in showing how the method of separation of variables is used and how the eigenvalues arise. The region is $0 \leq x \leq a$, $0 \leq y \leq b$ (where a and b are known positive constants). Inserting the expression $v(x, y) = f(x)g(y)$ into eq. (3.80), we obtain

$$f\ddot{g} + g\ddot{f} + k^2 fg = 0,$$

127

CHAPTER THREE

which can be rewritten as

$$\frac{\ddot{f}}{f} + k^2 = -\frac{\ddot{g}}{g} = \alpha^2,$$

where α is as yet unknown. $g(y)$ must involve trigonometric functions in order for the BC $g(0) = g(b) = 0$ to be valid. This implies that $\alpha = m\pi/b$, where m is an integer. Since $\ddot{f} + (k^2 - \alpha^2)f = 0$, we must have

$$k^2 - \left(\frac{m\pi}{b}\right)^2 = \left(\frac{n\pi}{b}\right)^2, \quad n \text{ an integer.}$$

Solving for k yields

$$k = \pi\sqrt{\frac{n^2}{a^2} + \frac{m^2}{b^2}} = k_{mn}. \tag{3.82}$$

The wave numbers k_{mn} are the eigenvalues for this case. It follows that the eigenfunctions $v_{mn}(x, y)$ are also functions of m and n. They are given by

$$v_{mn}(x, y) = A_{mn} \sin\frac{n\pi x}{a} + B_{mn} \sin\frac{m\pi y}{b}, \tag{3.83}$$

where A_{mn} and B_{mn} are constants determined by the ICs. It is clear that the BCs $v(0,0) = v(a, b) = 0$ are satisfied. Using the fact that $u(x, y, t) = v(x, y)w(t)$ and eqs. (3.81) and (3.83) and redefining the coefficients, we get

$$u(x, y, t) = \sum_{m,n} (a_{mn} \cos k_{mn}ct + b_{mn} \sin k_{mn}ct) \sin\left(\frac{n\pi x}{a}\right)\sin\frac{m\pi y}{b}. \tag{3.84}$$

The ICs in eq. (3.77) require that

$$f(x, y) = \sum_{m,n} a_{mn} \sin\frac{n\pi x}{a} \sin\frac{m\pi y}{b}$$

THE WAVE EQUATION

and

$$g(x, y) = c \sum_{m,n} k_{mn} b_{mn} \sin \frac{n\pi x}{a} \sin \frac{m\pi y}{b}.$$

The above expressions tell us that the a_{mn} and $ck_{mn}b_{mn}$ are the Fourier coefficients of $f(x, y)$ and $g(x, y)$, respectively. These Fourier coefficients are given by the following expressions:

$$a_{mn} = \frac{4}{ab} \int_0^a \int_0^b f(x, y) \sin \frac{n\pi x}{a} \sin \frac{m\pi y}{b} \, dx \, dy,$$

$$b_{mn} = \frac{4}{abck_{mn}} \int_0^a \int_0^b g(x, y) \sin \frac{n\pi x}{a} \sin \frac{m\pi y}{b} \, dx \, dy. \quad (3.85)$$

The normal functions of a given rectangle are

$$\sin \frac{n\pi x}{a} \sin \frac{m\pi y}{b}.$$

If $m = n = 1$ then u retains the same sign over the rectangle at a given time, vanishing at the ends only. But in any other case there are *nodal lines* running parallel to the coordinates. A nodal line is a line on the membrane where $u = 0$ (an equilibrium line). There are $m - 1$ nodal lines parallel to the x axis, their equations being

$$y = \frac{b}{m}, \frac{2b}{m}, \ldots, \frac{(m-1)b}{m}.$$

Similarly, the number of nodal lines parallel to the y axis is $n - 1$; they are

$$x = \frac{a}{n}, \frac{2a}{n}, \ldots, \frac{(n-1)a}{n}.$$

Square Boundary

It is clear that the radial frequency $\omega_{mn} = k_{mn}c$ (the frequency equals the wave number times the wave speed). For a square boundary we set $b = a$. Using eq. (3.82) the frequency becomes $\omega = \sqrt{2}c\pi/b$ (where we

CHAPTER THREE

suppressed the subscripts). We get the lowest tone or fundamental mode, which is

$$u(x, y, t) = \sin \frac{\pi x}{a} \sin \frac{\pi y}{b} \cos \omega t.$$

This tells us that the only nodes are the BCs.

Next, suppose for example that one of the numbers m, n is equal to 2 and the other to unity, yielding two types of vibrations. If they are in phase (or synchronous) the whole motion is

$$u(x, y, t) = \left(A \sin \frac{2\pi x}{a} \sin \frac{\pi y}{a} + B \sin \frac{\pi x}{a} \sin \frac{2\pi y}{a} \right) \cos \omega t.$$

A and B are the amplitudes. There are four cases to consider:
(1) If $B = 0$ then

$$u = A \sin \frac{2\pi x}{a} \sin \frac{\pi y}{a} \cos \omega t.$$

This indicates a vibrational mode with one node at $x = a/2$ and the other node at the boundaries $y = 0, a$.

(2) $A = 0$. The vibrational mode has nodes at $y = a/2$ and $x = 0, a$.

(3) $A = B$. Then, neglecting the time function, u is proportional to

$$\sin \frac{2\pi x}{a} \sin \frac{\pi y}{a} + \sin \frac{\pi x}{a} \sin \frac{2\pi y}{a}$$

$$= 2 \sin \frac{\pi x}{a} \sin \frac{\pi y}{b} \left(\cos \pi \frac{x}{a} + \cos \frac{\pi y}{a} \right).$$

This expression vanishes (causing nodes) for two cases:

(a) When

$$\sin \frac{\pi x}{a} = 0 \quad \text{or} \quad \sin \frac{\pi y}{a} = 0.$$

These two equations yield the edges or boundaries as the nodes (which happen to be the BCs).

(b) When

$$\cos \frac{\pi x}{a} + \cos \frac{\pi y}{a} = 0.$$

THE WAVE EQUATION

This case gives $x + y = a$, which is the node given by one diagonal of the square.

(4) $A = -B$. For this case the nodal lines are the edges of the square together with the other diagonal $x - y = 0$.

The next case in order of pitch is the first harmonic given by $m = n = 2$. The only type of mode to be considered is

$$u = \sin \frac{2\pi x}{a} \sin \frac{2\pi y}{a} \cos \omega t.$$

The nodes are determined by the equation

$$\sin \frac{\pi x}{a} \sin \frac{\pi y}{a} \cos \frac{\pi x}{a} \cos \frac{\pi y}{a} = 0.$$

The nodes are the edges of the square and the lines $x = a/2$, $y = a/2$.

Other cases occur for other values of m and n. The reader is referred to [29, vol. I p. 313]. (The above analysis is based on the work of Rayleigh in this reference.)

Circular Membrane

We now consider a drumhead, which is a stretched membrane fixed on a circular boundary. Let the radius of the membrane be R. We formulate the problem in polar coordinates (r, θ). Using the Laplacian in polar coordinates we get

$$u_{rr} + \frac{1}{r}u_r + \frac{1}{r^2}u_{\theta\theta} = \frac{1}{c^2}u_{tt}, \qquad 0 \le r \le R, \quad 0 \le \theta \le 2\pi, \quad t > 0,$$

$$\text{BC:} \quad u(R, \theta, t) = 0, \qquad 0 \le \theta \le 2\pi, \quad t > 0, \qquad (3.86)$$

$$\text{ICs:} \quad u(r, \theta, 0) = f(r, \theta), \qquad u_t(r, \theta) = g(r, \theta);$$

$$0 \le r \le R, \quad 0 \le \theta \le 2\pi.$$

This expresses the PDE in cylindrical coordinates with the appropriate BC and ICs. The ICs given by f and g are arbitrary.

We separate variables by setting

$$u(r, \theta, t) = v(r, \theta)w(t). \tag{3.87}$$

Inserting this expression into the above PDE yields

$$\frac{1}{v}\left(v_{rr} + \frac{1}{r}v_r + \frac{1}{r^2}v_{rr}\right) = \frac{1}{c^2}\left(\frac{\ddot{w}}{w}\right) = -k^2.$$

The time-dependent part of u becomes

$$w(t) = A \cos kct + B \sin kct, \tag{3.88}$$

while the PDE for v becomes

$$v_{rr} + \frac{1}{r}v_r + \frac{1}{r^2}v_{\theta\theta} + k^2 v = 0.$$

We separate variables in this PDE for v by setting $v(r,\theta) = \alpha(r)\beta(\theta)$. The PDE becomes

$$\beta\ddot{\alpha} + \frac{1}{r}\beta\dot{\alpha} + \frac{1}{r^2}\alpha\ddot{\beta} + k^2\alpha\beta = 0.$$

Separating variables, this becomes

$$\frac{r^2}{\alpha}\left(\ddot{\alpha} + \frac{1}{r}\dot{\alpha} + k^2\alpha\right) = -\frac{\ddot{\beta}}{\beta} = \lambda.$$

The ODE for β is

$$\ddot{\beta} + \lambda\beta = 0.$$

Setting $\lambda = n^2$, the solution for β is

$$\beta = A \cos n\theta + B \sin n\theta. \tag{3.89}$$

The ODE for α is

$$\ddot{\alpha} + \frac{1}{\rho}\dot{\alpha} + \left(1 - \frac{n^2}{\rho^2}\right)\alpha = 0, \quad \text{where } kr = \rho. \tag{3.90}$$

THE WAVE EQUATION

Equation (3.90) is called *Bessel's equation of index n*. Since this ODE is second order, the general solution is a linear combination of two linearly independent solutions. Solutions of Bessel's equation may be singular at the origin (blow up at $r = 0$). However, the theory of Bessel's equation tells us that, if n is an integer, then there exists exactly one independent solution of Bessel's equation that is finite at the origin. This denoted by $J_n(\rho)$. (Note that if the membrane consists of an annular region bounded by two circles then we must also include the Bessel function that is singular at the origin as a solution to α, since the region of the membrane does not include the origin.) In the case considered here, we get for α a set of values $\alpha_1, \alpha_2, \ldots, \alpha_n, \ldots$, each element of the set corresponding to a Bessel function of a particular index. Thus we have

$$\alpha_n(\rho) = J_n(\rho) = J_n(kr).$$

Since $v_n(r, \theta) = \alpha_n(r)\beta_n(\theta)$, it follows that

$$v_n(r, \theta) = J_n(kr)(A_n \cos n\theta + B_n \sin n\theta).$$

At this stage we invoke the BC at $r = R$, which is $v_n(R, \theta) = 0$ for all θ between 0 and 2π. This BC requires that the following set of equations be satisfied:

$$J_n(kR) = 0, \quad \text{for } n = 1, 2, \ldots. \tag{3.91}$$

For $n = 0$ we have $J_0(kR) = 0$, whose first three roots are given by $kR = \phi_{01} = 2.4048$, $\phi_{02} = 5.5201$, $\phi_{03} = 8.6537$. For $n = 1$ we have $J_1(kR) = 0$, whose first three roots (other than zero) are $kR = \phi_{11} = 3.8317$, $\phi_{12} = 7.015$, $\phi_{13} = 10.1735$. (See, for example, p. 318 of *C.R.C. Standard Mathematical Tables*, The Chemical Rubber Publishing Co., Cleveland, OH, 1963.) In general, we have $k = \phi_{nm}$. We thus see that the wave number k depends on the index n of the Bessel function and the index m that gives the ordered value of the root for a given n. The wavelength $\lambda = 2\pi/k = 2\pi/\phi_{nm}$. For a given ϕ_{nm}, the corresponding λ yields a node that is a circle centered at the origin. In order for these nodes to lie inside the membrane of radius R, we must have the inequality $\phi_{nm} > 2\pi$. We see that ϕ_{03}, ϕ_{12}, and ϕ_{13} satisfy this condition. The roots for a given n increase with the index m.

CHAPTER THREE

Using the above, we get for the nth mode of the solution

$$u_n(r,\theta,t) = \sum_{m=1} [J_n(k_{nm}r)(A_{nm}\cos n\theta + B_{nm}\sin n\theta)$$
$$\times (c_{nm}\cos k_{nm}ct + d_{nm}\sin k_{nm}ct)]. \quad (3.92)$$

The general solution for u for a circular membrane is the sum over the n modes, or $u(r,\theta,t) = \sum_{n=0} u_n(r,\theta,t)$. In detail, we have

$$u(r,\theta,t) = \sum_{nm} J_n(k_{nm}r)[A_{nm}\cos n\theta + B_{nm}\sin n\theta]$$
$$\times [c_{nm}\cos k_{nm}ct + d_{nm}\sin k_{nm}ct]. \quad (3.93)$$

Invoking the ICs $f(r,\theta)$ and $g(r\theta)$, we have

$$f(r,\theta) = \sum_{nm} J_n(k_{nm}r)[\alpha_{nm}\cos n\theta + \beta_{nm}\sin n\theta],$$
$$g(r,\theta) = c\sum_{nm} k_{nm}J_n(k_{nm}r)[C_{nm}\cos n\theta + D_{nm}\sin n\theta]. \quad (3.94)$$

We now show how to calculate the coefficients α_{nm}, β_{nm}, C_{nm}, and D_{nm} by using the appropriate orthogonality relations among the Bessel functions. Set

$$F_n(r) = \frac{1}{\pi}\int_0^{2\pi} f(r,\theta)\,d\theta. \quad (3.95)$$

Then we get

$$F_n(r) = \sum_{m=1}^{\infty} \alpha_{nm} J_n(k_{nm}r). \quad (3.96)$$

To solve for the α_{nm}s we need the following orthogonality condition on the Bessel functions:

$$\int_0^R rJ_n(k_{nm}r)\,dr = \tfrac{1}{2}\delta_{nm}R^2 J_{n+1}^2(k_{nm}R) = \delta_{nm}N_{nm}, \quad (3.97)$$

THE WAVE EQUATION

where δ_{nm} is the Kronecker delta defined by

$$\delta_{nm} = \begin{cases} 1 & \text{if } m = n, \\ 0 & \text{if } m \neq n. \end{cases}$$

We now multiply eq. (3.96) by $rJ_{nm}(k_{nm}r)$ and integrate with respect to r from 0 to R to obtain

$$\alpha_{nm} = \frac{1}{N_{nm}} \int_0^R r F_n(r) J_n(k_{nm}r)\, dr. \tag{3.98}$$

It is now clear how to obtain the other coefficients and hence solve the problem.

THREE-DIMENSIONAL WAVE EQUATION

The three-dimensional wave equation can be written as

$$c^2 \nabla^2 u - u_{tt} = 0, \quad \text{where } \nabla^2 = \frac{\partial^2}{\partial x^2} + \frac{\partial^2}{\partial y^2} + \frac{\partial^2}{\partial z^2}. \tag{3.99}$$

∇^2 is the three-dimensional Laplacian operator. For the IV problem in 3-space we append the following ICs:

$$u(x, y, z, 0) = f(x, y, z),$$
$$u_t(x, y, z, 0) = g(x, y, z), \tag{3.100}$$
$$t = 0, \quad \text{all } (x, y, z).$$

Spherical Symmetry

Before we attempt a general solution of this IV problem we consider the special case of spherical symmetry. x, y, z are tied up in the radius vector $r = \sqrt{x^2 + y^2 + z^2}$. Clearly, in the setting of wave propagation spherical symmetry means that a spherical wave is emitted as a progressing wave or shrinks as a regressing wave; the wave front at each instant of time is a spherical surface in three-dimensional space. Radial

CHAPTER THREE

symmetry is invoked, which means that the angular coordinate is missing. Therefore the displacement vector $u = u(r, t)$. The spherically symmetric wave equation is

$$u_{rr} + \frac{2}{r} u_r = c^{-2} u_{tt}. \tag{3.101}$$

It is clear that eq. (3.101) is the one-dimensional wave equation for ru. Therefore the solution for $u(r, t)$ is

$$u(r, t) = \frac{F(r - ct)}{r} + \frac{G(r + ct)}{r}, \tag{3.102}$$

where $F(r - ct)$ represents a progressing wave where the spherical wave surface expands, and $G(r + ct)$ a regressing wave where the wave surface contracts (at the rate of the wave speed c). F is an outgoing and G an incoming wave. The distance can be measured from an arbitrary point (ξ, η, ζ), so that $r^2 = (x - \xi)^2 + (y - \eta)^2 + (z - \zeta)^2$. Since (ξ, η, ζ) are parameters in the solution (where r is given as above), we can obtain a more general solution of the spherical wave equation by multiplying $f(r - ct)$ by the arbitrary function $h_1(\xi, \eta, \zeta)$ and $G(r + ct)$ by the arbitrary function $h_2(\xi, \eta, \zeta)$ and integrating over (ξ, η, ζ). Therefore, a more general solution of the spherical wave equation is given by

$$u(x, y, z, t) = \iiint h_1(\xi, \eta, \zeta) \frac{F(r - ct)}{r} d\xi \, d\eta \, d\zeta$$

$$+ \iiint h_2(\xi, \eta, \zeta) \frac{G(r + ct)}{r} d\xi \, d\eta \, d\zeta. \tag{3.103}$$

We now specialize to the following model: Consider an outgoing wave so that $G = 0$. Let $F(r - ct)$ be spread out over the spherical wave surface in an infinitely thin layer, so that $F(r - ct) = \delta(r - ct)$, where δ is a delta function. Then the solution for u becomes

$$u(x, y, z, t) = \iiint h_1(\xi, \eta, \zeta) \frac{\delta(r - ct)}{r} d\xi \, d\eta \, d\zeta. \tag{3.104}$$

We now use spherical polar coordinates as follows: Let (x, y, z) be the center of the spherical surface on which r is a constant (on a wave surface). Let (ξ, η, ζ) be the coordinates of a point on the surface

$r = $ const. Let (α, β, γ) be the angles r makes with (ξ, η, ζ), respectively. The new integration variables are $(r, \alpha, \beta, \gamma)$. The relationships between (x, y, z) and these integration variables are

$$\xi = x + r \cos \alpha,$$
$$\eta = y + r \cos \beta,$$
$$\zeta = z + r \cos \gamma,$$
$$\cos \alpha^2 + \cos \beta^2 + \cos \gamma^2 = 1.$$

It is clear that $(\cos \alpha, \cos \beta, \cos \gamma)$ defines a point on the unit sphere centered at (x, y, z). Let $d\Omega$ be the element of surface area on the unit sphere. Then the elementary volume $d\xi\, d\eta\, d\zeta$ is given by

$$d\xi\, d\eta\, d\zeta = r^2\, d\Omega\, dr.$$

Equation (3.104) becomes

$$u(x, y, z) = \iiint h_1(x + r \cos \alpha, y + r \cos \beta, z + r \cos \gamma) r\, dr\, d\Omega.$$

Using the property that the delta function $\delta(r - ct)$ picks out ct, eq. (3.104) becomes

$$u(x, y, z, t) = ct \iint h_1(x + ct \cos \alpha, y + ct \cos \beta, z \neq ct \cos \gamma)\, d\Omega.$$

Clearly, the ICs are $u(x, y, z, 0) = 0$, $u_t(x, y, z, 0) = 4\pi c h_1(x, y, z)$. This leads to the result that

$$u(x, y, z, t) = \frac{t}{4\pi} \iint h_1(x + ct \cos \alpha, y + ct \cos \beta, z + ct \cos \gamma)\, d\Omega \tag{3.105}$$

is the solution to the IV problem:

$$\text{PDE:}\quad c^2 \nabla^2 u - u_{tt} = 0,$$
$$\text{ICs:}\quad u(x, y, z, 0) = 0,\quad u_t(x, y, z, 0) = h_1(x, y, z). \tag{3.106}$$

CHAPTER THREE

To be concise, we let

$$\frac{1}{4\pi}\iint h_1(x + ct\cos\alpha, y + ct\cos\beta, z + ct\cos\gamma)\,d\Omega = \mu[h_1(x,y,z)].$$

(3.107)

It is clear that μ is a function of (x, y, z) and a functional of h_1. Then the solution of the above-mentioned IV problem becomes

$$u(x, y, z, t) = t\mu[h_1].$$ (3.108)

Now let us find the solution of the three-dimensional wave equation (3.99) for the following ICs:

$$u(x, y, z, 0) = f(x, y, z),$$
$$u_t(x, y, z, 0) = 0.$$ (3.109)

It is easily seen that the solution of this IV problem is

$$u(x, y, z, t) = \frac{\partial}{\partial t} t\mu[f(x, y, z)].$$ (3.110)

The solution of the IV problem where both ICs are not zero is given by

$$u(x, y, z, t) = \frac{\partial}{\partial t} t\mu[f(x, y, z)] + t\mu[h_1(x, y, z)].$$ (3.111)

A similar analysis of a regressing spherical wave of the form $G(r + ct)/r$ can be done. The term $\iint h_2(\xi, \eta, \zeta)[G(r + ct)/r]\,d\xi\,d\eta\,d\zeta$ is to be used in the right-hand side of eq. (3.103), instead of the first integral. For the regressing wave the arbitrary function $h_2(\xi, \eta, \zeta)$ replaces the function $h_1(\xi, \eta, \zeta)$ used for the progressing wave in the solution for the IV problem given by eq. (3.111).

Huygens' Principle

Equation (3.111) illustrates Huygens' principle, which can be explained as follows: Suppose an observer is stationed at a given field point (x, y, z) in the region where the ICs $f = h = 0$ (subscript suppressed on

THE WAVE EQUATION

h). The solution at the field point (x, y, z) at any time t is given by the integrals of f and h over a sphere of radius ct centered at that point. At $t = 0$ it is clear that the radius $r = ct = 0$, so that the solution is zero. As t increases the wave front (of radius ct) expands uniformly until it is large enough to penetrate the region where the ICs are not zero. Then the solution is not zero at the field point (x, y, z). As t becomes sufficiently large, the entire region in which the ICs are not zero will be engulfed by the spherical wave front, so that the solution remains zero thereafter. Figure 3.18 shows a projection of the spherical surfaces in two dimensions for various values of the radius ct; at ct_1 the wave front is too small to penetrate the region where the ICs are not zero, at ct_2 it is large enough to penetrate part of that region, then at ct_3 it is large enough to completely engulf the region. The important point is that, for three-dimensional spherical waveforms, the solution at any field point depends only on the values of the ICs on the wave surface.

Another way of looking at Huygens' principle is as follows: Every point on the progressing (advancing) spherical wave front is considered to be a source of secondary waves centered on the surface, which spread out with a velocity c as spherical wavelets. A later position of the wave front is given by the envelope of these secondary waves. This is shown in

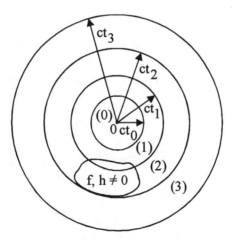

FIGURE 3.18. Two-dimensional projection of various spherical surfaces with a spherical wave source at 0. In region (0) ICs are zero. Region (1) partially engulfs the region where ICs are not zero. Region (2) completely engulfs the region where ICs are not zero. Region (3) engulfs the region where ICs are zero.

CHAPTER THREE

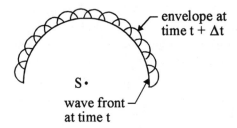

FIGURE 3.19. Projection of spherical waves emanating from source S. At time t, a series of wavelets (secondary sources) form an envelope, forming a new wave front at time $t + \Delta t$.

fig. 3.19, where the source S is at the center of the wave fronts. This figure shows a projection in the plane of spherical waves emanating from S. At time t the wave front emits secondary sources at each point. At $t + \Delta t$ this continuum of secondary sources forms an envelope on the new wave front. It is clear that we are dealing with a homogeneous medium so that the wave speed c is constant.

PROBLEMS

1. Show that $u = f(x - ct) + g(x + ct)$ satisfies the wave equation $c^2 u_{xx} - u_{tt} = 0$, and that all solutions of the wave equation must be of this form. Using eq. (3.10) show that the one-dimensional wave equation is given by (3.11).

2. Show that $u = f(x - ct)$ is a progressing wave by showing that an increase in x gives an increase in t in order for u to be constant along a C_+ characteristic. Show that $u = g(x + ct)$ is a regressing wave by showing that an increase in x produces a decrease in t in order for u to be constant along a C_- characteristic.

3. Discuss progressing and regressing solutions of the wave equation in the (x, t) plane and the (ξ, η) plane using characteristic theory, given the following initial pulse:

$$f(x) = \begin{cases} \sin 2\pi x & \text{for } 0 \le x \le \pi, \\ 0 & \text{for } x > \pi. \end{cases}$$

4. The vibrating string problem in the region $x > 0$ is given by eq. (3.14) for the case $L = \infty$.

   (a) Using D'Alembert's method show that the solution is

   $$u(x,t) = \begin{cases} \frac{1}{2}[f(x-ct) + f(x+ct)] + \frac{1}{2c}\int_{x-ct}^{x+ct} g(z)\,dz \\ \quad \text{for } x > ct, \\ \frac{1}{2}[f(ct+x) - f(ct-x)] + \frac{1}{2c}\int_{ct-x}^{ct+x} g(z)\,dz \\ \quad \text{for } 0 < x < ct. \end{cases}$$

   Hint: The progressing wave $F(x - ct)$ is valid only for $x - ct > 0$. We need an argument for F valid for $x < 0$. $u(0, t) = 0$ so that $F(-ct) = -G(ct)$. Set $r = -ct$; then $F(r) = -G(-r)$. G is known for a positive argument so that $-r > 0$ or $ct - x > 0$, which defines F for $0 < x < ct$.

   (b) Describe the solution in the (x, t) plane (plot t vs. x) using characteristic theory. Show that there is reflection from the boundary $x = 0$. Show that in the region $x < 0$ a characteristic stemming from a point on the negative x axis and intersecting the t axis carries initial data that cancel the initial data coming from a point on the positive x axis and intersecting the same point on the t axis, so that the BC $u(0, t) = 0$ is satisfied. Label all the appropriate characteristic lines and the various regions in the space $t > 0$.

5. Recast problem 4 into a corresponding problem in characteristic coordinates. Find the images of $t = 0$ and $x = 0$ in the characteristic plane. Label all the necessary characteristics and the various regions.

6. Solve problem 4 by the method of Laplace transforms.

7. Solve the following problem for $u(x, t)$:

   $$c^2 u_{xx} - u_{tt} = 0, \qquad x \geq 0, \quad t \geq 0,$$

   IC: $u(x, 0) = \begin{cases} 0 & \text{for } 0 < x_0, \ x > x_0 + 2L, \\ u_0 > 0 & \text{for } x_0 < x < x_0 + 2L, \end{cases}$

   $$u_t(x, 0) = 0,$$

   BC: $u(0, t) = 0.$

CHAPTER THREE

Using characteristic theory, plot and label the appropriate characteristics in the (x, t) plane and label the solution in the various regions. Consider two reflections off the t axis, one from a characteristic stemming from x_0, $t = 0$ and one stemming from $x - 0 + 2L$, $t = 0$.

8. Solve the following problem for $u(x, t)$ and discuss the solution during characteristic theory, labeling appropriate regions and characteristics:

$$c^2 u_{xx} - u_{tt} = 0, \quad 0 < t < \infty, \quad 0 < x < \infty,$$

$$\text{ICs:} \quad u(x, 0) = f(x), \quad u_t(x, 0) = g(x),$$

$$\text{BC:} \quad u_x(0, t) = 0, \quad t > 0.$$

Compare this problem to problem 4. Here the BC is that the gradient of u is zero rather than $u = 0$.

9. Solve problem 8 by the Laplace transform method and compare the solution to that obtained in problem 6.

10. Solve the following problem for $u(x, t)$ by the Laplace transform method:

$$c^2 u_{xx} - u_{tt} = 0, \quad 0 \leq x \leq L, \quad 0 \leq t < \infty,$$

$$\text{IC:} \quad u(x, 0) = f(x) = \begin{cases} \dfrac{u_0}{\xi} x, & 0 \leq x \leq \xi, \\ \dfrac{u_0}{L - x}, & \xi \leq x \leq L, \end{cases}$$

$$\text{BCs:} \quad u(0, t) = 0, \quad u(L, t) = 0.$$

This is the problem of the plucked harp string. Compare the solution you obtain with that given by eq. (3.32).

11. In transmission line theory, one considers the telegraph equation, which is the wave equation with two extra terms:

$$c^2 u_{xx} - u_{tt} - au_t - bu = 0.$$

Physically this equation represents the voltage across or current along a cable. The term au_t represents the dissipation of energy or damping of the waveform due to electrical resistance; the term bu represents the capacitance of the circuit. Show that if you set $u = v(t)w(x,t)$, insert this expression into the telegraph equation, and then set $2\dot{v} + av = 0$, you get the following PDE:

$$c^2 w_{xx} - w_{tt} + kw = 0,$$

where $k = b - (a^2/4)$. Then show that

$$u(x,t) = e^{-at/2} w(x,t).$$

Also show that if $k = 0$ then the general solution of the telegraph equation becomes

$$u(x,t) = e^{-at/2}[F(x - ct) + G(x + ct)].$$

This shows that the general solution of the telegraph equation under this condition consists of traveling progressing and regressing waves that are damped exponentially according to the damping factor a. These waves are called relatively undistorted, since they maintain their shape even though damped.

12. Solve the problem given by eq. (3.60) by the Laplace transform method.

13. Find the general solution of $c^2 u_{xx} - u_{tt} = axt + bx + dt$ where a, b, k are constants.

14. Solve the problem

$$c^2 u_{xx} - u_{tt} = ax^2, \qquad 0 \le x \le L, \quad 0 \le t < \infty,$$

ICs: $u(x,0) = x(L - x), \qquad u_t = 0,$

BCs: $u(0,t) = 0, \qquad u(L,t) = 0,$

where a is a constant. Transform the problem and solution to characteristic coordinates.

15. Consider the three-dimensional wave equation

$$c^2 \nabla^2 u = u_{tt}.$$

Prove that any function $f(\xi)$ that is twice continuously differentiable is a solution of this PDE providing that

$$\xi = \alpha x + \beta y + \gamma z \pm ct,$$

whose coefficients satisfy the relation

$$\alpha^2 + \beta^2 + \gamma^2 = 1.$$

It is clear that (α, β, γ) is a vector of unit magnitude. Therefore, prove that $u = f(\alpha x + \beta y + \gamma z \pm ct)$ is a general solution of the three-dimensional wave equation.

16. Let q be the distance of a point (x, y, z) in three-dimensional space from the plane $\alpha x + \beta y + \gamma z = 0$.

    (a) Prove that $q = \boldsymbol{\alpha} \cdot \mathbf{r} = \alpha x + \beta y + \gamma z$, where the vectors $\boldsymbol{\alpha} = (\alpha, \beta, \gamma)$ and $\mathbf{r} = (x, y, z)$. Prove that $\mathbf{q}$ is a vector normal to the plane $\boldsymbol{\alpha} \cdot \mathbf{r} = 0$.

    (b) Prove that, if u is a solution to the three-dimensional wave equation, then it is constant on a plane whose equation is $\alpha x + \beta y + \gamma z = 0$. Varying (α, β, γ) generates a family of characteristic surfaces which are plane waves (waves with planar wave fronts). The direction given by $\mathbf{q}$ is normal to the wave front and is in the direction of the traveling wave.

17. Show that

$$u = e^{ik(\mathbf{r} \cdot \mathbf{x})} e^{i\omega t}$$

represents a plane wave in three dimensions whose wave front is given by $\boldsymbol{\alpha} \cdot \mathbf{r} = 0$ and direction is given by the vector α where ω is the frequency and k the wave number, where $k = 1/\lambda = \omega/c$.

CHAPTER FOUR

Wave Propagation in Fluids

THIS CHAPTER is divided into two parts: part I deals with inviscid fluids where viscosity effects are neglected, so that there is no dissipation of thermal energy, Part II takes up this gap by treating viscous fluids.

PART I INVISCID FLUIDS

Here we apply the methods developed in chapter 2 to supersonic wave propagation in compressible fluids without viscosity. This field of one-dimensional gas dynamics or fluid mechanics is an excellent example of the interplay of the mathematics of PDEs discussed in chapter 2 (characteristic theory) and the physics of continuous media. We then treat the subject of two-dimensional steady flow using the same methods.

The physical model is this: Suppose we have a compressible gas such as air filling a long tube such that the motion of the gas is one dimensional. At any t the position of each gas particle is defined by a single space variable, which is the coordinate x of the axis measured from a given origin. We define a particle of gas as an element of gas occupying a differential volume in a continuous medium. This element has the same properties as the gas in the large. In postulating a continuous medium, we neglect the molecular structure of the gas as well as the field of fracture mechanics. Associated with each gas particle are the dependent variables such as the dynamic variables (particle velocity, acceleration, pressure, etc.) and thermodynamic variables (temperature, enthalpy, free energy, etc.), all of which are differentiable functions of space and time (x, t) (except across shock waves). There are two ways of representing the position of a gas particle in time and—space the Lagrangian and the Eulerian representations. These will now be described.

CHAPTER FOUR

Lagrangian Representation of One-Dimensional Compressible Gas Flow

Our first task is to identify the position of each particle in the tube at any time. There is no unique way of doing this. We shall choose the following scheme: Let the coordinate a be the initial coordinate of a particle, measured from the origin. a is called the *Lagrangian coordinate*. Let x be the distance from the same origin of a particle at time t. x is called the *Eulerian coordinate*. It is clear that the position of a particle (whose initial position is a) is x at time $t > 0$. It follows that $x = x(a, t)$, which means that x is a function of the independent variables (a, t). All the other dependent variables are also functions of (a, t). The IC is that at $t = 0$, $x(a, 0) = a$. Moreover, let f stand for any dependent variable. Then $f = f(x(a, t), t) = f(a, t)$. Kinematically speaking, we follow the motion of each particle down the tube. The continuous ensemble of these particle trajectories in the (a, t) plane describes the motion of the gas. We assume that our gas is adiabatic (no heat exchange with the environment) and isentropic (constant entropy) (except for shock wave propagation).

We now describe the three conservation laws of physics in the setting of the Lagrangian representation.

Conservation of Mass (Continuity Equation)

This law states that mass is neither created nor destroyed in the tube, meaning that the mass of each gas particle is invariant with time. Remember that we are dealing with a compressible gas so that the density changes with time. Therefore the differential volume occupied by each gas particle changes with time. To phrase this conservation law in quantitative terms, we assume the tube to be of unit cross-section. We consider the particles occupying the volume (a_1, a_2) at time $t = 0$. Since mass = (density) × (volume), the mass of the particles occupying this volume at $t = 0$ is $\int_{a_1}^{a_2} \rho(a, 0)\, da$. At $t > 0$ the mass of the same collection of particles is $\int_{x_1}^{x_2} \rho(x(a), t)\, dx$, since at t the volume interval is (x_1, x_2) and the volume element is dx (this volume element at $t = 0$ was da).

$$\int_{a_1}^{a_2} \rho(a, 0)\, da = \int_{x_1}^{x_2} \rho(x(a), t)\, dx = \int_{a_1}^{a_2} \rho(a, t) x_a\, da$$

since $dx = x_a \, da$. Assume that the initial density $\rho(a, 0) = \rho_0 =$ const. Divide this equation by the volume (a_1, a_2) and let $a_2 \to a_1 = a$; then the integral on the right-hand side of the above equation tends to ρx_a. Since the cross-sectional area of the tube is constant (unity), the conservation of mass for a particle element becomes

$$\rho_0 = \rho x_a. \tag{4.1}$$

Equation (4.1) is the expression for the conservation of mass for a one-dimensional compressible gas in the Lagrangian representation.

Conservation of Momentum

The conservation of linear momentum for a continuum is an application of Newton's second law, which states that the time rate of change of linear momentum of a gas particle is equal to the sum of the external forces acting on the particle. This is called the equation of motion for the continuum. (The conservation of angular momentum does not apply here since we have no rotational motion.) We assume that the only external force is due to the net pressure on the particle. Again we take a unit cross-section of volume (a_1, a_2). The pressure at the left and right ends of this volume segment is $p(a_1, t) = p_1$ and $p(a_2, t) = p_2$, respectively. The net force on this volume is $-[p(a_2, t) - p(a_1, t)] = -(p_2 - p_1)$ where we multiplied by the unit cross-sectional area. We used the convention that the pressure is a negative stress. Since the mass is invariant with respect to time, Newton's equation of motion tells us that the mass of the particle times its acceleration is equal to the net pressure force on the volume segment. The particle velocity v is x_t and the particle acceleration is $v_t = x_{tt}$. Integrating over the volume (x_1, x_2), Newton's equation of motion becomes

$$\int_{x_1}^{x_2} \rho(x(a, t), t) x_{tt} \, dx = -(p_2 - p_1).$$

In this equation the integration is with respect to x. We want to integrate with respect to the Lagrangian coordinate a. Since $x = x(a, t)$, the differential element becomes $dx = x_a \, da$, so that we can now

CHAPTER FOUR

integrate with respect to the Lagrange variable. We get

$$\int_{a_1}^{a_2} \rho x_{tt} x_a \, da = -(p_2 - p_1).$$

Using eq. (4.1) in the integrand we get

$$\int_{a_1}^{a_2} \rho_0 x_{tt} \, da = -(p_2 - p_1).$$

Dividing by $a_2 - a_1$ and letting $a_2 \to a_1 \equiv a$ gives

$$\rho_0 x_{tt} = \rho_0 v_t = -p_a. \tag{4.2}$$

Equation (4.2) is Newton's equation of motion for a one-dimensional gas acted on by a pressure force in the Lagrangian representation, which is the law of conservation of motion in this representation. We have three unknowns, p, v, ρ, to be solved for as functions of (a, t). But we have two equations: conservation of mass and momentum. We therefore need another equation. This is an *equation of state*, which relates the particle density to the net pressure. This is obtained from the energy equation that we now discuss.

Energy Equation

For a compressible adiabatic isentropic gas, pressure is a function of density only, $p = p(\rho)$. Let c be the velocity of a traveling wave in the gas; then c is the wave or phase velocity. Without going into details of fluid mechanics and thermodynamics, we now summarize some results that will be useful to us. c is given by

$$c^2 = \frac{dp}{d\rho}. \tag{4.3}$$

The adiabatic equation of state is

$$p = A\rho^\gamma, \qquad A = \frac{p_0}{\rho_0^\gamma}, \qquad \gamma = \frac{C_p}{C_v}, \tag{4.4}$$

where C_p and C_v are the specific heats at constant pressure and constant volume, respectively. For air the ratio of specific heats $\gamma = 1.4$. p_0, ρ_0 are defined at an equilibrium state, for example, at $t = 0$. It follows that c becomes the following function of density:

$$c^2 = \gamma A \rho^{\gamma-1}. \tag{4.5}$$

A gas whose equation of state is given by eq. (4.4) is called a *polytropic gas*. This equation of state is given in the Lagrangian representation since p, ρ, and c are functions of (a, t). Note that the same equation of state holds for the Eulerian representation, but p, ρ, and c are now functions of (x, t).

EULERIAN REPRESENTATION OF A ONE-DIMENSIONAL GAS

In this description of the motion of a one-dimensional gas, the space variable x is the independent variable (rather than the initial spatial distribution given by the coordinate a). An observer stationed at a distance x from the origin would see the particles (each identified by a particular value of a) flow by. The dependent variables (such as p, ρ, and v) associated with each particle are functions of the independent variables (x, t). Suppose we know the unsteady (time-varying) flow field in the tube; then at any time we know the position of each particle. This means that, in principle, we can plot the trajectories of the ensemble of particles in the (x, t) plane. Each trajectory is a curve of constant a (which identifies the particle). However, in the Lagrangian description we obtain an ensemble of trajectories in the (a, t) plane with x as the parameter that defines the position of a particle at t (constant x for each curve). If we desire we can cross-plot the (a, t) and (x, t) ensembles and obtain a set of curves in the (a, x) plane where t is the parameter constant on each curve.

We now express the three conservation laws in the Eulerian representation.

Conservation of Mass

Equation (4.1) tells us that $\rho x_\alpha = \rho_0 = $ const. This means that the total time derivative of the first term must equal zero. We make the fact that

CHAPTER FOUR

$\rho(dx_\alpha/dt) = \rho(dx/dt)_\alpha = \rho v_\alpha$ in the following calculation:

$$\frac{d\rho x_\alpha}{dt} = \rho v_\alpha + x_\alpha(v\rho_x + \rho_t) = \rho v_x x_\alpha + x_\alpha(v\rho_x + \rho_t) = 0.$$

This yields

$$\rho v_x + v\rho_x + \rho_t = 0. \tag{4.6}$$

Equation (4.6) is the continuity equation in the Eulerian representation. We put this into the more physically suggestive form

$$(\rho v)_x = -\rho_t. \tag{4.6a}$$

The left-hand side of this equation is the divergence of ρv. ρv is called the *flux of fluid*; it represents the mass of fluid flowing across a unit cross-sectional area per unit time. Equation (4.6a) is the one-dimensional representation of Gauss's theorem and is called the *divergence theorem*. It tells us that the net flux of fluid through a volume element depends only on the time rate of change of density, if there are no sources or sinks in the fluid. A decrease in density gives a positive divergence and an increase yields a negative divergence, hence the minus sign in the right-hand side.

Conservation of Momentum

Recall that, for a constant mass, Newton's equation of motion tells us that (mass) × (particle acceleration) = net pressure force, for our case. The acceleration is the total time derivative of the particle velocity with respect to time. Expanding by the chain rule gives

$$\frac{dv(x,t)}{dt} = \frac{dx}{dt}v_x + v_t = vv_x + v_t.$$

Recall that in chapter 2 we remarked that v_t is the part of the acceleration seen by an observer fixed at a particular x, and vv_x is the convective part, which represents the spatial distribution of the acceleration from x to $x + dx$ at a fixed time. Since the net pressure force on a fluid particle is $-p_x/\rho$, the equation of motion becomes

$$\rho(vv_x + v_t) = -p_x. \tag{4.7}$$

This is the law of conservation of linear momentum in the Euler representation The equation of state given by (4.4) still holds. Also, for an adiabatic gas, eq. (4.5) also holds, where p and ρ are functions of (x,t). Since p depends only on ρ we get

$$p_x = \frac{dp}{d\rho}\rho_x = c^2\rho_x,$$

so that eq. (4.7) becomes

$$\rho(vv_x + v_t) + c^2\rho_x = 0, \tag{4.8}$$

where c is a known function of ρ given by eq. (4.5).

In summary: equation (4.6) is the continuity equation and eq. (4.8) is the equation of motion for a polytropic gas in one dimension. It is customary to use the Euler representation for investigating the dynamic behavior of fluids, and the Lagrangian representation for the study of waves in solids. The reason is that it is easier to manipulate the boundary conditions in these two classes of media.

Solution by the Method of Characteristics: One-Dimensional Compressible Gas

We now come to the point where we can apply the methods in chapter 2. For convenience, we rewrite eqs. (4.6) and (4.8) in the following form:

$$L_1 = \rho v_x + v\rho_x + \rho_t = 0, \tag{4.6}$$

$$L_2 = \rho(vv_x + v_t) + c^2\rho_x = 0. \tag{4.8}$$

The expressions $L_1 = 0$ and $L_2 = 0$ bring out the fact that we have two equations that involve a linear combination of the partial derivatives $(\rho_x, \rho_t, v_x, v_t)$. We want an appropriate linear combination of L_1 and L_2 whose directional derivative is in the same direction as the directional derivatives of v and ρ. In chapter 2 we developed the theory of characteristics for a single first-order PDE in two independent variables based on the concept of the directional derivative. In this analysis we

CHAPTER FOUR

have, as mentioned, two dependent variables v and ρ, whose directional derivatives (tangent to the characteristic curves) must be in the same direction. In order to determine that linear combination of the four partial derivatives that is in this direction, we introduce the parameter λ, which determines the linear combination of L_1 and L_2. λ is fixed at each (x, t) point but will, in general, change from point to point. The linear combination is given by

$$L = L_1 + \lambda L_2. \tag{4.9}$$

We now form the directional derivatives of v and ρ by taking total derivatives with respect to t, and setting them equal to each other:

$$\left(\frac{dx}{dt}\right)v_x + v_t = \left(\frac{dx}{dt}\right)\rho_x + \rho_t = 0. \tag{4.10}$$

Using eqs. (4.6) and (4.8), L becomes

$$L = \rho(1 + \lambda v)v_x + \lambda \rho v_t + (v + \lambda c^2)\rho_x + \rho_t = 0. \tag{4.11}$$

In order for the directional derivative expressed by the linear combination L to be in the same direction as those of v and ρ, we must apply the following principle:

$$\text{coeff}(v_x) : \text{coeff}(v_t) = \text{coeff}(\rho_x) : \text{coeff}(\rho_t) = \frac{dx}{dt}.$$

This is an important rule. We reiterate that this set of ratios must be valid in order to allow that combination of the four derivatives to be in the same direction as the directional derivatives of v and ρ. Applying this rule to eqs. (4.9)–(4.11), we get

$$\frac{(1 + \lambda v)}{\lambda} = v + \lambda c^2 = \frac{dx}{dt}.$$

Solving this system for λ and dx/dt yields the values of λ and the corresponding characteristic ODEs:

$$\begin{aligned} &\text{for } \lambda = \frac{1}{c}, & \frac{dx}{dt} &= c + v: & C_+, \\ &\text{for } \lambda = -\frac{1}{c}, & \frac{dx}{dt} &= -(c - v): & C_-. \end{aligned} \tag{4.12}$$

The important result given by the system (4.12) is the fruit of characteristic theory applied to one-dimensional, unsteady, adiabatic, isentropic flow. It tells us the following: (1) For the linear combination given by $\lambda = 1/c$ the characteristic ODE is $dx/dt = c + v$, whose solution yields a one-parameter family of C_+ characteristic curves. (2) For $\lambda = -1/c$ the ODE is $dx/dt = -(c - v)$, yielding a C_- family of characteristics, since the slope is negative. In general, the ODEs are nonlinear. c is a known function of ρ, but ρ and v are unknown functions of (x, t).

We now construct the linear combinations L by using the values $\lambda = \pm 1/c$, which represent the two linear combinations of the four partial derivatives corresponding to the $C_\pm$ characteristics, respectively. Inserting eq. (4.12) into eq. (4.11) we obtain, after a little algebraic manipulation,

$$\lambda = \frac{1}{c}: \quad \left(\frac{dx}{dt}\right)\left[v_x + \frac{c\rho_x}{\rho}\right] + v_t + \frac{c\rho_t}{\rho} = 0,$$

$$C_+: \quad \frac{dx}{dt} = c + v, \tag{4.13a}$$

$$\lambda = -\frac{1}{c}: \quad \left(\frac{dx}{dt}\right)\left[v_x - \frac{c\rho_x}{\rho}\right] + v_t - \frac{c\rho_t}{\rho} = 0,$$

$$C_-: \quad \frac{dx}{dt} = -c + v, \tag{4.13b}$$

We consider eq. (4.13a). First we observe that the same functions of v and ρ appear in the differential forms $v_x + c\rho_x/\rho$ and $v_t + c\rho_t/\rho$. The only difference between these two expressions is that the first involves differentiation with respect to x and the second differentiation with respect to t. Riemann recognized this fact and introduced the function l whose partial derivatives are

$$l_x = \frac{c\rho_x}{\rho}, \quad l_t = \frac{c\rho_t}{\rho}, \quad \text{where } dl = \frac{c}{\rho}\,d\rho. \tag{4.14}$$

Inserting these partial derivatives of l into eq. (4.13a) gives

$$\left(\frac{dx}{dt}\right)[v_x + l_x] + [v_t + l_t] = 0.$$

CHAPTER FOUR

From this we immediately construct the function

$$f(v, \rho) = v + l, \quad \text{where } l(\rho) = \int_{\rho_0}^{\rho} \frac{c(\rho)\, d\rho}{\rho}. \quad (4.15)$$

This tells us that the directional derivative of $f(v, \rho)$ is in the same direction as those of v and ρ. Since the directional derivative $df/dt = 0$, it follows that $f = v + l = $ const on a C_+ characteristic. We can perform a similar analysis on eq. (4.13b) and arrive at the result that $v - l = $ const on a C_- characteristic. Reiterating, the important results are

$$v + l = \text{const} = r \quad \text{on } C_+, \qquad v - l = \text{const} = s \quad \text{on } C_-. \quad (4.16)$$

The constants r, s are called the Riemann invariants. On a given C_+ characteristic r is constant but s varies. The Riemann invariant r defines a C_+ characteristic that is generated by the Riemann invariant s. The converse is true for the C_- family of characteristics. These statements can be formulated analytically as follows: For the C_+ family of characteristics, let x and t be functions of s [$x = x(s)$, $t = t(s)$], where r is a constant defining a particular characteristic of that family. For the C_- family let $x = x(r)$, $t = t(r)$, where s defines a particular characteristic. As long as the determinant of the Jacobian of the mapping from the x, t to the r, s plane is not zero (there is unique inverse transformation) the characteristic ODEs take the form

$$\begin{aligned} x_s &= (c + v) t_s, & C_+, \\ x_r &= -(c - v) t_r, & C_-. \end{aligned} \quad (4.12a)$$

The solution to the first ODE yields the characteristic curves that generate the C_+ family of characteristics given by $x = x(s)$, $t = t(s)$, while the solution to the second ODE yields curves that generate the C_- family given by $x = x(r)$, $t = t(r)$.

Since $dp/d\rho = c^2$ we have $c\, d\rho/\rho = dp/\rho c$, so that the expression for l in eq. (4.15) becomes

$$l(p) = \int_{p_0}^{p} \frac{dp}{\rho c}. \quad (4.17)$$

WAVE PROPAGATION IN FLUIDS

The expression ρc is called the *acoustic impedance* and has important physical significance with respect to reflection and refraction of sound waves.

We now calculate the Riemann invariants for our polytropic gas defined by the equation of state for c given by eq. (4.5). Inserting this equation into eq. (4.15) gives

$$l = \frac{2c}{\gamma - 1}. \tag{4.18}$$

The Riemann invariants then become

$$r = v + \frac{2c}{\gamma - 1}, \quad s = v - \frac{2c}{\gamma - 1}. \tag{4.19}$$

This gives the following situation for our polytropic gas:

$$\frac{dx}{dt} = v + c \quad \text{on } C_+, \quad \text{where } v + \frac{2c}{\gamma - 1} = r,$$

$$\frac{dx}{dt} = -(c - v) \quad \text{on } C_-, \quad \text{where } v - \frac{2c}{\gamma - 1} = s. \tag{4.20}$$

The system given by eq. (4.20) is an example of a quasilinear system. The characteristic ODEs depend on the solution. The problem is best solved by a numerical method using a characteristic net. This method is beyond the scope of this book. However, we can obtain an explicit solution by linearizing the system, as shown below.

Linear Case

Let p' be the perturbed pressure and ρ' the perturbed density such that $p = p_0 + p'$, $\rho = \rho_0 + \rho'$. The perturbed pressure, density, and particle velocity are small in the following sense:

$$\left|\frac{v}{c}\right| \ll 1, \quad \left|\frac{\rho'}{\rho_0}\right| \ll 1, \quad \left|\frac{p'}{p_0}\right| \ll 1.$$

Inserting these conditions into eqs. (4.6) and (4.8) and neglecting the products of small terms, we get the linearized continuity equation and

CHAPTER FOUR

the linearized equation of motion, respectively:

$$L_1 = \rho_0 v_x + \rho'_t = 0, \qquad (4.21)$$

$$L_2 = \rho_0 v_t + c_0^2 \rho'_x = 0. \qquad (4.22)$$

Applying the method of characteristics to eqs. (4.21) and (4.22) results in straight-line characteristics in the (x,t) plane such that on C_+, $dx/dt = c_0$ and on C_-, $dx/dt = -c_0$. To see this we form L for our linear case:

$$L = L_1 + \lambda L_2 = \rho_0 v_x + \lambda \rho_0 v_t + \lambda c_0^2 \rho'_x + \rho'_t = 0. \qquad (4.23)$$

The conditions for $L = 0$ to be in the same direction as the directional derivatives of v and ρ are the appropriate ratios of the coefficients of the partial derivatives, as we described above. Applying these ratios yields

$$\frac{1}{\lambda} = \lambda c_0^2 = \frac{dx}{dt}.$$

These yield

$$\text{for } \lambda = \frac{1}{c_0}, \quad \frac{dx}{dt} = c_0 \text{ on } C_+,$$

$$\text{for } \lambda = -\frac{1}{c_0}, \quad \frac{dx}{dt} = -c_0 \text{ on } C_-. \qquad (4.24)$$

Inserting (4.24) into (4.23) gives the following results for $L = 0$:

$$L = \left(\frac{dx}{dt}\right)\left[v + \frac{c_0 \rho'}{\rho_0}\right]_x + \left[v + \frac{c_0 \rho'}{\rho}\right]_t = 0, \quad C_+: \quad \frac{dx}{dt} = c_0,$$

$$(4.25)$$

$$L = \left(\frac{dx}{dt}\right)\left[v - \frac{c_0 \rho'}{\rho_0}\right]_x + \left[v - \frac{c_0 \rho'}{\rho}\right]_t = 0, \quad C_-: \quad \frac{dx}{dt} = -c_0.$$

$$(4.26)$$

We set

$$l(\rho) = \int_{\rho_0}^{\rho'} \frac{c_0}{\rho'} d\rho' = c_0 \log\left(1 + \frac{\rho'}{\rho_0}\right) \approx c_0 \frac{\rho'}{\rho_0}, \quad (4.27)$$

where the approximation $c \approx c_0$ is used. Compare this result for the linear case with that for the quasilinear case given by eq. (4.15). The Riemann invariants for the linear case become

$$r = v + \frac{c_0 \rho'}{\rho_0}, \quad s = v - \frac{c_0 \rho'}{\rho_0}, \quad (4.28)$$

as seen by inserting these expressions for r and s in equations (4.25) and (4.26). For eq. (4.25), we get $dr/dt = 0$ and for (4.26), $ds/dt = 0$, which again shows that these expressions are the Riemann invariants for the linear case. As for the quasilinear case, r defines a C_+ characteristic each of which is generated by varying s, while s defines a C_- characteristic generated by r.

The characteristic net in the (x, t) plane consists of two families of straight parallel lines of slopes $\pm 1/c_0$ corresponding to the $C_\pm$ characteristics, respectively. For the Cauchy IV problem, if we are given ρ' and v (the ICs) along the IV curve $t = 0$, then we can easily solve this problem at any field point by constructing the Riemann invariants. The reason for the easy solution to this problem is that the characteristics are particularly simple and do not depend on the solution, as in the quasilinear case.

Two-Dimensional Steady Flow

We now investigate the properties of two-dimensional isentropic, adiabatic, irrotational flow. This is an excellent example of the application of the method of characteristics to second-order quasilinear PDEs.

Let the vector $\mathbf{q}$ be the particle velocity. For two-dimensional flow $\mathbf{q} = (u, v, 0)$, where u, v are the x, y components of $\mathbf{q}$, respectively. Both u and v are functions of x, y and independent of t, since the flow is assumed to be steady (independent of time). Irrotational flow means that $\text{curl}\,\mathbf{q} \equiv \nabla \times \mathbf{q} = \mathbf{0}$. For the two-dimensional case $\text{curl}\,\mathbf{q} = \mathbf{0}$ reduces to the single irrotational condition

$$v_x - u_y = 0. \quad (4.29)$$

CHAPTER FOUR

We now write the Eulerian representation of the conservation laws for steady two-dimensional flow. The continuity equation becomes

$$(\rho u)_x + (\rho v)_y = 0. \tag{4.30}$$

This may be put in another form by expanding the total derivative of ρ. We get

$$\frac{d\rho}{dt} = u\rho_x + v\rho_y, \quad \text{where} \quad \frac{dx}{dt} = u, \quad \frac{dy}{dt} = v. \tag{4.31}$$

Using this equation, the continuity equation becomes

$$\rho(u_x + v_y) = -\frac{d\rho}{dt}. \tag{4.32}$$

This tells us that the product of the density and the divergence of the particle velocity is equal to the negative of the total time rate of change of the density.

For our two-dimensional steady flow the Euler equations of motion become

$$\frac{du}{dt} = uu_x + vu_y = -\frac{p_x}{\rho}, \quad x \text{ direction},$$

$$\frac{dv}{dt} = uv_x + vv_y = -\frac{p_y}{\rho}, \quad y \text{ direction}. \tag{4.33}$$

The x and y components of the particle acceleration are given as the convective terms, which are quadratic terms representing the spatial distribution of the acceleration, since for the steady-state condition $u_t = v_t = 0$. For the linear approximation we neglect these terms.

We convert the partial derivatives of ρ to those of p by using the fact that $dp/d\rho = c^2$. To this end we rewrite eq. (4.33) as

$$uu_x + vu_y = -\frac{c^2 \rho_x}{\rho},$$

$$uv_x + vv_y = -\frac{c^2 \rho_y}{\rho}. \tag{4.34}$$

We now multiply the first equation of (4.34) by u and the second by v and add, obtaining

$$u^2 u_x + uv(u_y + v_x) + v^2 v_y = -\left(\frac{c^2}{\rho}\right)(u\rho_x + v\rho_y).$$

Using the continuity equation (4.30), the right-hand side of the above equation becomes $c^2(u_x + v_y)$. We thereby obtain

$$(c^2 - u^2)u_x - uv(u_y + v_x) + (c^2 - v^2)v_y = 0. \qquad (4.35)$$

c^2 is a known function of q^2. (This fact will be seen below when we investigate Bernoulli's law.) The irrotational condition (4.29) and eq. (4.35) are the two required first-order quasilinear PDEs for the two unknowns u, v, whose solution gives the velocity field in the (x, y) plane.

BERNOULLI'S LAW

To show the relationship between c and $\mathbf{q}$ for an irrotational flow field we digress and state some results for such a field. An irrotational field means that there exists a *velocity potential* ϕ such that grad $\phi = \mathbf{q}$. The flow consists of *streamlines*, which means that $\mathbf{q}$ is always tangent to a streamline. Along each streamline the following relationship between $\mathbf{q}$ and p holds:

$$\frac{1}{2}q^2 + \left(\frac{1}{\rho}\right)\nabla p = \text{const.} \qquad (4.36)$$

This is a form of the energy equation for potential flow. The constant holds on a streamline but may vary along different streamlines. The energy equation for an adiabatic gas given by Bernoulli's law takes the form

$$\mu^2 q^2 + (1 - \mu^2)c^2 = c_*^2, \qquad \mu^2 = \frac{\gamma - 1}{\gamma + 1}, \qquad (4.37)$$

where c_* called the critical sound speed. The critical sound speed allows us to ascertain the supersonic or subsonic character of the flow

CHAPTER FOUR

by comparing the flow speed q with c_*. This is seen by adding and subtracting q^2 and thereby putting eq. (4.37) into the following form:

$$q^2 - c_*^2 = (1 - \mu^2)(q^2 - c^2). \tag{4.37a}$$

Introducing the *Mach number* $M = q/c$, we deduce the following conditions from eq. (4.37a):

$$q > c_* \quad \text{if and only if } M > 1,$$
$$q < c_* \quad \text{if and only if } M < 1.$$

Clearly, $M > 1$ is supersonic flow and $M < 1$ is subsonic flow. $M = 1$ is called transonic flow.

In order to derive Bernoulli's law (4.37) we must present the necessary thermodynamic relations that enter into deriving the energy equation for an adiabatic gas. The first law of thermodynamics can be written as

$$di = \frac{dp}{\rho} + TdS.$$

Here i is the specific enthalpy (enthalpy per unit mass) defined by

$$i = e + \frac{p}{\rho},$$

where e is the specific internal energy, S the specific entropy, and T the absolute temperature. For an isentropic fluid $dS/dt = 0$ so that we get

$$\frac{di}{dt} = \frac{d}{dt}\left(\frac{p}{\rho}\right).$$

This can also be derived from the fact that the internal energy e is independent of time.

Euler's equation of motion in vector form for an external force due only to pressure is

$$\rho \frac{dq}{dt} = -\nabla p,$$

which can be written as

$$\frac{d}{dt}\left(\frac{1}{2}q^2 + \frac{p}{\rho}\right) = 0. \qquad (4.38)$$

Inserting the expression $d/dt(p/\rho) = di/dt$ into eq. (4.38) yields

$$\frac{d}{dt}\left(\frac{1}{2}q^2 + i\right) = 0. \qquad (4.39)$$

Integrating eq. (4.39) gives

$$\tfrac{1}{2}q^2 + i = \hat{q}^2. \qquad (4.40)$$

The physical significance of the integration constant $\hat{q}^2$ is that $\hat{q}$ is called the *limit speed*. It is constant along streamline but may be different along different streamlines. Equation (4.40) tells us that the particle speed q cannot exceed the limit speed $\hat{q}$.

Integrating the equation

$$di = \frac{dp}{\rho}$$

and using the adiabatic equation of state given by eq. (4.5) yields the following relationship between i and c^2:

$$i = \frac{c^2}{\gamma - 1}.$$

Substituting this expression for i into eq. (4.40) yields Bernoulli's law in the form given by eq. (4.37), which is what we set out to prove. Note that $c_* = \mu \hat{q}$.

Method of Characteristics Applied to Two-Dimensional Steady Flow

We are now in a position to obtain the two families of characteristics in the (x, y) plane and investigate their consequences. The structure we start with consists of the irrotational condition given by eq. (4.29) and

161

CHAPTER FOUR

the quasilinear PDE given by eq. (4.35). This is the system of equations we shall deal with in applying the method of characteristics. We now rewrite this system in a form similar to that used in our treatment of one-dimensional gas dynamics:

$$L_1 = v_x - u_y, \tag{4.41}$$

$$L_2 = (c^2 - u^2)u_x - uv(u_y + v_x) + (c^2 - v^2)v_y = 0. \tag{4.42}$$

We define the parameter λ involved in the linear combination of L_1 and L_2 by setting

$$L = \lambda L_1 + L_2.$$

The reason for this definition of L instead of $L = L_1 + \lambda L_2$ is the more convenient form that λ takes, namely, $\pm c\sqrt{q^2 - c^2}$ rather than its reciprocal. The equation $L = 0$ when written out becomes

$$L = (c^2 - u^2)u_x - (uv + \lambda)u_y - (uv - \lambda)v_x + (c^2 - v^2)v_y = 0.$$

Setting $dy/dx = z$, we use the rule of the ratio of coefficients. This gives

$$\text{coeff}(u_y) : \text{coeff}(u_x) = \text{coeff}(v_y) : \text{coeff}(v_x) = \frac{dy}{dx} = z, \tag{4.43}$$

$$\lambda_\pm = \pm c\sqrt{q^2 - c^2},$$

and

$$z_\pm = -uv \pm c \frac{\sqrt{q^2 - c^2}}{c^2 - u^2}. \tag{4.44}$$

It is easily seen that the values of $z_\pm$ are the roots of the quadratic equation

$$(c^2 - u^2)z^2 + 2uvz + (c^2 - v^2) = 0. \tag{4.45}$$

Equation (4.45) is the required characteristic ODE, which yields the two distinct families of characteristics $C_\pm$. It is a first-order second-degree ODE. Note that, in order for the roots to be real and unequal, we must have the inequality $q > c$, which establishes the fact that our system is supersonic. The character of the roots z characterizes the type of flow and the type of the quasilinear PDE. This is summarized in table 4.1.

TABLE 4.1
Classification of Roots z for Two-Dimensional Steady Flow

Nature of roots z	Type of PDE	$M = q/c$	Type of flow
Real unequal	Hyperbolic	> 1	Supersonic
Real equal	Parabolic	$= 1$	Transonic
Complex conjugates	Elliptic	< 1	Subsonic

SUPERSONIC VELOCITY POTENTIAL

We mentioned above that for irrotational flow a velocity potential ϕ exists such that

$$\nabla \phi = q, \quad u = \phi_x, \quad v = \phi_y.$$

It follows that the quasilinear PDE given by eq. (4.35) becomes

$$(c^2 - \phi_x^2)\phi_{xx} - 2(\phi_x \phi_y)\phi_{xy} + (c^2 - \phi_y^2)\phi_{yy} = 0. \quad (4.46)$$

This is a second-order quasilinear PDE whose generic form is given in chapter 2 [eq. (2.44)], where the lower-order terms are omitted. The method of characteristics applied to this PDE yields the same results as above, the characteristic ODE is still given by eq. (4.45), and so on. We can, of course, use the method described in chapter 2, "Another Method of Characteristics for a Second-Order Quasilinear PDE," where the system is transformed into a pair of first-order PDEs. The same results occur.

HODOGRAPH TRANSFORMATION

This transformation is a mapping of the flow equations from the x, y or physical space to the u, v or velocity space. This two-dimensional space is sometimes called the *hodograph plane*.

Jacobian as a Mapping Function from the Physical to the Hodograph Plane

The Jacobian transformation is a nonsingular mapping function (the determinant of the Jacobian is not zero) that allows us to map $(x, y) \rightleftarrows$

(u, v). We shall use the Jacobian to transform the flow equations (4.41) and (4.42) from the physical to the hodograph plane. We then obtain a system of first-order PDEs in the hodograph plane whose coefficients are known functions of u and v. Therefore, this hodograph transformation produces a linear system of PDEs—this gives us a great advantage. The system is then solved for $x = x(u, v)$, $y = y(u, v)$. The inverse mapping yields the flow field $u = u(x, y)$, $v = v(x, y)$.

To obtain the Jacobian mapping function we start by expanding the differentials dx and dy in terms of u and v:

$$dx = x_u \, du + x_v \, dv, \qquad dy = y_u \, du + y_v \, dv.$$

We first keep y constant, divide these equations by dx, and obtain a pair of algebraic equations for u_x and v_x:

$$1 = x_u u_x + x_v v_x,$$
$$0 = y_u u_x + y_v v_x.$$

The solution of this system is

$$u_x = \frac{y_v}{\det(J)}, \qquad v_x = -\frac{y_u}{\det(J)},$$

where the Jacobian mapping function and its determinant are

$$J = \begin{pmatrix} x_u & x_v \\ y_u & y_v \end{pmatrix}, \qquad \det(J) = x_u y_v - x_v y_u. \qquad (4.47)$$

We now keep x constant and divide the above differentials by dy and obtain a pair of equations for u_y and v_y:

$$0 = x_u u_y + x_v v_y, \qquad 1 = y_u u_y + y_v v_y.$$

The solution is

$$u_y = -\frac{x_v}{\det(J)}, \qquad v_y = \frac{x_u}{\det(J)}.$$

It is easily seen that j, the Jacobian of the mapping from the u, v to the x, y plane, is

$$j = \begin{pmatrix} u_x & u_y \\ v_x & v_y \end{pmatrix}, \qquad \det(j) = u_x v_y - u_y v_x = \frac{1}{\det(J)}. \qquad (4.48)$$

Summarizing the above results gives

$$\begin{pmatrix} u_x & u_y \\ v_x & v_y \end{pmatrix} = \det(j) \begin{pmatrix} y_v & -y_u \\ -x_v & x_u \end{pmatrix}. \qquad (4.49)$$

Since $\det(J) \neq 0$, it follows that $\det(j) \neq 0$. Equation (4.49) is the required hodograph transformation from the x, y to the u, v plane. It gives the solution for $u_x, \ldots, v_y$ in terms of $x_u, \ldots, y_v$, which can easily be read from the matrix equation (4.49). Applying these results to the flow equations (4.41) and (4.42) yields

$$-y_u + x_v = 0, \qquad (4.50)$$

$$(c^2 - u^2)y_v + uv(x_v + y_u) + (c^2 - v^2)x_u = 0. \qquad (4.51)$$

Equations (4.50) and (4.51) are the required flow equations in the u, v plane. As conjectured above, they are linear PDEs, since the coefficients are known functions of u, v. The importance of this transformation to a linear system of flow equations is that the characteristics in the hodograph or u, v plane are independent of the solution.

Characteristics in the Hodograph Plane

We shall obtain the two families of characteristics in the hodograph plane by the usual method of getting that linear combination of $x_u, \ldots, y_v$ which puts the directional derivatives in the characteristic direction dv/du in the u, v plane. To this end we multiply eq. (4.50) by λ, add eq. (4.51), and obtain

$$(c^2 - u^2)y_v + (uv + \lambda)x_v + (uv - \lambda)y_u + (c^2 - v^2)x_u = 0. \quad (4.52)$$

In order for the linear combination of partial derivatives to be in the same direction as the directional derivatives dx/du and dy/du, we apply the rule

$$\text{coeff}(x_v) : \text{coeff}(x_u) = \text{coeff}(y_v) : \text{coeff}(y_u) = \frac{dv}{du}$$

CHAPTER FOUR

and obtain

$$\left(\frac{dv}{du}\right)_{\pm} \equiv Z_{\pm} = \frac{uv \pm c\sqrt{(q^2 - c^2)}}{c^2 - v^2}, \tag{4.53}$$

$$\lambda_{\pm} = \pm c\sqrt{(q^2 - c^2)}. \tag{4.54}$$

Equation (4.53) are the ODEs for the slopes of the two families of characteristic curves in the u, v plane. We shall call these the $\Gamma_{\pm}$ families, relating to $Z_{\pm}$ and $\lambda_{\pm}$, respectively.

Geometric Interpretation of the Characteristics

We now give a geometric interpretation of the characteristics in the x, y plane. To this end we consider fig. 4.1, which shows the the $C_{\pm}$ characteristics and **q** drawn from the field point P. **q** makes the angle θ with the horizontal. c is drawn from A at the head of **q** normal to the

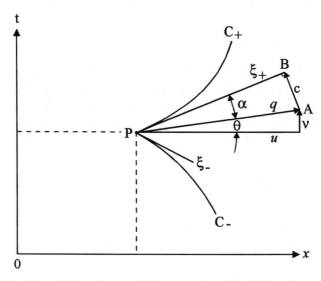

FIGURE 4.1. C_- and C_+ characteristics stemming from field point P. Particle velocity **q** stems from P. Sound speed is c; AB is normal to the tangent of C_+ at P.

tangent to C_+ and intersects the tangent at B. The tangents to $C_\pm$ make the angle α with $\mathbf{q}$. From the figure we get the relations

$$u = q\cos\theta, \qquad v = \sin\theta, \qquad \sin\alpha = \frac{c}{q} = \frac{1}{M}, \qquad z_\pm = \tan(\theta \pm \alpha).$$
(4.55)

α is called the *Mach angle*.

If the quadratic for z, given by eq. (4.45), is put in the form

$$c^2(1 + z^2) = (uz - v)^2,$$
(4.45a)

and if $z = \tan(\theta + \alpha)$ is inserted into the above equation, then we can obtain $c^2 = q^2 \sin^2\alpha$ by using some trigonometric relations.

Characteristics in the Hodograph Plane for an Adiabatic Gas

In [9] and [11] it was shown that the Γ characteristics in the hodograph plane for an adiabatic gas are epicycloids. It is convenient to repeat this analysis. We use Bernoulli's equation in the form

$$q^2[\mu^2 + (1 - \mu^2)\sin^2\alpha] = c_*^2,$$
(4.56)

where $c^2 = q^2\sin\alpha$ is inserted into eq. (4.37a).

We shall use eq. (4.56) to show that the Γ characteristics in the hodograph plane are *epicycloids* generated by the points on a circle of diameter $c_*[(1/\mu) - 1] = \hat{q} - c_*$ which rolls on the sonic circle of radius $\sqrt{u^2 + v^2} = c_*$.

Figure 4.2 shows the construction of the trajectory of a Γ characteristic. A circle C' of center O' and radius $QO' = r$ rolls without slipping on the circle C of radius $OQ = c_*$. The curve AP is a portion of the trajectory, the epicycloid, traced out by the point P fixed on the rolling circle. The point Q is the center of rotation so that QP is normal to the tangent to the trajectory at P. $PQ = \mathbf{q}$. Angle QPO turns out to be the Mach angle α. It also turns out that $QQ' = \frac{1}{2}(q' - c_*)$. These conjectures will now be proved. From the figure, we have angle $QO'P = \beta$ and angle $O'QP = (\pi - \beta)/2$, since triangle $QO'P$ is isosceles. Then angle

CHAPTER FOUR

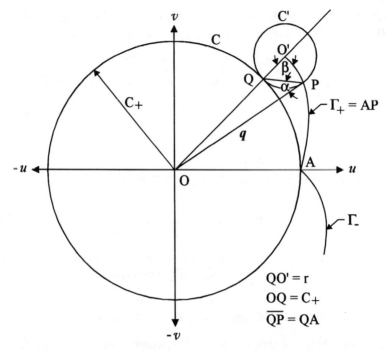

FIGURE 4.2. Construction of an epicycloid as a trajectory of the Γ_+ characteristic. Circle C' rolls on the circle C.

$OQP = (\pi + \beta)/2$. From the triangle $OO'P$ we obtain the relation

$$q^2 = (c_* + r)^2 + r^2 - 2r(c_* + r)\cos\beta.$$

Applying the law of sines to triangle OQP yields

$$\cos\frac{\beta}{2} = \left(\frac{q}{c_*}\right)\sin\alpha.$$

This gives

$$\cos\beta = 2\left(\frac{q}{c_*}\right)^2 \sin^2\alpha - 1.$$

168

We then get

$$q^2 = (c_* + r)^2 + r^2 - 2r(c_* + r)\left[2\left(\frac{q}{c_*}\right)^2 \sin^2 \alpha - 1\right].$$

This simplifies to

$$q^2\left[1 + 4\left(\frac{r}{c_*^2}\right)(c_* + r)\sin^2 \alpha\right] = (2r + c_*)^2.$$

Comparing this equation with eq. (4.56) gives the result

$$r = \left(\frac{c_*}{2}\right)(\mu^{-1} + 1) = \frac{1}{2}(\hat{q} - c_*). \tag{4.57}$$

This is the expression we set out to prove. The components of **q** are

$$u = (c_* + r)\cos \theta - r\cos\left[(c_* + r)\left(\frac{\theta}{r}\right)\right],$$
$$v = (c_* + r)\sin \theta - r\sin\left[(c_* + r)\left(\frac{\theta}{r}\right)\right]. \tag{4.58}$$

The system (4.58) gives the trajectory of a Γ_+ characteristic in the hodograph plane if we plot v versus u. The starting point of the rolling circle corresponds to $\theta = 0$ (point A in fig. 4.2). A similar construction can be made for Γ_-. Again referring to the figure, through each point of the annular ring $c_* < q < \hat{q}$ there pass two epicycloids (one for Γ_+ and one for Γ_-) so that this ring is covered with a net of two families of Γ characteristics in the hodograph plane.

Steady-state plane compressible flow offers a rich variety of special problems such as flow in nozzles, around corners, in jets, and around aerodynamic bodies. This is beyond the scope of this book. The reader is referred to the references on aerodynamics listed in the bibliography.

Shock Wave Phenomena

Introduction

The study of shock waves in air is important to the understanding of the flight of supersonic aircraft, and as a background for the study of detonation waves in explosives.

CHAPTER FOUR

Up to now we have been investigating one-dimensional unsteady compressible flow and two-dimensional steady compressible flow. We saw that the flow equations arose from the three conservation laws, where the energy equation assumed an adiabatic equation of state. All the dynamic and thermodynamic variables were assumed to be continuous everywhere in the medium, including across the wave fronts, which are the characteristic curves. Any disturbance in the initial data can only be transmitted along the characteristics with the speed of sound c. These families of characteristics represent wave fronts for very weak disturbances (sound waves), across which the dynamic and thermodynamic variables are continuous. This means that we are dealing with an adiabatic, isentropic flow such that there is a thermodynamically reversible transition of flow across the wave front.

The above assumptions do not fit the experimental facts for certain wave fronts. In 1870, Rankine (cited in [29, vol. II]) showed that no steady adiabatic process in which there are only pressure forces can represent a continuous change from one constant state to another in the region of a wave front of finite amplitude. He proposed instead that across this region a nonadiabatic process should occur, subject to the condition that heat may be transferred to the fluid particles but that no heat be received from the external environment. His condition satisfies the conservation of energy.

Riemann in 1860 developed the general theory of wave propagation making use of his Riemann invariants. He rediscovered and elaborated on the theory of shock waves, but made the incorrect assumption that the transition across the shock wave is adiabatic and reversible. In 1910, Rayleigh [29, vol. II] pointed out that an adiabatic reversible process would violate the principle of conservation of energy. He showed that the entropy must increase across a shock front. Other investigators such as Rankine (as mentioned above), Hugoniot, and Stokes showed the inconsistency of assuming an adiabatic and reversible flow across a shock.

Surfaces of Discontinuity

Before we embark on the systematic study of shock wave phenomena in one spatial dimension, we distinguish between two types of surfaces of discontinuity, *contact surfaces* and *shock fronts*.

A contact surface separates two parts of the medium without any flow across the surface. Therefore, this surface moves with the fluid and separates two zones of different densities and entropies. A contact surface is important in certain chemical processes where media of different densities and entropies occur.

A shock front is a surface of discontinuity across which gas flows. The side of the shock front through which the gas enters is called the front side of the shock while the other side is called the back side. We shall show that the shock front always moves with supersonic speed as observed from the front side, and with subsonic speed when observed from the back side. Since we are dealing with one dimension, the shock front is assumed to be a planar surface normal to the x axis, which is the direction of flow. In the (x, t) plane the shock trajectory is called a *shock line*.

Shock Tube

A shock tube is a long thin tube of gas in which a shock wave is initiated, usually at one end of the tube by a piston. Such a shock tube is used as a physical model for investigating one-dimensional shock wave phenomena. Instruments are installed that measure the dynamic and thermodynamic variables across the shock, the shock speed, etc.

Initially the gas in the tube is in a quiescent state where the dynamic and thermodynamic variables are constant and there is no motion. Thus, initially, $p = p_0$, $\rho = \rho_0$, $T = T_0$, and $\mathbf{q} = \mathbf{0}$, where $\mathbf{q} = (u, 0, 0)$. As time increases, it is clear that $p = p(x, t)$, $u = u(x, t)$, etc. The shock tube is represented by the positive x axis, and a piston is placed at the origin at $t < 0$. A shock front is generated by suddenly moving the piston to the right at $t = 0$ with a constant velocity U_p. Clearly, this is an idealization, for at $t = 0$ we assume a finite jump in the piston velocity, which means an infinite acceleration at $t = 0$ no matter how small the piston velocity. Figure 4.3 shows a plot of the piston velocity versus time, which is a step function.

Because of the assumption of infinite acceleration of the piston at $t = 0$, the motion of the gas particles in the neighborhood of $(x, t) = (0, 0)$ must be discontinuous. What happens is that initially a shock front forms at the right-hand side of the piston and eats into the quiet gas with a constant shock wave velocity $U > c$. In fig. 4.4 we show an advancing shock front in the shock tube. To the left of the moving piston there exists a rarefaction region (2) where the pressure is less than the ambient or quiescent pressure. The undisturbed or ambient region is designated by the subscript 0 and the disturbed region by the

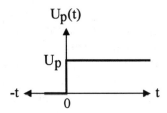

FIGURE 4.3. Plot of finite jump U_p in piston velocity $U_p(t)$ vs. time t.

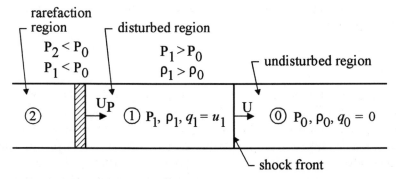

FIGURE 4.4. Advancing shock front of velocity U in a shock tube.

subscript 1. Figure 4.5 shows the trajectories of the piston, shock front, particle velocity (which is zero until disturbed by the shock front, whereupon it takes on the value of the piston velocity), and the C_+ characteristic. The situation depicted by this figure is an approximation that is valid in the neighborhood of the piston. A more exact solution can be obtained by solving the quasilinear flow equations with the appropriate boundary conditions valid in the wedge region bounded by the piston and the shock front. The solution will, in general, give a nonconstant c, yielding curved characteristics.

If we perform a series of experiments with decreasing piston velocities, we will obtain a sequence of shock fronts of decreasing supersonic values of U. In the limit as $U_p \to 0$, $U \to c_0$, the constant velocity of the sound wave. The *shock strength* is defined by the dimensionless ratio $(p - p_0)/p_0$. The pressure p in back of the shock front is greater than the ambient pressure p_0 and approaches p_0 from above as the piston velocity tends to zero, producing a sound wave or "sonic disturbance."

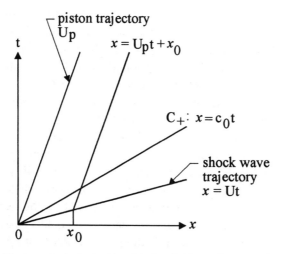

FIGURE 4.5. Various trajectories in the (x, t) plane for an advancing shock wave.

In this sense, a sound wave is called an *infinitely weak shock wave*.

There are three types of shock fronts:

(1) *Advancing shock front.* Suppose the ambient particle velocity $u_0 = 0$. The shock front advances into the quiescent region 0 with the velocity U, which will be shown to be supersonic as seen from the front side. The velocity of the shock front as observed from the disturbed region 1, the high-pressure side, is $U - u_1$, which turns out to be subsonic when observed from the back side.

(2) *Stationary shock front.* A shock front is stationary ($U = 0$) as seen by an observer riding with the front.

(3) *Receding shock front.* The shock front recedes into region 0 with velocity U, leaving behind a high pressure region 1 in which $u_1 = 0$. Such a receding shock is encountered as a shock front reflecting from a wall.

Qualitative Concepts and Analogy with Particle Motion

Before we embark on a quantitative treatment of shock phenomena, it is of interest to show how shock waves are produced by a finite system of particles, namely, moving cars on a highway. Suppose a column of

CHAPTER FOUR

cars is moving on a single-lane highway at uniform slow speed. A deceleration of one car will give the cars behind time to react, if they are not too close, so that no cars suddenly decelerate. The cars are said to be moving at less than the "sound speed" or "critical speed." Suppose the column speeds up so that if a car in front suddenly slows down the one behind is forced to suddenly slow down, and similarly for the ones behind. The speed at which this process just occurs is the sound speed. Clearly, it depends on how close the cars are, and their speed. Suppose the column moves faster than the sound speed. Then if the first car suddenly stops the second will crash into it, the third will crash into the second, etc. A receding shock wave will be produced, which travels back with a characteristic speed if the column were traveling with the same speed and the cars were equally spaced. Clearly, the trajectory of the shock front is the locus of points of intersection of the hit cars. Next, we can easily visualize an advancing shock front impinging on a quiescent region by considering a car moving with a speed greater than the critical speed and crashing into a uniform row of parked cars in neutral with their brakes off. The first car hit will set into motion the second car, etc., thus producing an advancing shock front. The locus of intersection of these successively hit cars is the trajectory of the shock front advancing into the quiescent region. We can represent each car by a mass coupled to neighboring masses by nonlinear springs. Thus a system of coupled nonlinear ODEs can be derived. The loci of points of collision are the mass-spring systems that give maximum spring compression.

This approach is merely illustrative. There is a discipline of traffic dynamics that makes use of the above model along with certain optimization principles. We shall not pursue this topic but will now investigate shock phenomena quantitatively.

Conservation Laws across the Shock Front

We reformulate the conservation laws in Eulerian form for a one-dimensional gas in order to obtain the jump or discontinuity conditions across the shock front. As usual, let x be the Eulerian coordinate and $a(x, t)$ be the Lagrangian coordinate. Suppose a particle of gas, identified by its Lagrangian coordinates, occupies the volume (a_0, a_1) at time t (a unit cross-sectional area is assumed). Therefore $a_i = a_i(t)$, $i = 0, 1$.

WAVE PROPAGATION IN FLUIDS

A discontinuity in the form of a shock front occupies the interior of this interval. This will be exploited below. We formulate the conservation laws for this volume of gas.

Conservation of Mass (Continuity Equation)

$\rho \, dx$ is the element of mass occupying the interval $(x, x + dx)$ at time t, where $a_0 \leq x \leq a_1$, $a_0 < a_1$. Since the mass occupying the interval (a_0, a_1) must be invariant with time, the law of conservation of mass becomes

$$\frac{d}{dt} \int_{a_0}^{a_1} \rho(x, t) \, dx = 0. \tag{4.59}$$

As seen, the integration is with respect to x for a fixed t. The end points, of course, depend on t.

Conservation of Momentum

The linear momentum of a gas particle occupying the interval $(x, x + dx)$ is $\rho u \, dx$, where u is the particle velocity. The time rate of increase of this momentum integrated over the interval (a_0, a_1) equals the net force on the column, which is assumed to be due only to the pressure gradient acting at the ends of the column. The conservation of momentum therefore becomes

$$\frac{d}{dt} \int_{a_0}^{a_1} \rho u \, dx = -(p_1 - p_0). \tag{4.60}$$

Conservation of Energy

To formulate the energy equation for the column of gas in the interval (a_0, a_1), we equate the time rate of increase of the total energy to the power input due to the pressure gradient at the ends of the interval. Let e be the specific internal energy (per unit mass) and let $\frac{1}{2} u^2$ be the kinetic energy per unit mass. The total energy per unit mass is the sum of the internal and the kinetic energies per unit mass. Also, let $p_i u_i$ be

CHAPTER FOUR

the power input due to the pressure at a_i for $i = 0, 1$. Then the energy equation becomes

$$\frac{d}{dt} \int_{a_0}^{a_1} \left(\rho e + \tfrac{1}{2} u^2 \right) dx = -(p_1 u_1 - p_0 u_0). \tag{4.61}$$

Entropy Change

Let S be the specific entropy (entropy per unit mass). Then the entropy of an element of gas in the interval $(x, x + dx)$ is $\rho S\, dx$, so that the total entropy of the column is

$$\int_{a_0}^{a_1} \rho S\, dx.$$

Since the shock front is in the interior of the column and the thermodynamic process involving flow across the front is irreversible, the entropy must increase in the column. This is represented by

$$\frac{d}{dt} \int_{a_0}^{a_1} \rho S\, dx > 0. \tag{4.62}$$

As already mentioned, for an infinitely weak shock or sound wave the thermodynamic process is revertible, so that there is no change in entropy across the wave front.

Jump Conditions

The integrals in the three conservation equations are of the form

$$I = \int_{a_0}^{a_1} F(x, t)\, dx. \tag{4.63}$$

It has been mentioned several times that an interior point of the interval (a_0, a_1) is the shock front. Call this point $x = \bar{x}$ where $a_0 < \bar{x} < a_1$. Because of the shock front there is a discontinuity in the integrand F. Let $F_\pm$ be the value of F as it approaches $\bar{x}$ from the right or left, respectively. Then $F_+ - F_- \neq 0$.

We now calculate dI/dt using the formula for differentiating an integral with respect to the parameter t:

$$\frac{dI}{dt} = \left(\frac{d}{dt}\right)\left[\int_{a_0}^{\bar{x}_-} + \int_{\bar{x}_+}^{a_1}\right] F(x,t)\, dx$$

$$= \left(\frac{d\bar{x}_-}{dt}\right)F(\bar{x}_-, t) - \left(\frac{da_0}{dt}\right)F(a_0, t) + \left(\frac{da_1}{dt}\right)F(a_1, t)$$

$$- \left(\frac{d\bar{x}_+}{dt}\right)F(\bar{x}_+, t) + \int_{a_0}^{a_1} F(x,t)_t\, dx. \quad (4.64)$$

This formula is valid no matter how small the interval is, so long as it contains $x = \bar{x}$. We now perform a limiting process by letting $a_1 - a_0 \to 0$ such that $a_0 \leq \bar{x} \leq a_1$. F_t is continuous across $\bar{x}$ so that the last integral in eq. (4.64) tends to zero. $F_- \to F(a_0, t) \equiv F_0$ and $F_+ \to F(a_1, t) \equiv F_1$. We also have $da_i/dt = u_i$ $(i = 0, 1)$, where u_i is the velocity at each end of the column and $\overline{dx/dt}$ is the velocity of the discontinuity point $\bar{x}$. This is the location of the shock front so that $\overline{dx/dt} = U$. Applying this limiting process to eq. (4.64) yields

$$\lim_{a_1 - a_0 \to 0}\left(\frac{dI}{dt}\right) = v_1 F_1 - v_0 F_0, \quad (4.65)$$

where

$$v_i = u_i - U \quad (i = 0, 1), \quad (4.66)$$

so that v_i is the velocity of each end point relative to the shock front velocity U. The jump conditions become

$$\frac{dI}{dt} = v_1 F_1 - v_0 F_0, \quad F = \begin{cases} \rho & \text{(cons. of mass)}, \\ \rho u & \text{(cons. of momentum)}, \\ \rho\left(e + \frac{1}{2}u^2\right) & \text{(cons. of energy)}. \end{cases}$$

$$(4.67)$$

Shock Conditions

Applying the jump conditions given by the values of F from eq. (4.67) to the conservation laws given by eqs. (4.59)–(4.61), we obtain the follow-

CHAPTER FOUR

ing shock conditions, which are finite jump conditions across the shock front:

$$\rho_1 v_1 = \rho_0 v_0 = m \quad \text{(cons. of mass)} \quad (4.68)$$

where m is the mass flux of fluid across the shock front;

$$(\rho_1 u_1)v_1 - (\rho_0 u_0)v_0 = -(p_1 - p_0) \quad \text{(cons. of momentum);} \quad (4.69)$$

$$\rho_1 v_1\left(e_1 + \tfrac{1}{2}u^2\right) - \rho_0 v_0\left(e_0 + \tfrac{1}{2}u^2\right) = -(p_1 u_1 - p_0 u_0)$$
$$\text{(cons. of energy);} \quad (4.70)$$

and

$$\rho_1 v_1 S_1 - \rho_0 v_0 S_0 > 0 \quad \text{(entnropy increase).} \quad (4.71)$$

For a shock front there is flow across the discontinuity surface so that $m \neq 0$. For a contact surface there is no mass flow so that $m = 0$.

We shall show that each of the three shock conditions, eqs. (4.69), (4.70), (4.71), can be written in a form involving only the relative velocities v_0, v_1. This means that the shock conditions are invariant with respect to a translation of the coordinate system, assuming a constant value of U. The shock conditions have the same form for an observer fixed in space or fixed on the shock front (stationary shock). To show this we use eq. (4.68), $u_i = v_i + U$, and write the momentum equation (4.69) in the form

$$mu_1 - mu_0 = m(v_1 + U) - m(v_0 + U) = m(v_1 - v_2) = -(p_1 - p_0).$$

Using the definition of m from eq. (4.68) we rewrite the above equation as

$$\rho_1 v_1^2 + p_1 = \rho_0 v_0^2 + p_0 = P, \quad (4.72)$$

where P is called the *total flux of momentum* across the shock front; it is clearly constant since it depends only on the ambient conditions.

The energy equation (4.70) can also be rewritten in terms of the relative velocities:

$$\frac{1}{2}(v_1^2 - v_0^2) + e_1 - e_0 = -(p_1 \tau_1 - p_0 \tau_0), \quad \tau_i = \frac{1}{\rho_i}. \quad (4.73)$$

WAVE PROPAGATION IN FLUIDS

τ_i is the specific volume for the region i. Recall that the specific enthalpy $i = e + p\tau$. Inserting this expression into eq. (4.73) gives Bernoulli's equation in the form

$$\tfrac{1}{2}v_1^2 + i_1 = \tfrac{1}{2}v_0^2 + i_0 = \tfrac{1}{2}\hat{q}^2. \tag{4.74}$$

We write down other forms of the Bernoulli equation using the conservation laws:

$$v_1^2 - v_0^2 = (p_0 - p_1)(\tau_0 + \tau_1) = -\Delta p(\tau_0 + \tau_1), \quad p_1 - p_0 = \Delta p, \tag{4.75}$$

$$\tfrac{1}{2}\Delta p(\tau_0 + \tau_1) = i_1 - i_0 = \Delta i. \tag{4.76}$$

Equation (4.76) tells us that the increase in specific enthalpy across the shock front is equal to the work done by the pressure gradient acting on the average specific volume. This equation may be put in the form

$$-\tfrac{1}{2}\Delta\tau(p_0 + p_1) = \Delta e. \tag{4.77}$$

Equation (4.77) tells us that the increase in the internal energy across the shock front is equal to the work done by the average pressure performing the compression $+\Delta\tau$. Equations (4.76) and (4.77) are important equations that hold across the shock front. They are called the *Hugoniot relations*. They will be discussed below for an adiabatic gas.

Mechanical Shock Conditions

The mechanical shock conditions are defined as those conditions across the shock front that depend only on the conservation of mass and momentum. Since they are independent of the energy equation or equation of state, they are valid for any continuous medium as long as there is flow across the wave front (as long as $m \neq 0$). It is easily seen that the following relations (which are given for reference) summarize

CHAPTER FOUR

the mechanical shock conditions:

$$
\begin{aligned}
& m = \rho_1 v_1 = \rho_0 v_0, \qquad \rho_1 v_1^2 + p_1 = \rho_0 v_0^2 + p_0 = P, \\
& m\,\Delta v = -\Delta p, \qquad m^2 = \frac{\Delta p}{\Delta \tau}, \qquad v_0 v_1 = \frac{\Delta p}{\Delta \rho}, \\
& \tau_0\,\Delta p = -v_0\,\Delta v, \qquad \tau_1\,\Delta p = -v_1\,\Delta v, \\
& \Delta\tau\,\Delta p = -(\Delta v)^2, \qquad (\tau_0 + \tau_1)(p_0 + p_1) = -(v_1^2 - v_0^2).
\end{aligned}
\qquad (4.78)
$$

Hugoniot Relations for an Adiabatic Gas

In order to investigate the shock conditions for a specific gas, in this case an adiabatic gas, we must invoke the conservation of energy or equation of state, in addition to the mechanical shock conditions. To this end we reformulate the Hugoniot equation given by eq. (4.77) by replacing p_1 and e_1 with the variables p and e, respectively, since the disturbed state given by region 1 is now considered a variable state. We now introduce the *Hugoniot function* $H(p, \tau, e)$, which we set equal to zero, obtaining

$$H(p, \tau, e) = e - e_0 + \tfrac{1}{2}(\tau - \tau_0)(p + p_0) = 0. \qquad (4.79)$$

This is the Hugoniot relation. It characterizes all admissible pairs of (p, τ) when values of (p_0, τ_0) are given in the undisturbed region, for a gas with a given equation of state defined by e, assumed to be a function of p and τ.

We now apply the Hugoniot relation to an adiabatic gas. The specific heat at constant volume C_v is defined in terms of e and the temperature T by $e = C_v T$. Our adiabatic gas is assumed to be a perfect gas, so that $p\tau = RT$, where R is the gas constant or Boltzmann constant. It is easily shown that $R = C_p - C_v$, where C_p is the specific heat at constant pressure. The above relations allow us to get e as a function of p and τ, thus obtaining

$$e = \frac{p\tau}{\gamma - 1}.$$

WAVE PROPAGATION IN FLUIDS

Inserting this expression for e into the Hugoniot relation (4.79) and simplifying yields

$$p(\tau - \mu^2\tau_0) - p_0(\tau_0 - \mu^2\tau) = 0, \qquad \mu^2 = \frac{(\gamma - 1)}{(\gamma + 1)}. \qquad (4.80)$$

Equation (4.80) is the Hugoniot relation for an adiabatic gas. For air under standard conditions, $\gamma = 1.4$ so that $\mu^2 = \frac{1}{6}$. For a given undisturbed condition (given p_0, τ_0), this Hugoniot relation generates a rectangular hyperbola. Figure 4.6 shows a plot of p versus τ (a Hugoniot curve) for an adiabatic gas. As τ decreases to its minimum value $\tau_m = \mu^2\tau$, p increases asymptotically to infinity, and as τ increases to its maximum value $\tau_M = (1/\mu^2)\tau$, p tends to zero.

Prandtl's Relation

Prandtl's relation is an elegant relation between the relative velocities v_0 and v_1 on both sides of the shock wave and the critical sound speed. It is given by the equation

$$v_0 v_1 = c_*^2, \qquad (4.81)$$

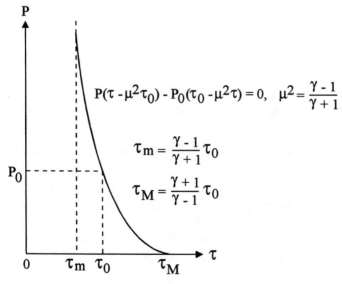

FIGURE 4.6. Hugoniot curve: plot of P vs. τ.

CHAPTER FOUR

where c_* is the critical sound speed. (Recall that the supersonic or subsonic character of the flow can be determined by comparing the flow speed q with the critical sound speed c_*.)

The reader can easily derive eq. (4.81) by applying Bernoulli's law in the form given by eq. (4.37) on both sides of the shock; set $q = v_0, v_1$, where $c = c_0, c_1$, respectively, using the expression $c_i^2 = \gamma p_i/\rho_i$ ($i = 0, 1$) and the mechanical shock condition given by eq. (4.72). The result is the equation

$$\frac{p_1 - p_0}{\rho_1 - \rho_0} = c_*^2. \tag{4.82}$$

This is left for the reader.

Comparing eq. (4.82) with one of the shock conditions given by eq. (4.78) yields eq. (4.81).

One of the important features of Prandtl's relation is that it allows us to go smoothly from a strong shock to a very weak shock or sound wave. For a very weak shock we have $v_0 = v_1$ approximately, so that Prandtl's relation becomes $v_0^2 = v_1^2 = c_*^2 = c_0^2$. The shock condition $v_0 v_1 = (p_1 - p_0)/(\rho_1 - \rho_0)$ becomes approximately $\Delta p/\Delta \rho = v_1^2 = v_0^2 = c_0^2$. But $dp/d\rho = c_0^2$, which means that there exists an infinitely weak shock wave that propagates with the velocity of sound in the undisturbed region. Since the shock strength is defined by $(p_1 - p_0)/p_0$, it follows that a sound wave is a shock wave of infinitely weak strength.

Shock Transition

We state four basic properties of a shock transition, that is, the relations governing the pressure, density, and entropy across the wave front.

(1) The increase of entropy across a shock front is third order in shock strength.

(2) The increase in pressure, density, and temperature across the shock front differs from reversible adiabatic changes of these quantities at most in the terms of third order in the shock strength.

(3) Shocks are compressive, which means that the pressure, density, and entropy increase across the shock front.

(4) The flow velocity relative to the shock front is supersonic at the front side (undisturbed side) and subsonic at the back side. This

statement clearly follows from the mechanical shock condition $(\tau_0 + \tau_1)/(p_1 - p_0) = (v_0 - v_1)(v_0 + v_1)$ and statement 3, thus showing that $v_0 > v_1$.

PART II VISCOUS FLUIDS

Here we investigate the effect of energy dissipation on fluid motion. This dissipation, which is due to viscosity, is the result of *thermodynamic irreversibility*, which means that heat is lost in going through a thermodynamic cycle (an equilibrium state, motion, back to the equilibrium state). The physical laws that govern the dynamics of a viscous fluid (as in any continuum) are the three conservation laws, mass, momentum, and energy, which will be taken up in some detail. But first we give an elementary discussion of viscosity as a background.

ELEMENTARY DISCUSSION OF VISCOSITY

We start with some elementary ideas about the behavior of viscous fluids. Consider the model of a viscous fluid in the (x, y) plane between two parallel horizontal plates a distance h apart. Let the lower plate be fixed and the upper plate move to the right (in the x direction) with a constant velocity U. An axial force P must be applied to the upper plate in order to move it. Experiment shows that P is approximately proportional to U. Mathematically, we have

$$P = \frac{\mu A U}{h}, \qquad (4.83)$$

where A is the area of each plate and μ is a positive constant called the *dynamic viscosity coefficient*.

If U is not too large, the fluid performs a parallel gliding motion called *laminar flow*, where there is no interchange of particles between neighboring layers or laminae. This means that the particle velocity is axial (in the x direction), and also depends on y and t, so that its transverse component is zero. As U gets larger the flow pattern becomes *turbulent*, so that there is mixing of particles among the neighboring laminae.

Since we assume laminar flow, the particle velocity $\mathbf{v} = (u, 0, 0)$, where $u = u(x, y, t)$. A thin film of fluid, called the *boundary layer*, adheres to the upper and lower plates. This is due to the roughness of the plates, which gives us the boundary conditions that the particle velocity at each plate is zero, meaning

$$u(x, 0, t) = 0, \quad u(x, h, y) = U. \tag{4.84}$$

To satisfy these BCs we must have

$$u = \frac{y}{h}U, \quad u_y = \frac{U}{h}. \tag{4.85}$$

This tells us that the velocity profile u is linear with respect to y, so that the velocity gradient u_y is constant for a given ratio U/h. Let τ_{xy} be the shear stress due to the applied force P. Since stress equal force per unit area, we have

$$\tau_{xy} = \frac{P}{A}.$$

This gives

$$\tau_{xy} = \mu u_y. \tag{4.86}$$

This important result tells us that the shear stress on a fluid between two parallel plates in slow relative motion is proportional to the velocity gradient u_y, the proportionality constant being the dynamic viscosity coefficient μ. This model is an example of *shear flow*.

If μ is independent of temperature and the velocity gradient, the fluid is said to be *Newtonian*. If otherwise, then the fluid is said to be *non-Newtonian*.

Sometimes it is convenient to use the *kinematic viscosity coefficient* ν defined by

$$\nu = \frac{\mu}{\rho}. \tag{4.87}$$

Conservation Laws

We now reintroduce the three conservation laws of physics for viscous fluids. In our treatment of one-dimensional gas flow we used the Lagrangian representation of the conservation laws. In discussing viscous fluids (in general, three-dimensional flow) it is more convenient to use the Euler representation of the conservation laws. Recall that in this representation the spatial coordinates are the independent variables. The observer is stationed at points in space and observes the flow field.

Conservation of Mass

The continuity equation arising from the conservation of mass in the Eulerian representation tells us that the divergence of the particle velocity is equal to the negative of the time rate of change of the density. We now derive this equation. Let $\mathbf{x} = (x, y, z)$ be the Eulerian or spatial coordinates. Let the particle velocity be $\mathbf{v} = (u, v, w)$, the stress tensor have the components τ_{ij} ($i, j = 1, 2, 3$), and let the density be ρ. These dynamic variables are tied together by the conservation laws and are to be solved for as functions of $(\mathbf{x}, t)$. We consider a volume V of fluid contained in a closed surface S. The mass of fluid in an elementary volume of fluid dV is $\rho\, dV$. The *mass flux density* of fluid is $\rho \mathbf{v}$. Its direction is that of the motion of the fluid, while its magnitude equals the mass of fluid flowing per unit time through a unit area normal to the velocity. The mass flux of fluid flowing across an element of surface dS bounding the volume V is $\rho \mathbf{v} \cdot d\mathbf{S}$. It is positive if the fluid is flowing across dS out volume and negative if the fluid is flowing into the volume. The total mass flux of fluid out of the volume V is

$$\int_S \rho \mathbf{v} \cdot d\mathbf{S}.$$

The decrease in the mass of fluid in volume V per unit time is

$$-\frac{\partial}{\partial t} \int_V \rho\, dV.$$

CHAPTER FOUR

Equating these two expressions gives

$$\int_S (\rho \mathbf{v}) \cdot d\mathbf{S} = -\frac{\partial}{\partial t} \int_V \rho \, dV.$$

The surface integral is transformed to a volume integral by the divergence theorem of Gauss, which is

$$\int_S (\rho \mathbf{v}) \cdot d\mathbf{S} = \int_V \nabla \cdot (\rho \mathbf{v}) \, dV. \tag{4.88}$$

We therefore have

$$\int_V [\nabla \cdot (\rho \mathbf{v}) + \rho_t] \, dV = 0$$

(reverting to subscripts for partial differentiation). Since this equation must be valid for any volume, however small (a differential volume), the integrand must vanish, yielding

$$\nabla \cdot (\rho \mathbf{v}) = (\rho u)_x + (\rho v)_y + (\rho w)_z = -\rho_t. \tag{4.89}$$

This is the *equation of continuity*. It is the mathematical description of the conservation of mass in the Eulerian representation. Another form of eq. (4.89) is obtained by expanding the expression $\nabla \cdot (\rho \mathbf{v})$. We obtain

$$\rho \nabla \cdot \mathbf{v} = -\mathbf{v} \cdot \nabla \rho - \rho_t = -\frac{d\rho}{dt}. \tag{4.90}$$

Conservation of Momentum

In deriving the equation of motion for a one-dimensional inviscid gas we used the Lagrangian representation. Here we use the Eulerian representation, which is more convenient for flow problems in three dimensions. We first derive *Euler's equations* (the equations of motion for an inviscid fluid) under the action of a pressure force, as a preparation for the derivation of the *Navier-Stokes equations* (the equations of motion for a viscous fluid).

Consider again a volume V bounded by a closed surface S. The total external force acting on V is equal to

$$-\int_S p\, dS = -\int_V (\nabla p)\, dV,$$

where we transformed the surface integral into the volume integral of grad p. (The minus sign is used since p is a compressive stress, which is, by definition, negative.) This equation is valid for any volume, however small. Therefore, the fluid surrounding any volume element dV exerts a force $-(\text{grad } p)\, dV$ on that element. In other words, a force $-\text{grad } p$ acts on a unit volume of the fluid. Euler's equations are then obtained by equating this force to the product of the mass per unit volume (density ρ) and the particle acceleration dv/dt. This gives

$$\rho \frac{dv}{dt} = -\nabla p.$$

The acceleration can be expanded as

$$\frac{dv}{dt} = \frac{\partial v}{\partial t} + dx\,\frac{\partial v}{\partial x} + dy\,\frac{\partial v}{\partial y}\, dz\,\frac{\partial v}{\partial z}$$

$$= \frac{\partial v}{\partial t} - (\mathbf{v}\cdot\nabla)v.$$

As mentioned in chapter 1, "Wave Equation for a Bar," the first term on the right-hand side represents the time rate of change of the velocity at a fixed point in space, while the second term represents the difference in velocity between two points a distance dr apart at a fixed time, the convective term. Using the above expanded form for the acceleration, Euler's equations of motion become (in vector form)

$$v_t + (\mathbf{v}\cdot\nabla)v = -\frac{1}{\rho}\nabla p. \tag{4.91}$$

(We reverted to subscript notation for partial differentiation.)

The Navier-Stokes equations can be obtained by adding to the pressure gradient on the right-hand side of the Euler equations (4.91) the divergence of the shear stresses that cause viscous drag on the fluid. Reverting to the x_i coordinate system, we let τ_{ij} be the (ij)th compo-

CHAPTER FOUR

nent of the stress tensor τ. The shear components are those for which $i \neq j$ and the normal components are the diagonal elements of the stress matrix $(i = j)$. The stress term is described in detail in chapter 5. The pressure $p = -\tau_{11} = -\tau_{22} = -\tau_{33}$. We incorporate this pressure into the stress tensor τ, which thus contains the normal and the shear stresses. We now use the divergence of this stress tensor to generalize the Euler equations for a viscous fluid. We obtain

$$\rho v_t + \rho(\mathbf{v} \cdot \nabla)v = \nabla \cdot \tau. \qquad (4.92)$$

It is clear that, for an inviscid fluid, there exist only normal stresses, which are equal to the negative of the pressure gradient. For a viscous fluid we must, in addition, relate the shear stresses to the gradient of the velocity vector $\mathbf{v}$ if the motion is small enough. This follows experimentally, for the internal friction causing viscous flow occurs only when different laminae of fluid particles move with different velocities so that there is relative motion between the neighboring laminae. We must subtract the rotational part of the velocity since rotation does not cause internal friction. It turns out that the (ij)th component of the stress tensor is

$$\tau_{ij} = -p\delta_{ij} + \eta\left(\frac{\partial v_i}{\partial x_j} + \frac{\partial v_j}{\partial x_i} - \frac{2}{3}\delta_{ij}\frac{\partial v_k}{\partial x_k}\right) + \zeta\delta_{ij}\frac{\partial v_k}{\partial x_k}. \qquad (4.93)$$

(See [22, pp. 47–48] and [21, p. 571], for a more complete treatment.) Note that in the tensor form of eq. (4.93) summation is performed over the double index k. The positive constants η and ζ are the coefficients of viscosity. ζ is the additional viscosity coefficient that occurs in a compressible fluid.

The divergence of the stress tensor is given by

$$\nabla \cdot \tau = \frac{\partial \tau_{ij}}{\partial x_j} = -\nabla p + \eta\left(\frac{\partial^2 v_i}{\partial x_j \partial x_j} + \frac{\partial}{\partial x_i}\frac{\partial v_j}{\partial x_j} - \frac{2}{3}\frac{\partial}{\partial x_i}\frac{\partial v_k}{\partial x_k}\right)$$

$$+ \zeta\frac{\partial}{\partial x_i}\frac{\partial v_k}{\partial x_k}$$

$$= \eta\frac{\partial^2 v_i}{\partial x_j \partial x_j} + \left(\zeta + \frac{1}{3}\eta\right)\frac{\partial}{\partial x_i}\frac{\partial v_k}{\partial x_k}. \qquad (4.94)$$

It is clear that

$$\frac{\partial v_k}{\partial x_k} \equiv \nabla \cdot \mathbf{v}, \qquad \frac{\partial^2 v_i}{\partial x_j \, \partial x_j} \equiv \nabla^2 v.$$

We can therefore write the equation of motion for a viscous fluid in vector form as

$$\rho\left(\frac{\partial v}{\partial t} + (\mathbf{v} \cdot \nabla)v\right) = -\nabla p + \eta \nabla^2 v \left(\zeta + \frac{1}{3}\eta\right)\nabla(\nabla \cdot \mathbf{v}). \quad (4.95)$$

For an incompressible fluid $\nabla \cdot v = 0$, so that eq. (4.95) becomes

$$\frac{\partial v}{\partial t} + (\mathbf{v} \cdot \nabla)v = -\frac{1}{\rho}\nabla p + \nu \nabla^2 v, \qquad \nu = \frac{\eta}{\rho}, \quad (4.96)$$

where ν is the kinematic viscosity coefficient. Equation (4.96) is called the Navier-Stokes equation. The (ij)th component of the stress tensor for an incompressible fluid is

$$\tau_{ij} \equiv p\delta_{ij} + \eta\left(\frac{\partial v_i}{\partial x_j} + \frac{\partial v_j}{\partial x_i}\right). \quad (4.97)$$

If the particle velocity is small we may neglect the convective term $(\mathbf{v} \cdot \nabla)v$ the Navier-Stokes equation, which becomes linear:

$$\frac{\partial v}{\partial t} = -\frac{1}{\rho}\nabla p + \nu \nabla^2 v. \quad (4.98)$$

(In this analysis we omitted the external force F, for simplicity. This may easily be added to the right-hand side of the Navier-Stokes equation.) Equation (4.98) has an interesting interpretation. The first term on the right-hand side expresses the variation of v due to the pressure gradient. It has the same form as in the case of an inviscid fluid. The term $\nu \nabla^2 v$ is due to the viscosity of the fluid. If we neglect the pressure gradient, we have

$$\frac{\partial v}{\partial t} = \nu \nabla^2 v.$$

CHAPTER FOUR

If we replace v by the temperature T, this equation becomes Fourier's unsteady heat conduction equation (a parabolic PDE). Since the heat conduction equation represents a dissipation of heat, the above equation represents the dissipation of the particle velocity due to viscous damping.

BOUNDARY CONDITIONS AND BOUNDARY LAYER

We now consider the BCs on the Navier-Stokes equation, since these are necessary to solve any specific problem in viscous flow. Consider a viscous fluid exhibiting laminar flow over a stationary flat plate. There are always molecular forces of attraction between the fluid particles and the plate, so that a thin layer of fluid adjacent to the surface adheres to it (the fluid at the surface is at rest). A velocity gradient normal to the plate occurs, forming a layer called the boundary layer. The fluid is viscous in the boundary layer and inviscid outside the boundary layer. (The study of boundary layer theory is very important in the investigation of flow over airfoils, for example. We shall not pursue this field here.) The BC is that $\mathbf{v} = \mathbf{0}$ on the surface. This means that both the normal and tangential components of the particle velocity vanish on the boundary. This is in contrast to an inviscid fluid, where only the normal component vanishes, meaning that there is slip at the surface.

A delicate point now arises: As the roughness of the plate gets less the fluid becomes less viscous—the kinematic viscosity coefficient ν becomes smaller. As long as this coefficient remains positive, however small, there is no slip at the surface, so that the BC remains $\mathbf{v} = \mathbf{0}$ and there is a boundary layer, however small. But if we set $\nu = 0$, a dramatic change occurs in the boundary condition, to a nonzero tangential component of the velocity at the surface, giving slip, and the flow becomes inviscid. The surface of the plate changes from being rough, however slightly, to being infinitely smooth. The mathematical situation is this: As long as $\nu > 0$ the Navier-Stokes equation (4.96) is valid. But, if we set $\nu = 0$, this equation reduces to Euler's equation (4.91). Comparing these two equations, we see that Euler's equation is first order in the partial derivatives while the Navier-Stokes equation is second order since it contains the Laplacian of $\mathbf{v}$, namely, the term $\nu \nabla^2 v$. Since the number of BCs depends on the order of the PDE, a boundary condition is lost in going from the Navier-Stokes to Euler equation. What happened to that BC?

Before we attempt an answer, consider a very simple example: The ODE for the displacement x of a damped harmonic oscillator is $\ddot{x} + \gamma \dot{x} + \omega x = 0$, where γ is the attenuation constant and ω is the frequency. As long as ω remains nonzero we have damped oscillations. But if we set $\omega = 0$ then the ODE reduces to $\dot{x} + \gamma x = 0$. The ODE is reduced from second order to first order so that we go from two to one initial condition. (We are dealing with an ODE instead of a PDE, so that we consider initial conditions instead of boundary conditions.) The ODE changes from a damped harmonic oscillator (under certain conditions) to a nonoscillatory ODE representing damping. Again, what happened to the missing IC and why the sudden change in the character of the ODE?

The problem of loss of a boundary condition due to a decrease in the order of the PDE is solved by the *method of singular perturbations*, which involves an *asymptotic expansion* in terms of ν. This procedure is beyond the scope of this book.

Energy Dissipation in a Viscous Fluid

The conservation of energy states that the sum of the potential and kinetic energy of a system is constant. This principle is not valid for a viscous fluid, since the presence of viscosity results in the dissipation of energy.

We now calculate the energy dissipation for an incompressible viscous fluid. We first calculate the rate of change of the kinetic energy and then use the Navier-Stokes equation to bring in the viscous shear stress. The total kinetic energy T is

$$T = \tfrac{1}{2}\rho \int_V v^2 \, dV,$$

where, as usual, the integration is taken over the volume V of the fluid. Since the fluid is assumed to be incompressible, the density is constant. The time derivative of the kinetic energy per unit volume is

$$\frac{1}{2}(\rho v^2)_t = \frac{\rho}{2}(\mathbf{v} \cdot \mathbf{v})_t = \rho \mathbf{v} \cdot \mathbf{v}_t.$$

CHAPTER FOUR

We use the condition that for an incompressible fluid the (ij)th component of the divergence of the viscosity shear stress-tensor reduces to

$$\tau_{ij,j} = \eta \nabla^2 v_i.$$

From the Navier-Stokes equation (4.96) we solve for v_t. We get

$$v_t = -(\mathbf{v} \cdot \nabla)v - \mathbf{v} \cdot \nabla p + \tau_{ij,j}.$$

Using this equation we obtain the following expression for the rate of change of the kinetic energy per unit volume:

$$\left(\frac{1}{2}\rho v^2\right)_t = -\rho(\mathbf{v} \cdot \nabla)v - \mathbf{v} \cdot \nabla p + v_i \tau_{ij,j}$$

$$= -\rho(\mathbf{v} \cdot \nabla)\left(\frac{1}{2}v^2 + \frac{p}{\rho}\right) + \nabla \cdot (\mathbf{v} \cdot \tau) - \tau_{ij} v_{i,j},$$

where $\mathbf{v} \cdot \tau$ denotes the vector (since it is the scalar product of a vector and a tensor) whose components are $v_i \tau_{i,j}$. Since the fluid is incompressible we have $\nabla \cdot \mathbf{v} = 0$ and the above expression becomes

$$\left(\frac{1}{2}\rho v^2\right)_t = -\nabla \cdot \left[\rho \mathbf{v}\left(\frac{1}{2}v^2 + \frac{p}{\rho}\right) - \mathbf{v} \cdot \tau\right] - \tau_{ij} v_{i,j}. \quad (4.99)$$

The term $\rho v(\frac{1}{2}v^2 + p/\rho)$ is the energy flux due to the transfer of fluid mass and is the same as in an ideal fluid (since it does not involve viscosity). The term $\mathbf{v} \cdot \tau$ is the energy flux due to viscosity. The term $\tau_{ij} v_{i,j}$ is the transfer of viscous energy and is the scalar product of the momentum flux and the velocity gradient.

To obtain the total rate of change of kinetic energy we integrate eq. (4.99) over the volume V and use the divergence theorem. We get

$$T_t = \frac{\partial}{\partial t} \int_V \frac{1}{2} \rho v^2 \, dV$$

$$= -\int_S \left[\rho \mathbf{v}\left(\frac{1}{2}v^2 + \frac{p}{\rho}\right) - \mathbf{v} \cdot \tau\right] \cdot d\mathbf{S} - \int_V \tau_{ij} v_{i,j} \, dV. \quad (4.100)$$

The first term of the right-hand side gives the rate of decrease in kinetic energy of the fluid in the volume V due to the energy flux across the bounding surface S. The second term expresses the decrease in kinetic energy due to viscous dissipation. We now let the volume tend to infinity. There is no fluid flow across the infinite bounding surface so that the surface integral vanishes. (For a finite volume the surface integral also vanishes since the normal component of the velocity across the surface is zero.) Equation (4.100) becomes

$$T_t = -\int \tau_{ij} v_{i,j} \, dV.$$

For an incompressible fluid the tensor τ_{ij} is given by eq. (4.97) (subtracting the pressure term). It follows that

$$\tau_{ij} v_{i,j} = \eta v_{i,j}(v_{i,j} + v_{j,i}) = \frac{\eta}{2}(v_{i,j} + v_{j,i})^2.$$

The energy dissipation in an incompressible viscous fluid then becomes

$$T_t = -\frac{\eta}{2} \int (v_{i,j} + v_{j,i})^2 \, dV. \qquad (4.101)$$

Since the dissipative energy leads to a decrease in the rate of change of kinetic energy and the integral in the above expression is positive, we have a proof that the viscosity coefficient η is positive.

Wave Propagation in a Viscous Fluid

We are now in a position to treat wave propagation in an incompressible viscous fluid. To this end we take as our model the oscillations of a body immersed in an incompressible viscous fluid. The vibrating body produces waves in the fluid which we wish to examine.

We start with a simple example. We consider an incompressible viscous fluid bounded by an infinite flat plate in the $z = 0$ plane. The region of the fluid is the half space $z > 0$. We let the plate oscillate with a frequency ω in the x direction (the x, y or horizontal plane contains the plate). Let U be the velocity of the oscillating plate. We assume that U is time harmonic. It is convenient to use complex variable notation. We have

$$U = \mathrm{Re}(U_0 e^{i\omega t}), \qquad U_0 = A e^{i\delta}, \qquad (4.102)$$

CHAPTER FOUR

where Re indicates the real part, A is the amplitude of U, and δ is the phase angle. The velocity of the fluid $\mathbf{v} = (u,v,w)$ must satisfy the BC at $z = 0$, which is

$$u(x,y,0,t) = U, \qquad v(x,y,0,t) = 0, \qquad w(x,y,0,t) = 0. \qquad (4.103)$$

It is clear from symmetry that $\mathbf{v}$ depends spatially only on z. Also from symmetry $v = 0$. The continuity equation for an incompressible fluid is $u_x + v_y + w_z = 0$. Since $w_z = 0$, using the BC that $w = 0$ throughout the fluid, then $\mathbf{v} = (u,0,0)$ where $u = u(z,t)$. This is also seen physically: Since laminar flow is assumed it is clear that in every plane $z = \text{const}$, the fluid velocity is a constant (depending on z) and is in the x direction, In addition, the convective or nonlinear term of the Navier-Stokes equation (4.96), $(\mathbf{v} \cdot \nabla)v$, becomes zero. Thus the Navier-Stokes equation becomes the linear equation (4.98). Moreover, since the pressure p depends only on z and t and the z component of the Navier-Stokes equation gives $p_z = 0$, it follows that the pressure is constant everywhere in the fluid. The Navier-Stokes equation therefore reduces to

$$u_t = \nu u_{zz}. \qquad (4.104)$$

The reader will recognize this equation as Fourier's one-dimensional unsteady heat conduction equation if u were the temperature. Since u is the fluid velocity we conclude that, for this example, the fluid behaves in the same dissipative manner as that of heat transfer by conduction.

We seek time-harmonic solutions of eq. (4.104) of the form

$$u = u_0 e^{i(\zeta z - \omega t)}, \qquad (4.105)$$

where ζ is the wave number. Inserting this equation into eq. (4.104) and factoring out the exponentials yields

$$\zeta^2 = i\frac{\omega}{\nu}.$$

The roots of this quadratic are the two values of the wave number:

$$\zeta_\pm = \pm(1+i)\beta, \qquad \beta = \sqrt{\frac{\omega}{2\nu}}. \qquad (4.106)$$

Inserting this expression for the wave number into eq. (4.105) yields

$$u = u_0 e^{-\beta z} e^{i(\beta z - \omega t)}.$$

β is positive so that the factor $e^{-\beta z}$ decays exponentially with distance z from the plate. Note that we neglected the root $\zeta_- = -(1 + i)$; this leads to the factor $e^{\beta z}$, which is an unstable solution since it gives an exponential increase in z.

Using the BC given by eq. (4.102) the solution for the fluid velocity becomes

$$u = A e^{i\delta} e^{-\beta z} e^{i(\beta z - \omega t)}, \qquad \beta = \sqrt{\frac{\omega}{2\nu}}. \qquad (4.107)$$

[It is clear that we must take the real part of eq. (4.107).] Equation (4.07) tells us that the oscillating plate produces a transverse progressing wave, since the fluid velocity u is normal to the direction of wave propagation (z direction). The important property of the wave is that it is rapidly damped as we proceed away from the plate. In fact, over a distance of one wavelength, the amplitude diminishes by a factor of $e^{2\pi} \approx 540$. The *depth of penetration* of the wave is defined as the distance z away from the plate over which the amplitude falls off to e^{-1} of its value at the boundary. Looking at the solution it is clear that the

$$\text{depth of penetration} = \beta^{-1} = \sqrt{\frac{2\nu}{\omega}}. \qquad (4.108)$$

In the case of air, $\nu = 0.13$ so that $\beta^{-1} = 0.21 T^{1/2}$, where T is the period of oscillation in seconds.

The phase velocity c of the transverse wave is obtained in the usual way by setting the exponent $i(\beta z - \omega t)$ equal to a constant and then forming $dz/dt = c$, yielding

$$c = \sqrt{2\nu\omega} \qquad (4.109)$$

We now calculate the shear stress τ_{xz} acting on the oscillating plate due to the viscous flow. We have

$$\tau_{xz} = \eta u_z = i\zeta \eta u = i(1+i)\beta \eta u_0 e^{i(\zeta z - \omega t)}$$

$$= (i-1)\sqrt{\frac{\omega \eta \rho}{2}}\, e^{i(\zeta z - \omega t)}$$

upon using eqs. (4.105) and (4.106). The shear stress on the plate is obtained by setting $z = 0$, yielding

$$\tau_{xz} = (i-1)\sqrt{\frac{\omega \eta \rho}{2}}\, U_0 e^{-i\omega t} \quad \text{at } z = 0. \tag{4.110}$$

This tells us that there is a phase difference of $\pi/4$ between the simple harmonic motion of the plate and the shear stress on the plate due to friction.

We now calculate the time average of the energy dissipated due to viscous damping. The power lost per unit area of the plate is equal to the mean value of the product of the shear stress τ_{xz} and the fluid velocity u, which is $\tfrac{1}{2} U_0^2 \sqrt{\omega \eta \rho / 2}$.

Oscillating Body of Arbitrary Shape

The more general case of viscous flow around an oscillating body of arbitrary shape is now considered under certain conditions. For such a body the nonlinear or convective term $(\mathbf{v} \cdot \nabla)v$ in the Navier-Stokes equation is not necessarily zero. What are the conditions under which we can neglect this nonlinear term?

To answer this question we take the curl of both sides of the linear Navier-Stokes equation (4.96) and get

$$(\operatorname{curl} v)_t = \nu \nabla^2 \operatorname{curl} v \tag{4.111}$$

since curl(grad p) = 0. Equation (4.111) tells us that curl v (the vorticity) satisfies the three-dimensional heat conduction or diffusion equation. This means that the vorticity exponentially decays as we go away from the body.

Similarity Considerations and Dimensionless Parameters; Reynolds' Law

In fluid mechanics we can obtain important results by simple physical considerations based on certain similarity principles, where dimensionless parameters come into play. Dimensionless parameters such as the Reynolds number are introduced, allowing us to compare the viscous flow of bodies of geometrically similar types having the same Reynolds number. Two bodies are geometrically similar in a flow field if the same single linear dimension can be used to describe the body. (This will be described in more detail below.)

We shall investigate the steady incompressible flow of a solid body immersed in a viscous fluid. Let U be the free-stream velocity (the constant velocity at infinity). What are the physical parameters that specify a given flow? The flow is governed by the Navier-Stokes (N-S) equations, eqs. (4.96), with the appropriate boundary conditions. The N-S equations tell us that the kinematic viscosity coefficient ν and the dependent variables $\mathbf{v}$, p, and ρ govern the flow. The flow depends on the shape and dimensions of the immersed body. The shape is known, so that the body essentially depends on a linear dimension L (for example, the diameter of a sphere or the length of an arbitrary body in the direction of its motion). Recall that ν has dimensions L^2/T and U dimensions L/T (where T stands for time). The only dimensionless combination of L, U, and ν is UL/ν (or its reciprocal or any function of them). This gives us a motivation for defining the *Reynolds number R* as

$$R = \frac{UL}{\nu} = \frac{\rho UL}{\eta}, \qquad (4.112)$$

where η is the dynamic viscosity coefficient.

Suppose we want to compare two different viscous flows, one of an arbitrary body of length L with a velocity v and the other of a sphere of radius r and a free-stream velocity U. Let the ratio v/U be given by

$$\frac{v}{U} = f\left(\frac{L}{r}, R\right), \tag{4.113}$$

where f is a dimensionless vector function of the dimensions of the bodies and the Reynolds number. From this formula we see that for the same Reynolds number, the ratio of the velocities is the same function of the ratio of the lengths L/r. These flows are similar, and are said to obey *Reynolds law of similarity*, discovered by Osborne Reynolds in 1883. This law also says that two flows having the same Reynolds number have geometrically similar streamlines for nonturbulent flow.

We shall give a physical interpretation of the Reynolds number R, by showing that R is the ratio of the inertial to the viscous force acting on the body. Observe that the left-hand side of the Navier Stokes equation (4.96) gives the inertial force $\rho \dot{v}$ while the right-hand side gives the viscous force $\eta \nabla^2 v$. We therefore obtain

$$\frac{\text{inertial force}}{\text{viscous force}} = \frac{\rho \dot{v}}{\eta \nabla^2 v}.$$

The numerator can be expressed as a function of the physical quantities ρ, U, and L. We therefore use the nonlinear convective term $\rho \dot{v} = \rho(\mathbf{v} \cdot \mathbf{\nabla})v$ as the inertial force. A typical component is $\rho u u_x$. Therefore the dimensions of the inertial force are $\rho U^2 L^{-1}$. A typical term of the viscous force is ηu_{xx}, whose dimensions are $\eta U/L^2$. Combining these dimensional forces yields eq. (4.112), showing that the Reynolds number is dimensionless.

For the asymptotic case of infinite Reynolds number the flow becomes inviscid. In aerodynamics the air has little viscosity so that high Reynolds numbers are used to calculate the flow around an airfoil. At the other end of the spectrum, when the Reynolds number approaches zero, the flow becomes infinitely viscous and the inertial force can be neglected.

The Navier-Stokes equations cannot be solved by analytical methods except for special asymptotic cases of large or small Reynolds numbers. However, with the development of high-speed digital computers a whole host of numerical methods have been developed to solve special cases in great detail. But these special cases are not very helpful in giving us a deeper insight into the more general aspects of viscous flow. One approach is to use the method of asymptotic expansions involving Taylor series expansions for the Reynolds number. Such an approach gives a good insight into the changeover of boundary conditions from inviscid to viscous flow and from laminar to turbulent flow. These considerations are beyond the scope of this book.

We now turn to the case of *unsteady flow* in connection with dimensional analysis. These flows are characterized by an additional parameter τ, which is some time interval that determines how the flow changes with time. For instance, for oscillatory flow we may take τ to be the period of oscillation. These unsteady flows are of interest to us in connection with the phenomenon of wave propagation. In order to investigate such flows we define the dimensionless *Strouhal number S* by

$$S = \frac{U\tau}{L}. \tag{4.114}$$

In comparing two unsteady flows, similar motion occurs when the numbers R and S have the same values for each flow.

Poiseuille Flow

We consider the steady-state flow of an incompressible, viscous fluid in a pipe or cylinder. This is called *Poiseuille flow*.

Let the x axis be the axis of the cylinder and let the flow be in the x direction so that the particle velocity $\mathbf{v} = (u, 0, 0)$, where $u = u(y, z)$. This means that each cross-section x-const has the same velocity distribution. This also means that $u_x = 0$, so that the continuity equation div $v = 0$ is satisfied identically. If we write out the Navier-Stokes equation (4.96) in extended form, we easily see that

$$p_y = p_z = 0.$$

Thus, $p = p(x)$, meaning that the pressure is also constant over each cross-section $x = $ const. The Navier-Stokes equation becomes

$$\frac{dp}{dx} = \eta \nabla^2 u = \eta(u_{xx} + u_{yy}). \tag{4.115}$$

CHAPTER FOUR

The left-hand side of this equation is a function of x [since $p = p(x)$] and the right-hand side is a function of y and z. It follows that $dp/dx = $ const. Therefore the pressure gradient may be written as

$$\nabla p = \frac{\Delta p}{L}, \tag{4.116}$$

where Δp is the pressure difference between the ends of the pipe of length L. It follows that the velocity distribution $u = u(y, z)$ can be found by solving the BV problem

$$u_{xx} + u_{yy} = \text{const}$$

in the region bounded by the closed curve C representing the circumference of a cross-section of the cylinder. The BC is $u = 0$ on C.

We now specialize to a cylinder of length L and circular cross-section of radius R. The Navier-Stokes equation becomes

$$\frac{1}{r}\frac{d}{dr}\left(r\frac{du}{dr}\right) = -\frac{\Delta p}{\eta L}. \tag{4.117}$$

We obtain u by integrating this equation:

$$u = -\left(\frac{\Delta p}{4\eta L}\right)r^2 + a \log r + b,$$

where a and b are constants to be determined from the BC, which is no slip or $u(R) = 0$. Since $\log r$ has a singularity at $r = 0$, we must have $a = 0$. We obtain

$$u(r) = -\left(\frac{\Delta p}{4\eta L}\right)(R^2 - r^2). \tag{4.118}$$

Equation (4.118) tells us that the velocity distribution across any cross-section of the pipe is parabolic in r, being a maximum at $r = 0$ and zero at $r = R$.

We can now determine the *discharge* Q, defined as the mass of fluid flowing past a unit cross-section per unit time. A mass of fluid equal to $2\pi\rho r u \, dr$ flows past an element $2\pi r \, dr$ of the cross-sectional area per unit time. This yields

$$Q = 2\pi\rho \int_0^R r u \, dr.$$

Upon using eq. (4.118) and integrating, we obtain

$$Q = \left(\frac{\pi \Delta p}{8\nu L}\right) R^4. \tag{4.119}$$

Equation (4.119) is called *Poiseuille's formula*. It says that the discharge of fluid through a circular cross-section is proportional to the fourth power of the radius.

In biophysics, Poiseuille flow has important applications to the field of hemorheology, which is the study of blood flow. However, since a blood vessel is a flexible tube rather than a rigid pipe, more complicated BCs must be used to take into account the tube's flexibility. Also, the vessel is not straight so that the axis of the tube curves in space. Moreover, arterial flow is pulsatile so that we no longer have steady flow; the particle acceleration must be considered in the Navier-Stokes equations. However, the above treatment is the basis for more complicated models.

Stokes' Flow

In the discussion of the Reynolds number it was pointed out that small values of the Reynolds number yield highly viscous flow with small velocity. This type of flow has applications in the materials sciences and rheology in particular in the fields of plastics, settling of suspensions, and colloids. Sir George Stokes (one of the developers of the Navier-Stokes equations) studied this type of flow. He observed that for very small Reynolds numbers he could neglect the convective term $(\mathbf{v} \cdot \text{grad})\mathbf{v}$, thus linearizing the Navier-Stokes equations. This linearization, as we have seen, leads to the study of highly viscous fluids exhibiting slow motion. To fix his ideas, he investigated the incompressible steady-state flow of a highly viscous fluid around a sphere having a rough surface.

The steady-state Navier-Stokes equation neglecting the nonlinear term is

$$\nabla p = \eta \nabla^2 \mathbf{v} \tag{4.120}$$

subject to the incompressibility condition

$$\nabla \cdot \mathbf{v} = 0. \tag{4.121}$$

CHAPTER FOUR

Equations (4.120) and (4.121) can be solved for the velocity and pressure distributions, thus completely defining the flow field. In investigating such a flow over a sphere, we must append the BCs of zero slip, $v = 0$, on the surface of the sphere and uniform motion with a velocity U at infinity.

Our physical model is a sphere fixed in the fluid whose free-stream velocity is U. We interpret v as the perturbed velocity of the fluid around the sphere. To formulate the problem we use a spherical coordinate system (r, θ, ϕ), where r is the radius vector, θ is the polar angle, and ϕ is the azimuth angle. Figure 4.7 shows the spherical coordinate system. $r =$ const generates the sphere, $\theta =$ const generates a cone with the origin at the vertex, and $\phi =$ const generates a portion of the plane whose horizontal coordinate is the x axis and whose vertical coordinate makes the angle ϕ with respect to the y axis.

Let the angular velocity be ω. Then curl $\mathbf{v} = 2\omega$. Taking the curl of eq. (4.120) and using the definition of ω we get

$$\nabla^2 \omega = 0. \tag{4.122}$$

This means that the components of ω are harmonic functions. It is evident from symmetry that the vortex lines are circles in plane normal to the x axis with centers on the x axis. Vortex lines are described in

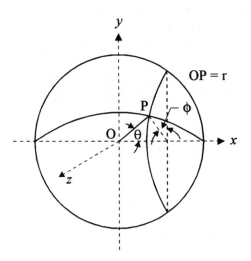

FIGURE 4.7. Spherical coordinates (r, θ, ϕ).

chapter 8. It is clear that the axis of $\boldsymbol{\omega}$ is in the x direction (the free-stream velocity U is in the x direction).

We write $\boldsymbol{\omega}$ in spherical coordinates. Let $(\mathbf{i},\mathbf{j},\mathbf{k})$ be unit vectors in the (x, y, z) directions. $\boldsymbol{\omega}$ is in the $(\mathbf{j},\mathbf{k})$ plane. We have

$$\boldsymbol{\omega} = -\mathbf{j} f(r, \theta) \sin \phi + \mathbf{k} f(r, \theta) \cos \phi, \qquad (4.123)$$

where $f(r, \theta)$ is to be determined. Equation (4.122) becomes, in spherical coordinates,

$$\frac{\partial}{\partial r}\left(r^2 \frac{\partial f}{\partial r}\right) + \left(\frac{1}{\sin \theta}\right) \frac{\partial}{\partial \theta}\left(\sin \theta \frac{\partial f}{\partial \theta}\right) - \frac{f}{\sin^2 \theta} = 0. \qquad (4.124)$$

$f(r, \theta)$ may be written as a product of a function of r and a function of θ. The strength of a vortex filament is zero for $\theta = 0$ and π and a maximum for $\theta = \pi/2$. Therefore, f is of the form

$$f(r, \theta) = g(r) \sin \theta.$$

Inserting this expression for f into eq. (4.124) gives the following ODE for g:

$$\frac{d}{dr}\left(r^2 \frac{dg}{dr}\right) - 2g = 0.$$

Assuming a solution of the form $g = r^n$ is easily seen that $g = r$ or r^{-2}. Since $\boldsymbol{\omega}$ must vanish as r becomes infinite, we must have $g = r^{-2}$. $\boldsymbol{\omega}$ then becomes

$$\boldsymbol{\omega} = -\mathbf{j}\frac{A}{r^2} \sin \theta \sin \phi + \mathbf{k}\frac{A}{r^2} \sin \theta \cos \phi$$

$$= A\left(-\mathbf{j}\frac{z}{r^3} + \mathbf{k}\frac{y}{r^3}\right) = A \operatorname{curl}\left(\frac{\mathbf{i}}{r}\right), \qquad (4.125)$$

where A is an arbitrary constant. We have used the transformation $y = r \sin \theta \sin \phi$, $z = r \sin \theta \cos \phi$. Since $\boldsymbol{\omega} = \frac{1}{2} \operatorname{curl} \mathbf{v}$, the particle velocity becomes

$$\mathbf{v} = \frac{2A}{r}\mathbf{i} + \nabla \Psi, \qquad (4.126)$$

CHAPTER FOUR

where Ψ is any scalar function of the coordinates. We may add this last term since we recall that $\nabla \times \nabla \Psi = 0$. Let $v = (v_r, v_\theta, v_\phi)$ be the components of v in the direction of increasing (r, θ, ϕ). We use the continuity equation (4.121) in spherical coordinates to obtain

$$\nabla^2(A \cos \theta + \Psi) = \nabla^2(\Phi) = 0, \quad \text{where } A \cos \theta + \Psi = \Phi. \tag{4.127}$$

This equation tells us that Φ is a solution of Laplace's equation. We have

$$\Psi = \left(-A + Br + \frac{C}{r^2}\right) \cos \theta.$$

Then eq. (4.126) becomes

$$\mathbf{v} = \mathbf{i}\left(\frac{2A}{r}\right) - A\nabla \cos \theta + B\nabla(r \cos \theta) + C\nabla\left(\frac{1}{r^2}\right) \cos \theta. \tag{4.128}$$

We put this equation in component form and obtain

$$u_r = \left(B + \frac{2A}{r} - \frac{2C}{r^3}\right) \cos \theta,$$

$$u_\theta = -\left(B + \frac{A}{r} + \frac{C}{r^3}\right) \sin \theta.$$

The constants A, B, C can be determined as follows: To satisfy the BC at infinity we must have

$$u_r(\infty, \theta) = U, \quad u_\theta(\infty, \theta) = 0.$$

This yields $B = U$. Let R be the radius of the sphere. The BC at the surface is zero slip, so that

$$u_r(R, \theta) = 0, \quad u_\theta(R, \theta) = 0.$$

From these BCs we obtain $A = -(3/4)RU$, $C = -(1/4)R^3U$. The velocity components then become

$$u_r = \left(1 - \frac{3}{2}\frac{R}{r} + \frac{1}{2}\frac{R^3}{r^3}\right)U\cos\theta,$$

$$u_\theta = -\left(1 - \frac{3}{4}\frac{R}{r} - \frac{1}{4}\frac{R^3}{r^3}\right)U\sin\theta. \quad (4.129)$$

From (4.125) the magnitude of ω becomes

$$\omega = -\frac{3}{4}\left(\frac{R}{r^2}\right)U\sin\theta. \quad (4.130)$$

We have thus solved the kinematic portion of the problem, namely, the velocity field for the incompressible flow of a viscous fluid over a sphere.

The next step is the dynamic part of the problem: to calculate the stress on the surface of the sphere. We immediately obtain the mean pressure from the steady-state Navier-Stokes equation (4.96), by using eq. (4.128) with the BCs for A, B, C:

$$p = -\frac{3}{2}\eta RU\left(\frac{\cos\theta}{r^2}\right) + p_0, \quad (4.131)$$

where p_0 is the pressure at infinity (atmospheric pressure). This pressure distribution tells us that the pressure is a maximum at the leading edge of the sphere ($\theta = 0$), a minimum at the trailing edge ($\theta = \pi$), and zero where $\theta = \pi/2$ and $(3/2)\pi$. We observe the same distribution of the r component of the velocity, u_r. u_θ is zero at the leading and trailing edges of the sphere and has a maximum amplitude at $\theta = \pi/2$ and $(3/2)\pi$. From symmetry, the ϕ components of all the dependent variables are zero, so this is actually a two-dimensional problem.

To calculate the stress distribution (stress tensor) on the surface of the sphere, we first need to obtain the strain tensor. It will simplify the analysis if we use a right-handed two-dimensional rectangular coordinate system where the x axis is in the direction of increasing r and the y axis is normal to it in the direction of increasing θ. (For more on the stress and strain tensors the reader is referred to chapter 5.)

CHAPTER FOUR

The strain tensor is represented by the symmetric matrix

$$\varepsilon = \begin{pmatrix} \varepsilon_{xx} & \varepsilon_{xy} \\ \varepsilon_{yx} & \varepsilon_{yy} \end{pmatrix},$$

where by symmetry $\varepsilon_{xy} = \varepsilon_{yx}$. The diagonal elements of the strain tensor represent the normal strain, the off-diagonal elements represent the shear components. In fluid mechanics, instead of using the strain tensor we use the *strain rate tensor*, defined by ε_t. To this end we define the displacement vector $\mathbf{d} = (\xi, \eta, \zeta)$, which represents the displacement of a fluid particle from its equilibrium configuration. For a particle in equilibrium $\mathbf{d} = \mathbf{0}$. Since the problem is two dimensional, we have the z component $\zeta = 0$. ε and ε_t are functions of (x, y, t). The elements of the strain matrix are

$$\varepsilon_{xx} = \frac{\partial \xi}{\partial x}, \qquad \varepsilon_{xy} = \frac{1}{2}\left(\frac{\partial \xi}{\partial y} + \frac{\partial \eta}{\partial x}\right),$$

$$\varepsilon_{xy} = \varepsilon_{yx}, \qquad \varepsilon_{yy} = \frac{\partial \eta}{\partial y}.$$

The particle velocity $\mathbf{v} = \mathbf{d}_t$ (reverting to subscripts for partial derivatives) so that the components of the strain rate tensor become

$$\varepsilon_{xx,t} = u_x, \qquad \varepsilon_{xy,t} = \tfrac{1}{2}(u_y + v_x),$$
$$\varepsilon_{yy,t} = v_y, \qquad \text{where } v = (u, v, 0). \tag{4.132}$$

Turning to the stress tensor τ, in matrix form it becomes

$$\boldsymbol{\tau} = \begin{pmatrix} \tau_{xx} & \tau_{xy} \\ \tau_{yx} & \tau_{yy} \end{pmatrix}.$$

This tensor is also represented by a symmetric matrix. The normal components are given by the diagonal elements, the components of the shear stress are given by the off-diagonal elements. Since the unit vector $\mathbf{i}$ is in the direction of the outward normal to the surface of the sphere, the normal component of τ is the scalar product of r. Thus

$$\tau_n = \mathbf{i} \cdot \boldsymbol{\tau} = i\tau_{xx} + j\tau_{xy}. \tag{4.133}$$

These components of the stress tensor are related to the components of the strain rate tensor by

$$\tau_{xx} = -p + 2\eta u_x, \qquad \tau_{xy} = \eta(u_y + v_x), \qquad (4.134)$$

where η is the dynamic viscosity coefficient. Upon using eq. (4.129) and remembering that we are using a Cartesian coordinate system, we obtain

$$u_x = \frac{3}{2}\left(\frac{R}{r^2}\right)\left[1 - \left(\frac{R}{r}\right)^2\right]U\cos\theta,$$

$$u_y + v_x = -\frac{3}{2}\left(\frac{R^3}{r^4}\right)U\sin\theta. \qquad (4.135)$$

The BC on the spherical surface is $r = R$ so that we get

$$u_x = 0, \qquad u_y + v_x = -\frac{3}{2R}U\sin\theta \qquad \text{for } r = R. \qquad (4.136)$$

Inserting eq. (4.136) into eq. (4.134) and using eq. (4.131) yields

$$\tau_{xx} = -p = \frac{3\eta}{2R}U\cos\theta - p_0,$$

$$\tau_{xy} = -\frac{3\eta}{2R}U\sin\theta. \qquad (4.137)$$

τ_{xx} is the normal stress in the plane containing r in the direction normal to the surface. If it is positive (tension) then it is directed away from the surface, if negative (compression) then it is directed inward. For example, if $\theta = \pi/2$ then the normal component $\tau_{xx} = 0$, so that the only stress is that due to atmospheric pressure acting inward. τ_{xy} is the shear stress in the plane normal to r in the direction tangent to the surface.

We now calculate $\mathbf{F}$, the resultant force on the surface. It is clear that, by symmetry, $\mathbf{F}$ is in the direction of the fluid motion at infinity

CHAPTER FOUR

given by the free-stream velocity **U**. By integrating with respect to θ over the surface of the sphere, we obtain the following expression for **F**:

$$\mathbf{F} = \int_0^\pi \tau_{xx} \cos\theta \, 2\pi R^2 \sin\theta \, d\theta - \int_0^\pi \tau_{xy} \sin\theta \, 2\pi R^2 \sin\theta \, d\theta$$

$$= 6\pi\eta RU. \tag{4.138}$$

Equation (4.138) is called *Stokes' law*. The drag of the viscous fluid on the sphere is the force due to the resistance of the flow over the sphere. This drag depends only on the motion of the fluid relative to the sphere.

The solution we have just obtained for the dynamic portion of Stokesian flow does not satisfy the BC at infinity where $v = 0$ even though the Reynolds number is very small. To see this, we estimate the magnitude of the nonlinear term $(\mathbf{v} \cdot \text{grad})\mathbf{v}$ in the Navier-Stokes equation. If this term is neglected we get $\nabla p = \eta \nabla^2 \mathbf{v}$. At large distances from the sphere, the order of magnitude of the nonlinear term is $U^2 R/r^2$. The linear term in the Navier-Stokes equation is of the order $\eta RU/\rho r^3$. The condition that the linear term dominates the nonlinear term is given by the inequality

$$\frac{\eta RU}{\rho r^3} \gg \frac{U^2 R}{r^2}.$$

This inequality holds only at distances r from the sphere such that

$$r \ll \frac{\eta}{\rho U}.$$

Clearly, in the limiting case of zero Reynolds number, r is unbounded. But this case means inviscid flow or slip on the spherical surface, which leads to D'Alembert's paradox of no drag, so that the upper limit is finite.

OSEEN APPROXIMATION

C. W. Oseen, the Swedish physicist, in 1910 corrected Stokes' law in that it did not satisfy the BC at infinity. His approach was to take into account the nonlinear term but to linearize it by recognizing that at large distances from the sphere, **v** is approximately **U**, so that we can use the approximation $\mathbf{v} \cdot \nabla \approx \mathbf{U} \cdot \nabla$ and thus obtain the following approximation to the Navier-Stokes equation:

$$\rho(\mathbf{U} \cdot \nabla)v = -\nabla p + \eta \nabla^2 v. \tag{4.139}$$

This reduction of the nonlinear to the linear Navier-Stokes equation is called the *Oseen approximation*. We shall merely present Oseen's result; the details are given in [21, p. 608]. The more exact equation is

$$F = 6\pi\eta R U \left[1 + \left(\frac{3\rho R U}{8\eta} \right) \right]. \tag{4.140}$$

Example. We consider the case of a sphere of radius R and density σ falling under the action of a gravitational force at a very small Reynolds number in a still viscous fluid of density ρ and viscosity coefficient η. The two external forces acting on the sphere are

(1) the force of gravity given by $\frac{4}{3}R^3\sigma g$
(2) the buoyancy force given by $\frac{4}{3}R^3\rho g$

When a steady-state or equilibrium condition is reached, the difference between these two forces must be set equal to the drag force (we use Stokes' law):

$$\tfrac{4}{3}R^3 g(\sigma - \rho) = 6\pi\eta R U.$$

From this equation we solve for the steady-state velocity acquired by the sphere and obtain

$$U = \frac{2(\sigma - \rho)R^2 g}{9\eta}. \tag{4.141}$$

If the density σ of the sphere is greater than that of the fluid then U

CHAPTER FOUR

is positive and is directed downward, and conversely if the density of the sphere is less than that of the fluid. In either case U is constant.

Problems

1. If the fluid moves radially and the velocity $u = u(r, t)$, prove that the equation of continuity is

$$u\rho_r + \frac{\rho}{r^2}(r^2 u)_r = -\rho_t.$$

2. If σ is the cross-sectional area of a stream filament, prove that the equation of continuity is

$$(\rho\sigma v)_s = -(\rho\sigma)_t,$$

where v is the particle velocity and ds is an element of arc length of the filament. See chapter 8, p. 6 for the definition of a stream filament.

3. If $S(r, t) = 0$ is the equation of a surface which always consists of the same fluid particles, show that, after an infinitesimal time δt, $S(r + v\,\delta t, t + \delta t) = 0$, and deduce that

$$S_t + (v\nabla)S = 0.$$

4. The equation of state for an ideal or perfect gas is $pV = RT$. Derive the following equations:

$$c_0 = \sqrt{\gamma RT},$$

$$\frac{T}{T_0} = 1 - \frac{1}{2}(\gamma - 1)\left(\frac{v}{c_0}\right)^2 = 1 - \frac{\gamma - 1}{\gamma + 1}\left(\frac{v}{c_*}\right)^2.$$

The gas constant $R = 8.314 \times 10^7$ ergs/deg mole.

5. Show that, for an adiabatic gas, $\rho/\rho_0 = (c/c_0)^{2/(\gamma-1)}$. Making use of the Riemann invariant l show that the particle velocity $dx/dt = u = 2/(\gamma - 1)(c - c_0)$. Let the C_+ characteristic have the equation $x = c_0 t$.

Derive expressions for $u(x, t, c_0)$ and $x(t, c_0)$. Answer:

$$\frac{dx}{dt} = u = \frac{2}{\gamma + 1}\left(\frac{x}{t} - c_0\right),$$

$$x = -\frac{2}{\gamma - 1}c_0 t + \frac{\gamma + 1}{\gamma - 1}c_0 t_0 \left(\frac{t}{t_0}\right)^{2/(\gamma+1)}.$$

6. Let $u = u(x, y)$ be the steady-state velocity. The PDE $u_{xx} + xu_{yy} = 0$ is known as *Tricomi's equation*. Show that this is a mixed problem in the sense that it is elliptic for $x > 0$, parabolic for $x = 0$, and hyperbolic for $x < 0$.

7. Consider the case $\gamma = 3$. Show that the general solution of the one-dimensional isentropic compressible flow problem is given by the following set of equations:

$$x = (v + c)t + F(v + c), \qquad x = (v - c)t + G(v - c),$$

where F and G are arbitrary functions of their arguments. Interpret these solutions in terms of traveling waves and their relation to characteristic theory.

8. Consider two-dimensional steady-state subsonic flow. Show that, if we use the transformation

$$x = \bar{x}, \qquad y = \sqrt{1 - M^2}\,\bar{y},$$

where M is the Mach number, then the velocity potential ϕ satisfies Laplace's equation.

9. Prove Prandtl's relation given by eq. (4.81).

10. Show that $C_p = [\gamma/(\gamma - 1)]R$ for a perfect gas.

11. (a) show that $q = \sqrt{[2\gamma/(\gamma - 1)]R(T_0 - T)}$ for a perfect gas, where T_0 is the *stagnation temperature*, the temperature where $q = 0$.

(b) Show that the limit or maximum speed $\hat{q} = \sqrt{[2\gamma/(\gamma - 1)]RT_0}$.

(c) Show that the speed of sound at the stagnation temperature is $c_0 = \sqrt{\gamma RT_0}$.

CHAPTER FOUR

12. Prove the following shock relations for a polytropic gas:

$$\frac{T_0}{T_1} = \frac{\rho_1}{\rho_0} = \frac{p_1 + \mu^2 p_0}{p_0 + \mu^2 p_1},$$

$$(u_1 - u_0)^2 = (p_1 - p_0)^2 \frac{(1 - \mu^2) T_0}{p_1 + \mu^2 p_0},$$

$$\frac{p_1}{p_0} = (1 + \mu^2) M_0^2 - \mu^2,$$

$$\frac{u_1 - u_0}{c_0} = (1 - \mu^2)\left(\frac{U - u_0}{c_0} - \frac{c_0}{U - u_0}\right).$$

13. Determine the frictional force f on each of two parallel plates a distance h apart between which there is a layer of viscous fluid, when one plate is fixed and the other oscillates in its own plane. *Hint*: Seek a solution of the form

$$u(x, t) = (A \sin kx + B \cos kx) e^{-i\omega t}.$$

Let the BCs be $u(0, t) = u_0 e^{i\omega t}$, $u(h, t) = 0$. Show that the velocity is

$$u(x, t) = u_0 e^{i\omega t} \frac{\sin k(h - x)}{\sin kh}.$$

Show that the frictional force on the oscillating plate is

$$f_0 = -\mu ku \cot kh$$

and the frictional force on the fixed plate is

$$f_h = \mu ku \operatorname{cosec} kh.$$

14. Determine the frictional force on an oscillating plate covered by a layer of viscous fluid of thickness h, the upper surface being free. *Hint*: The BC at the solid plate is oscillatory; at the free surface it is $\tau_{xy} = \mu v_x = 0$ at $x = h$. Show that the velocity is

$$v(x, t) = u_0 e^{i\omega t} \frac{\cos k(h - x)}{\cos kh}$$

and the frictional force is

$$f = \mu(v_x)_{x=0} = \mu k u_0 e^{i\omega t} \tan kh.$$

CHAPTER FIVE

Stress Waves in Elastic Solids

Introduction

In this chapter we shall investigate the propagation of waves in elastic media. Since stresses are set up in the material under dynamic loads, we shall call propagating waves in elastic media stress waves. A good historical introduction to elasticity is given in [24], and a brief but more up-to-date historical introduction is given in [11].

In previous chapters it was pointed out that the field equations of physics (yielding the dynamic and thermodynamic variables) for a given continuous medium arise from the three conservation equations: conservation of mass, momentum, and energy. For an elastic medium these equations of motion are called the *Navier equations*. From them, a rich variety of stress waves is obtained (unlike waves in fluids, which are rather limited). The first two conservation laws are common to all media, while the energy equation defines a particular medium through the equation of state. For example, we saw that for an inviscid fluid the equation of state is adiabatic, relating pressure to density. When we deal with an elastic solid we are concerned with the dynamic variables: stress and strain. The energy equation for an elastic solid yields a relation between the stress and strain. For small-amplitude deformations there is a linear relation between the stress and strain, which is given by Hooke's law: stress is proportional to strain for the one-dimensional case. For two and three dimensions the stress tensor is a linear function of the strain tensor. For large-amplitude deformations, there is a more complicated nonlinear relation between stress and strain. A knowledge of wave propagation in elastic solids (which we call stress wave propagation) is very important to engineers in many fields (mechanical, civil, aeronautical, etc.), where an investigation of the dynamic effects of various loads on engineering materials is considered.

CHAPTER FIVE

Fundamentals of Elasticity

In order to properly introduce the field equations of elasticity, we review some fundamental concepts in elasticity. In this section we shall investigate the nature of deformation, strain, stress, stress-strain relations, equations of motion for the stress components, and equations of motion for the displacement.

Deformation

An elastic material is a deformable, continuous medium which suffers no energy loss when its deformed state returns to the equilibrium state. The deformation of any medium is a purely geometric concept. Since strain is derived from deformation, strain is also a geometric concept. We recall from chapter 1 that a continuum is a medium that has a continuous distribution of matter in the sense that its molecular and crystalline structure is neglected. This means that we can define mathematically a differential volume element dV that has the same continuous properties as the material in the large. This concept of a continuum is based on an averaging process, where we take advantage of the large number of molecules in a differential volume element to smear out the effects of individual molecules.

To describe a deformable continuum we consider two states of the medium (or body):

(1) The undeformed or equilibrium configuration or state
(2) The deformed configuration

Any two neighboring points P_1, P_2 in the body in its undeformed state under a deformation suffer the transformation $P_1 \to P_1', P_2 \to P_2'$. The distance between the two undeformed points changes in the deformed state. If $P_1'P_2' < P_1P_2$ that part of the body undergoes a compression; if $P_1'P_2' > P_1P_2$ we have tension. Clearly, if there is no change we have the equilibrium state. We use the Lagrange representation. A point in the undeformed state is given by the coordinates $\mathbf{a} = (a_1, a_2, a_3)$, where $\mathbf{a}$ is the radius vector. In the deformed state, that point goes into the Eulerian coordinates given by the radius vector $\mathbf{x}$, where $\mathbf{x} = (x_1, x_2, x_3)$. The displacement vector $\mathbf{u}$ is defined by $\mathbf{u} = \mathbf{x} - \mathbf{a}$, so that $P_i' - P_i = u_i$ $(i = 1, 2)$. Since we use the Lagrange representation, we have $\mathbf{x} = \mathbf{x}(\mathbf{a}, t)$, $\mathbf{u} = \mathbf{u}(\mathbf{a}, t)$, and all the dynamic and thermodynamic variables are functions of $(\mathbf{a}, t)$. We have thus defined deformation (compression or tension) in terms of the displacement vector.

Strain Tensor

For the equilibrium state, let $\mathbf{a} = (a, b, c)$ and for the deformed state let $\mathbf{x} = (x, y, z)$. In the equilibrium state a particle occupies the volume element $dV_a = da\,db\,dc$. In the deformed state that particle occupies the volume element $dV_x = dx\,dy\,dz$. A deformation is given by the transformation $\mathbf{a} \to \mathbf{x}$, where $x = (a, b, c)$, $y = y(a, b, c)$, $z = z(a, b, c)$. There is a unique inverse transformation. The relation between these two volume elements is

$$dV_x = \det(J)\, dV_a, \qquad (5.1)$$

where $\det(J)$ is the determinant of J, the Jacobian of the transformation $\mathbf{a} \to \mathbf{x}$. J is also called the *mapping function* and is given by the matrix

$$J = \frac{(x, y, z)}{(a, b, c)} = \begin{pmatrix} x_a & x_b & x_c \\ y_a & y_b & y_c \\ z_a & z_b & z_c \end{pmatrix}. \qquad (5.2)$$

The principle of conservation of mass is given by

$$\rho_x\, dV_x = \rho_a\, dV_a.$$

The *compression ratio* R is defined by

$$R = \frac{\rho_x}{\rho_a} = [\det(J)]^{-1}. \qquad (5.3)$$

We develop the strain tensor (in the form of a matrix) by considering a curve C_a in the body in the undeformed state. Under the mapping $\mathbf{a} \to \mathbf{x}$, this curve maps into the curve C_x in the deformed state. The two curves are composed of the same particles. The column matrix $d\mathbf{a}$ has components da, db, dc, while the row matrix is the transpose $d\mathbf{a}^*$. We have $d\mathbf{a} = (da, db, dc)^*$, and similarly for $d\mathbf{x}$. Let ds_a be an element of arc length of C_a and let ds_x be an element of arc length of C_x under

CHAPTER FIVE

the transformation $d\mathbf{a} \to d\mathbf{x}$ which is given by

$$d\mathbf{x} = J\,d\mathbf{a}. \tag{5.4}$$

The magnitude of ds_a is the square root of the expression

$$(ds_a)^2 = (d\mathbf{a})^*(d\mathbf{a}).$$

Similarly, we have

$$(ds_x)^2 = (d\mathbf{x})^*(d\mathbf{x}).$$

The transpose of eq. (5.4) is $(d\mathbf{x})^* = (d\mathbf{a})^*J^*$. Using this expression and eq. (5.4), we get

$$(ds_x)^2 = (d\mathbf{x})^*(d\mathbf{x}) = (d\mathbf{a})^* J^* J (d\mathbf{a}). \tag{5.5}$$

Suppose the transformation $\mathbf{a} \to \mathbf{x}$ has the property that for every curve C_a all arc lengths are unchanged in being transformed to the corresponding curve C_x. It follows that $(d\mathbf{x})^*(d\mathbf{x}) = (d\mathbf{a})^*(d\mathbf{a})$, so that $J^*J = E$, the 3×3 identity matrix. This means that $\mathbf{J}$ is the *rotation matrix* yielding a rigid-body rotation (no deformation), where $\det(\mathbf{J}) > 0$. Since the measure of strain can be considered as a deviation from a pure rotation, we may take the expression

$$J^*J - E$$

as twice the three-dimensional strain tensor. We have

$$\varepsilon = \tfrac{1}{2}(J^*J - E), \tag{5.6}$$

where ε is the strain tensor. The reason for the factor of $1/2$ in eq. (5.6) is seen when we derive the linear approximation of ε. For this approximation we have $|(ds_x - ds_a)/ds_a| \ll 1$, so that

$$\frac{(ds_x)^2 - (ds_a)^2}{(ds_a)^2} = \frac{(ds_x - ds_a)(ds_x + ds_a)}{ds_a} \approx \frac{2\,ds_a(ds_x - ds_a)}{ds_a}$$

$$= 2\frac{ds_x - ds_a}{ds_a}.$$

Let ε_L be the linear strain tensor. We have

$$\frac{ds_x - ds_a}{ds_a} = \left(\frac{da}{ds_a}\right)^* \varepsilon_L \left(\frac{da}{ds_a}\right), \qquad (5.7)$$

where eq. (5.6) is used for the linear strain tensor.

Strain as a Function of Displacement

We now obtain the strain matrix ε as a function of the displacement vector $\mathbf{u}$, where $\mathbf{u} = (u, v, w)$. We have $\mathbf{u} = \mathbf{x} - \mathbf{a}$, the components being $u = x - a$, $v = y - b$, $w = z - c$. Instead of calculating the right-hand side of eq. (5.6) and setting $\mathbf{x} = \mathbf{u} - \mathbf{a}$ to obtain ε, we introduce $\mathbf{K}$, the Jacobian of $\mathbf{u}$ with respect to $\mathbf{a}$, so that $\mathbf{K} = (u, v, w)/(a, b, c)$. This is the matrix

$$K = \begin{pmatrix} u_a & u_b & u_c \\ v_a & v_b & v_c \\ w_a & w_b & w_c \end{pmatrix}. \qquad (5.8)$$

It follows that $\mathbf{J} = \mathbf{K} + \mathbf{E}$, so that the strain tensor (in matrix form) becomes

$$\varepsilon = \tfrac{1}{2}(K + K^*) + KK^* = \varepsilon_L + \varepsilon_N. \qquad (5.9)$$

The linear part of this is given by the matrix

$$\varepsilon_L = \begin{pmatrix} u_a & \tfrac{1}{2}(u_b + v_a) & \tfrac{1}{2}(u_c + w_a) \\ \tfrac{1}{2}(v_a + u_b) & v_b & \tfrac{1}{2}(v_c + w_b) \\ \tfrac{1}{2}(w_a + u_c) & \tfrac{1}{2}(w_b + v_c) & w_c \end{pmatrix}. \qquad (5.10)$$

The determination of the elements of the nonlinear part of the strain tensor ε_N is left to a problem. The diagonal elements of ε represent the pure or normal components of the strain, while the off-diagonal elements are the components of the shear strain. It is clear that the strain tensor ε is represented by a symmetric matrix (a matrix that is equal to its transpose). This is true for both the linear and nonlinear parts. The proof is left to a problem.

CHAPTER FIVE

Note that the concept of a tensor was not yet defined A tensor has certain defining properties with respect to specific transformations, which are useful in certain other definitions of strain and in relativity theory. These properties are not relevant here. However, we shall discuss below the strain tensor in terms of its action on a strained surface of the elastic body.

Recall that we defined strain by considering a curve in our body and its deformation either by stretching (tension) or by compression. An element of the curve may be considered as the distance ds_a between two neighboring points in the undeformed body, which maps (under the Jacobian transformation) into the distance ds_x between the same points under deformation. There are other definitions of strain, but this one is suitable for our purpose. To get an understanding of how strain acts on an elastic body, we show how the strain tensor acts on the three orthogonal faces of a cube in (a, b, c) (undeformed) space to deform it into a parallelepiped in the (x, y, z) (deformed) space. The unstrained cube is oriented such that the normal to each face is in a coordinate direction. In the tensor notation, each element of the tensor is represented by two indices i, j. Let the ijth element of the strain tensor be given by ε_{ij}, where $i, j = 1, 2, 3$, so that there are nine elements. Since this tensor is symmetric, we have $\varepsilon_{ij} = \varepsilon_{ji}$, meaning that each element is invariant with respect to an interchange of i and j. This gives six distinct elements. The correspondence between this notation and the elements of the strain matrix is $\varepsilon_{11} = u_a$, $\varepsilon_{12} = \frac{1}{2}(u_b + v_a), \ldots, \varepsilon_{33} = w_c$. We now consider how each element of the strain tensor acts on the undeformed cube. The index i represents the normal to the face of the cube in the i direction, while j gives the direction of the strain. (Clearly, i or $j = 1, 2, 3$ gives the a, b, c directions, respectively.) Suppose the elements of ε are represented in a matrix array. For $i = j$ we have the diagonal elements, which give the direction of the strain normal to the cube surfaces. ε_{11} is the normal component of the strain on the cube surface whose normal is in the a direction, etc. For $i \neq j$ we have the off-diagonal elements, which yield the shear strains. ε_{12} is the shear strain in the b direction on the cube surface whose normal is in the a direction. By symmetry (symmetric tensor) this shear strain is also equal to the strain in the a direction acting on the cube surface whose normal is in the b direction. Note that the components of a vector are given by a single index. The matrix form of the strain tensor may be considered as composed of row vectors or column vectors.

A more complete discussion of strain, including its geometric properties, principal-axes transformations, curvilinear and generalized coordinates, etc., is given in [24] and it is treated from a more modern vector point of view in [32] and [26], where nonlinear deformations are treated in a matrix notation. (We have modeled our discussion on strain after [26].) The treatment given here is sufficient for an understanding of stress wave propagation in solids.

Stress Tensor

To develop the stress tensor we consider a body occupying a volume V enclosed by a surface S imbedded in the elastic medium. We define the *stress vector* **T** as the surface force per unit area acting at a point P on the surface. Let dS be the element of surface area at P. The force **T** dS represents the action of the material at P directed into the volume V. If a unit normal vector ν is drawn on the surface element at P directed outward (to the positive side of dS), then **T** ds represents the action of the part of the body lying on the positive side of S upon the negative side. It follows from Newton's third law of mechanics that the action of the material lying on the negative side of the normal upon that on the positive side is $-\mathbf{T}\,ds$. Figure 5.1 shows the action of the stress vector at a point P on the surface. **T** depends on two parameters: the orientation of the surface given by its normal ν and the direction of **T**. In order to display these two parameters, we define the stress tensor by the notation T^ν; the direction of the stress vector is given by **T** and the superscript identifies the orientation of the surface element in terms of its normal.

The great French mathematical physicist August Cauchy enunciated the following stress principle: *The state of stress at any point in the body is completely and uniquely characterized by specifying the stress tensor at that point*. The proof of this principle is given in any treatise on elasticity (e.g., [11, 288 p.]).

Let T^i be the stress vector in the x_i direction, $i = 1, 2, 3$. Let e_i be the unit vector in the x_i direction. We resolve T^i into its components along the coordinate axes and obtain

$$T^i = e_j \tau_{ij}, \qquad (5.11)$$

where we used the Einstein summation convention by summing over the repeated index j. τ_{ij} is the ijth component of the stress tensor, which

CHAPTER FIVE

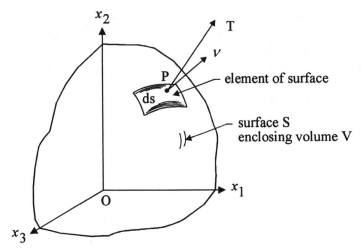

FIGURE 5.1. Action of stress vector T on point P on the surface. ν is the normal to the surface at P.

can be represented by the corresponding component of its matrix representation. The subscript i gives the orientation of the surface element dS. This is given by the scalar product $d\mathbf{S} \cdot \mathbf{e}_i$. The subscript j identifies the direction of this component of the stress tensor. For example, τ_{23} is the component of the stress tensor in the x_3 direction acting on the element of surface whose outward-drawn normal is in the x_2 direction. Like the strain tensor, the stress tensor is symmetric.

Stress-Strain Relations

For an elastic solid under small deformations, the stress tensor is a linear function of the strain tensor. This means that, in general, each component of the stress tensor is a linear combination of all the components of the strain tensor. This result can be obtained by expanding each component of the stress tensor in a Taylor series in the strain components about the equilibrium configuration and keeping only the linear terms. We simplify the notation for the strain components by setting

$$\varepsilon_{11} = e_1, \quad \varepsilon_{22} = e_2, \quad \varepsilon_{33} = e_3, \quad \varepsilon_{12} = e_4,$$

$$\varepsilon_{23} = e_5, \quad \varepsilon_{13} = e_6.$$

We use a similar notation for the stress components τ_{ij} in terms of the t_is.

The linear expansion of the stress in terms of the strain becomes

$$t_i = c_{ij} e_j, \qquad i, j = 1, 2, \ldots, 6 \tag{5.12}$$

summed over j. This may be reformulated into the matrix equation

$$t = Ce, \tag{5.13}$$

where t is a 6×1 column matrix, C is a 6×6 nonsingular matrix (has a unique inverse), and e is a 6×1 column matrix. C contains all the elastic constants of the material. Inverting eq. (5.13) yields

$$e = Et, \tag{5.14}$$

where E is the *elastic modulus* of the material. Clearly, the matrix E is the inverse of C.

The *strain-energy density function* W was introduced by the English mathematical physicist Green in 1839 in the *Transactions of the Cambridge Philosophical Society*. W is a consequence of the first law of thermodynamics and is a scalar potential. The reader is referred to [33, p. 81]. W is represented as a quadratic function of the strain in the matrix form

$$W = \tfrac{1}{2} e^* C e \tag{5.15}$$

or in the extended form

$$W = \tfrac{1}{2} c_{ij} \varepsilon_i \varepsilon_j \tag{5.16}$$

summed over i and j from 1 to 6. Since W is a scalar it is symmetric in i and j, which means that C is a symmetric matrix; this reduces the number of elements from 36 to 21.

Since the t_is are given by eq. (5.12), we get the fundamental relationship between the strain-energy density function and the stress:

$$\frac{\partial W}{\partial e_i} = t_i. \tag{5.17}$$

CHAPTER FIVE

Hooke's Law for an Isotropic Material

An isotropic material is defined as one whose elastic constants C are invariant with respect to any rotation of the reference frame. J. L. Davis [11, p. 293] showed by use of similarity transformations that repeated rotations finally show that only two elastic constants are involved in Hooke's law for an isotropic medium. (See also [32, p. 62].) These elastic constants λ and μ are called the *Lamé constants*.

The trace (sum of the diagonal elements) of the strain matrix Θ is called the *dilatation*. It is a measure of the relative change in volume of the body due to a compression or dilation. It is given by

$$\Theta = \varepsilon_{11} + \varepsilon_{22} + \varepsilon_{33}. \tag{5.18}$$

Hooke's law can be expressed by the set of equations

$$\tau_{ij} = \lambda\,\delta_{ij}\Theta + 2\mu\varepsilon_{ij}, \tag{5.19}$$

where the Kronecker delta $\delta_{ij} = 1$ for $i = j$ and 0 or $i \neq j$. Let Φ be the trace of the stress matrix so that

$$\Phi = \tau_{11} + \tau_{22} + \tau_{33}. \tag{5.20}$$

We have

$$\Phi = (3\lambda + 2\mu)\Theta. \tag{5.21}$$

The inverse of eq. (5.19) is

$$\varepsilon_{ij} = \frac{\lambda\delta_{ij}}{2\mu(3\lambda + 2\mu)}\Phi + \frac{\tau_{ij}}{2\mu}. \tag{5.22}$$

The Lamé constants λ, μ can he expressed in terms of the two elastic constants *Young's modulus E* and *Poisson's ratio σ*. It can be shown that

$$\lambda = \frac{E\sigma}{(1 + \sigma)(1 - 2\sigma)}, \quad \mu = \frac{E}{2(1 + \sigma)},$$
$$\sigma = \frac{\lambda}{2(\lambda + \mu)}, \quad E = \frac{\mu(3\lambda + 2\mu)}{\lambda + \mu}. \tag{5.23}$$

Equations of Motion for the Stress

Consider a continuum occupying a volume V enclosed by a surface S. In the interior of V we consider a volume element dV enclosed by a surface element of area dS that is acted on by external forces whose resultant induces an inertial force due to the acceleration of the body. Let $\mathbf{x} = (x_1, x_2, x_3)$ be a coordinate system fixed in the body, $\mathbf{u} = (u_1, u_2, u_3)$ be the displacement vector describing the deformation, $\mathbf{v} = \mathbf{u}_t = (u_{1,t}, u_{2,t}, u_{3,t}) = (v_1, v_2, v_3)$ be the particle velocity, $\mathbf{a} = \mathbf{v}_t = (v_{1,t}, v_{2,t}, v_{3,t}) = (a_1, a_2, a_3)$ be the particle acceleration, and ρ be the density of the deformed body. The inertial force is $\rho \mathbf{a}\, dV$. Let $\rho \mathbf{F}\, dV$ be the resultant of the body forces acting on dV, where $\mathbf{F}$ is the specific resultant force (force per unit volume), and let $\mathbf{T}\, dS$ be the resultant of the surface forces on dS. The vector sum of these forces must equal the inertial force. This gives the equation of motion for the particle occupying dV, in vector form,

$$\mathbf{F}\rho\, dV + \mathbf{T}\, dS = \rho \mathbf{a}\, dV.$$

We integrate the first term over the volume V and the second term over the surface S surrounding V and express the result in tensor form for the ith equation,

$$\int_V \rho F_i\, dV + \oint_S \tau_{ij} \eta_j\, dS = \int_V \rho v_{i,t}\, dV,$$

where the surface integral is represented as a contour integral over the closed surface S, and the component of the stress tensor τ normal to S is given in terms of the sum of the components of the scalar product $\boldsymbol{\tau} \cdot \boldsymbol{\nu}$. The surface integral is now transformed to a volume integral by the divergence theorem (see, for example, [11, p. 76]), giving

$$\oint_S \tau_{ij} \nu_j\, dS = \int_V \tau_{ij,j}\, dV,$$

where, expanding the tensor notation, we have

$$\tau_{ij,j} = \frac{\partial \tau_{i1}}{\partial x_1} + \frac{\partial \tau_{i2}}{\partial x_2} + \frac{\partial \tau_{i3}}{\partial x_3}.$$

CHAPTER FIVE

The ith equation of motion becomes

$$\int_V (\rho F_i + \tau_{ij,j})\, dV = \int_V \rho v_{i,t}\, dV.$$

Clearly, this is a system of three equations. Since we have a continuum, if the region of integration shrinks, in the limit, to an interior point, we obtain the differential form of the equations of motion, which we put in the tensor representation

$$\tau_{ij,j} = \rho(-F_i + v_{i,t}). \tag{5.24}$$

This is the ith equation of motion, which represents the conservation of linear momentum. These equations of motion are therefore Newton's equations of motion for a continuum. (Recall that, in the tensor notation, we sum over the double subscript j.) These are the linearized equations of motion, since the nonlinear terms for the particle acceleration, which represent the convective terms, are neglected. Since they are linear the Lagrangian and the Eulerian representations are the same; the difference between these two representations appears only in the nonlinear terms. These equations of motion are valid for any continuous medium (solid, liquid, gas) since they do not invoke the energy equation that defines the material. Clearly, these equations of motion give an incomplete description of the physical situation, since we have three equations and six components of the stress tensor τ and three components of the displacement vector **u**. The additional requisite equations are given from the conservation of energy, which supplies the *constitutive equations*; these we take as Hooke's law for an isotropic elastic medium. We investigate this in the next section.

NAVIER EQUATIONS OF MOTION FOR THE DISPLACEMENT

In this section we derive the linear equations of motion for an elastic medium in terms of the components of the displacement vector by using Hooke's law. These equations are called the Navier equations. In many problems in elasticity it is more convenient to obtain the equations of motion for the displacement vector and then derive the stress field from the definition of strain and Hooke's law.

We obtain the linear strain in tensor form from eq. (5.10) by using the x coordinate system instead of the Lagrangian variables **a** and the displacement vector $\mathbf{u} = u(x)$ The ijth component of the linear strain tensor is $\varepsilon_{ij} = u_{i,j}$. Since the strain tensor is symmetric, we have

$$\varepsilon_{ij} = \tfrac{1}{2}(u_{i,j} + u_{j,i}). \tag{5.25}$$

This takes care of the off-diagonal elements, which are the components of the shear strain. Inserting eq. (5.25) into eq. (5.19) gives the stress tensor as a function of the spatial derivatives of **u**:

$$\tau_{ij} = \lambda\,\delta_{ij}\Theta + \mu(\varepsilon_{ij} + \varepsilon_{ji}). \tag{5.26}$$

To use the equations of motion (5.24) for the stress components we must differentiate each τ_{ij} with respect to x_i in eq. (5.26) and sum over i. Thus

$$\tau_{ij,i} = \mu\nabla^2 u_i + (\lambda + \mu)\Theta_i. \tag{5.27}$$

The first term in the right-hand side of eq. (5.27) involves the Laplacian operator ∇^2 operating on u_i, since $\varepsilon_{ij,i} = u_{i,jj} = \nabla^2 u_i$. The dilatation Θ may be written as

$$\Theta = \varepsilon_{ii} = u_{i,i} = \nabla \cdot \mathbf{u}.$$

The ith equation of motion (5.24) becomes (in tensor form)

$$\mu\nabla^2 u_i + (\lambda + \mu)\Theta_i = \rho(-F_i + u_{i,tt}). \tag{5.28}$$

The set of equations (5.28) are the Navier equations. They are three coupled second-order PDEs for the displacement vector **u**. To elucidate the tensor equation (5.28) we write it out in extended form using (x, y, z) coordinates and $\mathbf{u} = (u, v, w)$:

$$\mu\nabla^2 u + (\lambda + \mu)\Theta_x = \rho(-X + u_{tt}),$$
$$\mu\nabla^2 v + (\lambda + \mu)\Theta_y = \rho(-Y + v_{tt}), \tag{5.29}$$
$$\mu\nabla^2 w + (\lambda + \mu)\Theta_z = \rho(-Z + w_{tt}),$$

CHAPTER FIVE

where the dilatation can be written as

$$\Theta = \nabla \cdot \mathbf{u} = u_x + v_y + w_z$$

and $\mathbf{F} = (X, Y, Z)$. We now write the Navier equation (5.29) as a single vector equation. Let $\mathbf{i}, \mathbf{j}, \mathbf{k}$ be unit vectors in the x, y, z directions, respectively. We multiply the first equation of (5.24) by $\mathbf{i}$, the second by $\mathbf{j}$, and the third by $\mathbf{k}$, and add. We get

$$\mu \nabla^2 \mathbf{u} + (\lambda + \mu) \nabla \Theta = \rho(-\mathbf{F} + \mathbf{u}_{tt}). \tag{5.30}$$

Using the well-known vector identity

$$\nabla^2 \mathbf{u} = \operatorname{grad} \operatorname{div} \mathbf{u} - \operatorname{curl} \boldsymbol{\psi} \tag{5.31}$$

where $\boldsymbol{\psi} = \operatorname{curl} \mathbf{u}$, eq. (5.30) becomes

$$-\mu \operatorname{curl} \boldsymbol{\psi} + (\lambda + 2\mu) \operatorname{grad} \Theta = \rho(-\mathbf{F} + \mathbf{u}_{tt}). \tag{5.32}$$

$\boldsymbol{\psi}$ is called the rotation vector. Note that the vector forms of the Navier equations are independent of the coordinate system. From the form given by eq. (5.32) we now derive two types of waves: (1) longitudinal waves involving the wave equation for the dilatation, and (2) transverse waves, involving the rotation vector.

(1) We take the divergence of each term of eq. (5.32) and use the fact that $\operatorname{div} \operatorname{curl} \boldsymbol{\psi} = 0$. We obtain

$$\frac{(\lambda + 2\mu)}{\rho} \nabla^2 \Theta = \Theta_{tt} - \nabla \cdot \mathbf{F}. \tag{5.33}$$

Equation (5.33) is the vector form of the three-dimensional nonhomogeneous wave equation for Θ, the nonhomogeneous term being the one involving $\mathbf{F}$. We set

$$\frac{\lambda + 2\mu}{\rho} = c_L^2. \tag{5.34}$$

It will be seen below that c_L is the longitudinal wave velocity.

(2) Next we get the equation of motion for $\boldsymbol{\psi}$ by taking the curl of each term in eq. (5.30). We use the fact that $\operatorname{curl} \operatorname{grad} \Theta = 0$ and obtain

$$\frac{\mu}{\rho} \nabla^2 \boldsymbol{\psi} = \boldsymbol{\psi}_{tt} - \operatorname{curl} \mathbf{F}. \tag{5.35}$$

We set

$$\frac{\mu}{\rho} = c_T^2, \qquad (5.36)$$

where $\mathbf{c}_T$ is the velocity of the rotational vector, which, we shall show, is the velocity of a transverse wave.

Propagation of Plane Elastic Waves

We simplify the Navier equations by treating plane waves in an infinite isotropic elastic medium. A *plane wave* is defined as one whose wave front is a planar surface normal to the direction of the propagating wave. The displacement vector $\mathbf{u} = (u, v, w)$ is a function of one coordinate which we take as x (and, of course, t). All derivatives with respect to y and z in eq. (5.29) are zero. For simplicity we set $\mathbf{F} = 0$. The dilatation becomes $\Theta = u_x$, so that $\Theta_x = u_{xx}$. Making use of the definitions of the wave velocities given by eqs. (5.34) and (5.36), the Navier equations (5.29) become

$$c_L^2 u_{xx} = u_{tt},$$
$$c_T^2 v_{xx} = v_{tt}, \qquad (5.37)$$
$$c_T^2 w_{xx} = w_{tt}.$$

The first equation of (5.37) is the one-dimensional wave equation for the component u in the direction of wave propagation x. This gives a longitudinal wave, so that c_L is indeed the longitudinal wave velocity. The second equation of (5.37) is the wave equation for v. The direction of wave propagation is still x, so that $v = v(x,t)$ gives particle vibrations in the plane of the wave front (the y direction), which is normal to x. This means that we have a transverse wave, and c_T is indeed the transverse wave velocity. Similarly for the third equation, where $w(x,t)$ shows transverse wave propagation. From their definitions it is seen that the velocity of longitudinal waves is always greater than that of transverse waves.

CHAPTER FIVE

GENERAL DECOMPOSITION OF ELASTIC WAVES

From the definitions of c_L and c_T we can write the vector Navier equation (5.30) in the form

$$c_T^2 \nabla^2 \mathbf{u} + (c_L^2 - c_T^2)\,\text{grad}\,(\text{div}\,\mathbf{u}) = \mathbf{u}_{tt}, \tag{5.38}$$

where, again, $\mathbf{F} = 0$. We now split eq. (5.38) into two vector equations by decomposing the displacement vector $\mathbf{u}$ as follows:

$$\mathbf{u} = \mathbf{u}_L + \mathbf{u}_T. \tag{5.39}$$

We shall show that $\mathbf{u}_L$ represents a longitudinal wave and $\mathbf{u}_T$ a transverse wave. A longitudinal wave is rotationless, which means that

$$\text{curl}\,\mathbf{u}_L = 0. \tag{5.40}$$

It follows from vector analysis that a scalar function of space and time exists such that

$$\mathbf{u}_L = \text{grad}\,\phi, \tag{5.41}$$

where ϕ is called the scalar potential. On the other hand, $\mathbf{u}_T$ satisfies the equation

$$\text{div}\,\mathbf{u}_T \equiv \mathbf{\nabla} \cdot \mathbf{u}_T = 0. \tag{5.42}$$

This clearly means that a transverse wave suffers no change in volume (an *equivoluminal wave*) but is rotational, so that

$$\mathbf{u}_T = \text{curl}\,\mathbf{\psi}, \tag{5.43}$$

where $\mathbf{\psi}$ is the vector potential.

Using eqs. (5.41) and (5.43), eq. (5.39) becomes

$$\mathbf{u} = \text{grad}\,\phi + \text{curl}\,\mathbf{\psi}. \tag{5.44}$$

This tells us that the displacement vector can be decomposed into an irrotational vector and an equivoluminal vector.

We now insert eq. (5.39) into (5.38), take the divergence of each term, and use eq. (5.42). We obtain

$$\text{div}(c_L^2 \nabla^2 \mathbf{u}_L - \mathbf{u}_{L,tt}) = 0.$$

Since curl $\mathbf{u}_L = 0$ and div( ) = 0, it follows that the terms in parentheses are also zero, yielding

$$c_L^2 \nabla^2 \mathbf{u}_L = \mathbf{u}_{L,tt}. \tag{5.45}$$

Equation (5.45) is the vector wave equation for the displacement representing longitudinal wave (irrotational waves), since the wave velocity is c_L. This means that, in general, each component of $\mathbf{u}$ satisfies the three-dimensional wave equation. Since $\mathbf{u}_L = \text{grad } \phi$, it is clear that the scalar potential ϕ also satisfies the wave equation with the same wave velocity.

Similarly, inserting eq. (5.39) into (5.38), taking the curl of each term, and using the fact that curl $\mathbf{u}_L = 0$ we obtain

$$\text{curl}(c_T^2 \nabla^2 \mathbf{u}_T - \mathbf{u}_{T,tt}) = 0.$$

Since div( ) = 0, it follows that

$$c_T^2 \nabla^2 \mathbf{u}_T - \mathbf{u}_{T,tt} = 0. \tag{5.46}$$

Equation (5.46) is the vector wave equation for u_T, whose solutions yield transverse, equivoluminal, rotational waves. It follows that the vector potential ψ also satisfies this wave equation.

Sometimes it is easier to solve the wave equations in terms of the scalar and vector potentials. Then the displacement can be obtained from eq. (5.44). In extended form we get

$$u = \phi_x + \psi_{z,y} - \psi_{y,z},$$
$$v = \phi_y + \psi_{x,z} - \psi_{z,x},$$
$$w = \phi_z + \psi_{y,x} - \psi_{x,y}, \tag{5.47}$$

$$\psi = (\psi_x, \psi_y, \psi_z), \quad \psi_{z,y} \equiv \frac{\partial \psi_z}{\partial y}, \dots$$

CHARACTERISTIC SURFACES FOR PLANAR WAVES

The wave front of a planar wave is a plane normal to the direction of wave propagation. Let the normal to the wave front be $\boldsymbol{\nu} = (l, m, n)$, where l, m, n are the direction cosines of $\boldsymbol{\nu}$. Let the radius vector $\mathbf{r} = \mathbf{i}x + \mathbf{j}y + \mathbf{k}z$ be directed from the origin to any point on the wave front. The scalar product $\mathbf{r} \cdot \boldsymbol{\nu}$ is the projection of $\mathbf{r}$ on the wave front.

CHAPTER FIVE

All solutions of the wave equations (5.45) and (5.46) are of the form

$$u = f(\phi_\pm),$$
$$v = G(\phi_\pm),$$
$$w = H(\phi_\pm), \qquad (5.48)$$
$$\phi_\pm = xl + ym + zn \mp ct = \mathbf{r} \cdot \mathbf{v} \mp ct,$$

where F, G, H are arbitrary functions of the argument $\phi_\pm$, which is called the phase of the wave. ϕ_+ represents a progressing wave and ϕ_- a regressing wave. It is also true that there are no solutions that are not of the form given by eq. (5.48). If we set $\phi_+ = \mathbf{r} \cdot \mathbf{v} - ct = \text{const}$, we obtain a characteristic surface; this is a planar wave front that progresses into the medium in the direction normal to the wave front with a wave velocity equal to c. Setting $\phi_- = lx + my + nz + ct = \text{const}$ yields a characteristic surface that regresses in the direction opposite to that of the normal to the wave front. Inserting the components of $\mathbf{u}$ given by eq. (5.48) into eq. (5.29) (setting $X = Y = Z = 0$) yields the following quadratic equation for c^2:

$$(\rho c^2 - \mu)(\rho c^2 - \lambda - 2\mu) = 0. \qquad (5.49)$$

The roots of eq. (5.49) are $c^2 = c_L^2, c_T^2$. This substantiates the fact that there are two planar waves, one longitudinal and the other transverse.

TIME-HARMONIC SOLUTIONS AND REDUCED WAVE EQUATIONS

We now investigate *time-harmonic* solutions to the wave equations for longitudinal and transverse waves. These are waves whose time-dependent parts are of the form $e^{\pm i\omega t}$. This means we can separate the solutions of the wave equations (5.45) and (5.46) into the product of a spatial and a time-dependent part. For simplicity we consider mononochromatic waves, characterized by a single frequency ω (in radians per sec). To extend the treatment of time-harmonic solutions to a discrete frequency spectrum, we can synthesize waveforms by Fourier series expansions. For a continuous frequency spectrum we use the Fourier integral or other appropriate integral transforms.

We write the displacement vector **u** in the form

$$\mathbf{u} = \text{Re}[U(x, y, z)e^{\pm i\omega t}], \tag{5.50}$$

where Re[ ] is the real part of the bracket. **u** stands for either a longitudinal or a transverse wave. Substituting eq. (5.50) into eqs. (5.45) and (5.46) and factoring out the exponentials yields

$$\nabla^2 U_L + k_L^2 U_L = 0,$$
$$\nabla^2 U_T + k_T^2 U_T = 0, \tag{5.51}$$

where k_L and k_T are the wave numbers for the longitudinal and transverse waves, respectively, and are given by

$$k_L = \frac{\omega}{c_L}, \quad k_T = \frac{\omega}{c_T}. \tag{5.52}$$

[Note that the wave numbers have the dimension of reciprocal length $(2\pi/\lambda)$.] The equations (5.51) are called the *reduced wave equations* or *Helmholtz equations* for U_L and U_T, respectively. The reduced wave equation is of the elliptic type. Let U stand for U_L or U_T. Solutions of the reduced wave equation $\nabla^2 U + k^2 U = 0$ are of the form

$$U = A e^{i(2\pi/\lambda)(l_1 x + l_2 y + l_3 z)}. \tag{5.53}$$

Inserting eq. (5.53) into the reduced wave equation yields

$$\frac{4\pi^2}{\lambda^2}(l_1^2 + l_2^2 + l_3^2) = k^2.$$

Since the wave number $k = 2\pi/\lambda$, it follows that

$$l_1^2 + l_2^2 + l_3^2 = 1.$$

$l_1 = l, l_2 = m, l_3 = n$ are the direction cosines of $\boldsymbol{\nu}$, the normal to the wave front. Therefore eq. (5.53) can be rewritten as

$$U = e^{i(2\pi/\lambda)(\mathbf{r}\cdot\boldsymbol{\nu})}. \tag{5.53a}$$

Combining this solution of the reduced wave equation with the time-dependent solution given by eq. (5.50) yields the time-harmonic solutions

$$\mathbf{u} = \bar{\mathbf{u}}[e^{(2\pi i/\lambda)(\mathbf{r}\cdot\boldsymbol{\nu} \mp ct)}], \tag{5.54}$$

CHAPTER FIVE

where $c = \omega/k$ (clearly, the real part of the right-hand side is taken for calculations). The minus sign in the phase gives a progressing traveling wave and the plus sign a regressing wave. ω is the same for both a longitudinal and a transverse waveform, while c, λ, and k depend on the waveform. Written out in component form, this yields

$$u = A_L e^{(2\pi i/\lambda_L)(\mathbf{r}\cdot\mathbf{v} \mp c_L t)} + A_T e^{(2\pi i/\lambda_T)(\mathbf{r}\cdot\mathbf{v} \mp c_T t)},$$

$$v = B_L e^{(2\pi i/\lambda_L)(\mathbf{r}\cdot\mathbf{v} \mp c_L t)} + B_T e^{(2\pi i/\lambda_T)(\mathbf{r}\cdot\mathbf{v} \mp c_T t)}, \qquad (5.55)$$

$$w = C_L(\ ) + C_T(\ ).$$

(Clearly, one separates out the progressing and regressing waves, in practice.) The time-harmonic solutions expressed by eqs. (5.54) and (5.55) are special cases of the general solution given by eq. (5.48). Note that we have the same phase of the progressing and regressing waves, given by $\phi_\pm = xl + ym + zn \mp ct$, so that planes of constant phase yield the traveling wave fronts.

SPHERICALLY SYMMETRIC WAVES

Spherically symmetric waves are produced in three-dimensional space from a point source. For simplicity we consider the wave equation for the generic scalar $f(r, t)$, which may stand for the components of $\mathbf{u}$, the scalar potential ϕ, or the components of the vector potential $\boldsymbol{\psi}$. Setting $r^2 = x^2 + y^2 + z^2$ and neglecting angular dependence, we get the wave equation for f in spherical coordinates:

$$c^2 \left[f_{rr} + \left(\frac{2}{r}\right) f_r \right] = f_{tt}. \qquad (5.56)$$

We get time-harmonic solutions of eq. (5.56) by setting

$$f(r, t) = g(r) e^{-i\omega t}. \qquad (5.57)$$

This is the method of separation of variables. Inserting eq. (5.57) into eq. (5.58) yields

$$\ddot{g} + \left(\frac{2}{r}\right)\dot{g} + k^2 g = 0, \qquad k = \frac{\omega}{c}, \qquad \dot{g} = \frac{dg}{dr}, \dots.$$

This ODE for g may be rewritten as

$$(r\ddot{g}) + k^2 rg = 0. \tag{5.57a}$$

Note that eq. (5.57) is a second-order ODE for g where the first-derivative term is absent. Setting $w = rg$ gives $\ddot{w} + k^2 w = 0$, so that the general solution of eq. (5.57a) is

$$g(r) = \left(\frac{1}{r}\right) e^{\pm ikr}. \tag{5.58}$$

Multiplying this solution by the time-dependent part given by eq. (5.57) yields time-harmonic solutions of the spherical wave equation as a linear combination of terms of the form

$$\left(\frac{1}{r}\right) e^{ik(r-ct)}, \quad \left(\frac{1}{r}\right) e^{ik(r+ct)}. \tag{5.59}$$

The first expression of eq. (5.59) represents an outgoing attenuated spherical wave (emanating from the point source), while the second expression represents an incoming wave (from infinity where the amplitude is zero) going toward the source. Note that there is a singularity at the source. The phase of the wave is $\phi_\pm = k(r \mp ct)$. Setting the phase ϕ_+ equal to a constant and varying time generates an outgoing spherical wave front, and setting ϕ_- equal to a constant generates an incoming wave; these are the characteristic surfaces.

We may obtain a more general solution to the spherical wave equation by writing it as

$$c^2 (rf)_{rr} = (rf)_{tt}.$$

Therefore the general solution for the spherical wave equation is

$$f(r,t) = \left(\frac{1}{r}\right) F(r - ct) + \left(\frac{1}{r}\right) G(r + ct), \tag{5.60}$$

where F and G are arbitrary functions of their arguments to be specifically determined by the boundary and initial conditions. The

CHAPTER FIVE

argument of F is the phase $r - ct$ and thus represents an outgoing wave, while the argument of G is $r + ct$ and represents an incoming wave. All solutions of the spherical wave equation are of the form given by eq. (5.60), and there are no solutions that are not of those forms.

LONGITUDINAL WAVES IN A BAR

We consider the propagation of longitudinal waves in a one-dimensional bar. Let its longitudinal axis be the x axis. The displacement vector $\mathbf{u} = \mathbf{u}(u(x,t), 0, 0)$. Since we have plane wave propagation, eq. (5.37) is the equation of motion, where $v = w = 0$; the one-dimensional wave equation for the displacement is

$$c_L^2 u_{xx} = u_{tt}. \tag{5.37a}$$

It is instructive to rederive the one-dimensional wave equation for the bar from first principles. This will yield the correct expression for the wave velocity. Let the constant area of the bar be A. Consider an element of the bar at point x of length δx. The compressive stress σ at x is directed to the right, while the compressive stress $\sigma + \sigma_x \delta x$ at $x + \delta x$ is directed to the left. (One can also use a tensile stress—the directions are then reversed.) The particle acceleration is u_{tt}. Applying Newton's second law of motion gives

$$\rho A \, \delta x \, u_{tt} = A \sigma_x \, \delta x, \tag{5.61}$$

where ρ is the density of the bar. Let the one-dimensional linear strain be ε where $\varepsilon = u_x$. Let σ be the one-dimensional stress. Hooke's law tells us that $\sigma = E\varepsilon$, where E is the Young's modulus of the bar. Using these expressions, eq. (5.61) becomes

$$c_L^2 u_{xx} = u_{tt}, \quad \text{where } c_L^2 = \frac{E}{\rho}. \tag{5.62}$$

Solution of this wave equation gives the propagation of longitudinal waves in the bar. The general solution of eq. (5.62) is

$$u(x,t) = F(x - ct) + G(x + ct) \tag{5.63}$$

(where we left off the subscript of the longitudinal wave velocity c). F and G are arbitrary functions of their arguments. F represents progre-

STRESS WAVES IN ELASTIC SOLIDS

seing waves, and G regressing waves. The specific forms of F and G are obtained from the appropriate boundary and initial conditions.

We now derive a linear relation between the stress and the particle velocity at any point in the bar. For simplicity we consider a progressing wave $u = F(x - ct)$. Differentiating both sides with respect to x gives

$$F'(x - ct) = u_x = \varepsilon = \frac{\sigma}{E}.$$

We now differentiate both sides of F with respect to t and obtain

$$-cF'(x - ct) = u_t.$$

These expressions yield

$$\sigma = -(\rho c) u_t. \tag{5.64}$$

The term ρc is the *characteristic impedance* of the bar. The electrical analogy tells us that stress is analogous to voltage and particle velocity is analogous to current. It follows that the impedance is analogous to resistance. For wave propagation into two or more media the impedance is a measure of the relative amplitudes of stress waves.

Finite Bar

We consider a bar of length L. A compressive stress pulse σ is initiated at the front end ($x = 0$), producing an incident wave. As a first example let the back end ($x = L$) be a *free surface*, meaning that $\sigma(L, t) = 0$. Let the displacement due to the stress pulse (progressing wave) be given by $u_1 = F(ct - x)$. The reason for taking this expression for the waveform is that the displacement is zero until $x = ct$ and is positive for $ct > x$. Let the displacement due to the reflected pulse be $u_2 = G(ct + x)$ for a regressive wave. The stress produced by these two pulses is $Eu_{1,x}$ and $Eu_{2,x}$, so that the resultant stress is $Eu_{1,x} + Eu_{2,x}$. At the reflected end of the bar we have

$$F'(ct - L) - G'(ct + L) = 0.$$

Two results follow: (1) The waveforms of the incident and reflected waves are the same but the negative of each other. This means the incident compressive wave is reflected as a tensile wave to satisfy the free BC at $x = L$. Similarly, an incident tensile wave is reflected as a

235

CHAPTER FIVE

compressive wave. (2) At $x = L$ the displacements add, so that we have $u_1 + u_2 = 2F(L - ct)$.

As our second example we consider the same setup but let the BC at $x = L$ be rigid. This means that there is no motion of the bar at $x = L$. $u_1(L, t) - u_2(L, t) = 0$. The stresses produced by the incident and reflected pulses add up at the fixed boundary to give twice the stress. It follows that an incident compressive wave is reflected as a compressive wave of the same waveform, and an incident tensile wave is reflected as a tensile wave.

This one-dimensional treatment of a bar is approximate. The reason is that, in deriving the one-dimensional wave equation, it was assumed that all plane transverse sections (cross-sections) remain plane during the passage of the stress waves, and the stress acts uniformly over each section. However, the longitudinal expansions and contractions of sections of the bar will necessarily result in lateral contractions and expansions. This lateral motion will result in a nonuniform distribution of stress across each section of the bar, thus distorting the original planar cross-sections.

The ratio of lateral to longitudinal strain is given by Poisson's ratio. To understand this statement we give a simple example. Suppose a constant stress T acts axially on the bar. This component of the stress tensor is $\tau_{xx} = T$, where $T > 0$ in tension and $T < 0$ in compression. All the other components of the stress tensor vanish. It follows from eq. (5.22) with an obvious change in index notation) that all the components of the shear strain vanish, while the normal components of the strain tensor are related to those of the stress tensor by

$$\varepsilon_{xx} = \frac{(\lambda + \mu)T}{\mu(3\lambda + 2\mu)} = \frac{T}{E}, \quad \varepsilon_{yy} = \varepsilon_{zz} = \frac{-\lambda T}{2\mu(3\lambda + 2\mu)} = -\frac{\nu}{E}T,$$

where ν is Poisson's ratio. The ratio of lateral to longitudinal strain is given by Poisson's ratio ν. Specifically,

$$\frac{\varepsilon_{yy}}{\varepsilon_{xx}} = \frac{\varepsilon_{zz}}{\varepsilon_{xx}} = \frac{-\lambda}{2(\lambda + \mu)} = -\nu.$$

This shows that a longitudinal wave traveling in a bar distorts the transverse planar sections so that a one-dimensional treatment of wave

STRESS WAVES IN ELASTIC SOLIDS

propagation in a bar or cylinder is inadequate, and we must therefore resort to a three-dimensional approach.

CURVILINEAR ORTHOGONAL COORDINATES

The Navier equations or equations of motion for the displacement are more tractable in cylindrical than in Cartesian coordinates. We therefore digress in this section by investigating the transformations from Cartesian to curvilinear orthogonal coordinates and then specialize to the transformation to cylindrical coordinates.

Let (q_1, q_2, q_3) be coordinates of a point in any system. (x, y, z) will be functions of these coordinates, so that $x = x(q_1, q_2, q_3)$, $y = y(q_1, q_2, q_3)$, $z = z(q_1, q_2, q_3)$, and the inverse transformation is unique. (dx, dy, dz) are now expanded in terms of (dq_1, dq_2, dq_3):

$$dx = \frac{\partial x}{\partial q_1} dq_1 + \frac{\partial x}{\partial q_2} dq_2 + \frac{\partial x}{\partial q_3} dq_3,$$

with similar expansions for dy and dz. Let ds be an element of length. It follows that $ds^2 = dx^2 + dy^2 + dz^2$. Using the above expansions for dx, dy, and dz, we get

$$ds^2 = g_{11} dq_1^2 + g_{22} dq_2^2 + g_{33} dq_3^2 \\ + 2g_{12} dq_1 dq_2 + 2g_{13} q_1 q_2 + 2g_{23} q_2 q_3,$$

where

$$g_{ij} = g_{ji} = \frac{\partial x}{\partial q_i}\frac{\partial x}{\partial q_j} + \frac{\partial y}{\partial q_i}\frac{\partial y}{\partial q_j} + \frac{\partial z}{\partial q_i}\frac{\partial z}{\partial q_j}, \quad i,j = 1,2,3.$$

g_{ij} are called the *metric coefficients*.

If we consider the length ds_1 that corresponds to a change from q_1 to $q_1 + dq_1$, we have $ds_1 = (g_{11})^{1/2} dq_1$. In a similar manner, we get $ds_2 = (g_{22})^{1/2} dq_2$, $ds_3 = (g_{33})^{1/2} dq_3$. To shorten the notation we set $h_1 = (g_{11})^{1/2}$, $h_2 = (g_{22})^{1/2}$, $h_3 = (g_{33})^{1/2}$. For an orthogonal system of coordinates the surfaces $q_1 = \text{const}$, $q_2 = \text{const}$, $q_3 = \text{const}$ intersect each other at right angles.

CHAPTER FIVE

A vector function **F** in the **q** orthogonal system can be expressed by

$$\mathbf{F} = \mathbf{a}F_1 + \mathbf{b}F_2 + \mathbf{c}F_3,$$

where $(\mathbf{a}, \mathbf{b}, \mathbf{c})$ are the unit vectors along the coordinate axes in the **q** system and (F_1, F_2, F_3) are the components of **F** along these axes. The vector functions of **F** are

$$\operatorname{div} \mathbf{F} = \frac{1}{h_1 h_2 h_3}\left[\frac{\partial(h_2 h_3 F_1)}{\partial q_1} + \frac{\partial(h_3 h_1 F_2)}{\partial q_2} + \frac{\partial(h_1 h_2 F_3)}{\partial q_3}\right]$$

and

$$\operatorname{curl} \mathbf{F} = \frac{\mathbf{a}}{h_2 h_3}\left[\frac{\partial(h_3 F_3)}{\partial q_2} - \frac{\partial(h_2 F_2)}{\partial q_3}\right] + \frac{\mathbf{b}}{h_3 h_1}\left[\frac{\partial(h_1 F_1)}{\partial q_3} - \frac{\partial(h_3 F_3)}{\partial q_1}\right]$$
$$+ \frac{\mathbf{c}}{h_1 h_2}\left[\frac{\partial(h_2 F_2)}{\partial q_1} - \frac{\partial(h_1 F_1)}{\partial q_2}\right].$$

Finally, let V be a scalar. The gradient of V is given by

$$\operatorname{grad} V = \frac{\mathbf{a}}{h_1}\frac{\partial V}{\partial q_1} + \frac{\mathbf{b}}{h_2}\frac{\partial V}{\partial q_2} + \frac{\mathbf{c}}{h_3}\frac{\partial V}{\partial q_3}.$$

(For the derivation of these expressions see any advanced book on vector analysis.)

We now consider cylindrical coordinates (r, θ, z) where r is the radius vector, θ is the angle between the radius vector and the x axis, and z is the axis of the cylinder (the x, y Cartesian coordinates are in a plane normal to the z axis). We now calculate the metric coefficients (h_1, h_2, h_3) for this coordinate system. From elementary analytic geometry we have $ds_1 = dr$, $ds_2 = r\,d\theta$, $ds_3 = dz$, so that $h_1 = 1$, $h_2 = r$, $h_3 = 1$. Let the components of the displacement vector **u** be u_r, u_θ, u_z in the r, θ, and z directions, respectively. We use the metric coefficients for cylindrical coordinates in the above expressions for div **F**, curl **F**, and grad V. The dilatation is div **u** and is given by

$$\operatorname{div} \mathbf{u} = \Theta = \frac{1}{r}\left[\frac{\partial(ru_r)}{\partial r} + \frac{\partial u_\theta}{\partial \theta} + \frac{\partial ru_z}{\partial z}\right]. \tag{5.65}$$

STRESS WAVES IN ELASTIC SOLIDS

Our aim is to use the metric coefficients for the cylindrical coordinate system along with the above expressions for the vector and scalar functions to cast the Navier equations in the form given by eq. (5.32) into cylindrical coordinates [since eq. (5.32) can be used in any coordinate system]. Since we have the gradient and divergence in cylindrical coordinates it remains only to get the curl in the same coordinate system. Let the components of the rotation vector Ψ be $(\Psi_r, \Psi_\theta, \Psi_z)$ in the r, θ, and z directions respectively. These components become in cylindrical coordinates:

$$2\Psi_r = \frac{1}{r}\frac{\partial u_z}{\partial \theta} - \frac{\partial u_\theta}{\partial z},$$

$$2\Psi_\theta = \frac{\partial u_r}{\partial z} - \frac{\partial u_z}{\partial r}, \quad (5.66)$$

$$2\Psi_z = \frac{1}{r}\left[\frac{\partial(r u_\theta)}{\partial r} - \frac{\partial u_r}{\partial \theta}\right]$$

THE NAVIER EQUATIONS IN CYLINDRICAL COORDINATES

Using eqs. (5.65) and (5.66), the Navier equations (5.32) become in extended form

$$c_L^2 \frac{\partial \Theta}{\partial r} + c_T^2 \left[-\frac{1}{r}\frac{\partial \Psi_z}{\partial \theta} + \frac{\partial \Psi_\theta}{\partial z}\right] = \frac{\partial^2 u_r}{\partial t^2},$$

$$c_L^2 \frac{1}{r}\frac{\partial \Theta}{\partial \theta} + c_T^2 \left[-\frac{\partial \Psi_r}{\partial z} + \frac{\partial \Psi_z}{\partial r}\right] = \frac{\partial^2 u_\theta}{\partial t^2}, \quad (5.67)$$

$$c_L^2 \frac{\partial \Theta}{\partial z} + c_T^2 \left[-\frac{1}{r}\frac{\partial(r\Psi_\theta)}{\partial r} + \frac{1}{r}\frac{\partial \Psi_r}{\partial \theta}\right] = \frac{\partial^2 u_z}{\partial t^2}.$$

For longitudinal waves without rotation, the rotation vector Ψ vanishes and eq. (5.67) reduces to the wave equation for the dilatation Θ. For

CHAPTER FIVE

transverse waves, Θ vanishes and eq. (5.67) becomes the wave equation for Ψ.

RADIALLY SYMMETRIC WAVES

We now investigate the propagation of radially symmetric waves in a solid infinite cylinder of radius a. Some authors ([24, 20], etc.) take the following approach: They solve the Navier equations (5.67) for the dilatation and the rotation components. From them they obtain the displacement and rotation. Then they apply the results to an infinite solid cylinder with free-surface BCs.

The approach to be used here is to take advantage of the fact that an elastic medium has no internal friction, so that scalar and vector potentials exist. We shall make use of this fact by solving the wave equations for these potentials in cylindrical coordinates, and from these solutions obtain the displacement vector, the stress components, etc.

Radially symmetric waves in a solid cylinder are symmetric about the cylinder axis z, so that the angular displacement component vanishes and the other components do not depend on the angle θ. Each particle of the cylinder oscillates in the (r, z) plane. It turns out that the rotation vector has a nonzero angular component that is independent of θ. We consider an infinite train of time-harmonic waves of a single frequency along a solid infinite cylinder such that the displacement is a simple harmonic function of z. The cylindrical coordinates are (r, θ, z), where θ is the angle that the radius vector r makes with the x axis. (The x, y coordinates are in a plane normal to the cylinder axis z.) The displacement vector $\mathbf{u} = (u_r, 0, u_z)$, where the components are functions of (r, z, t). From eq. (5.66) we see that the rotation vector $\Psi = (0, \Psi_\theta, 0)$, which is also a function of (r, z, t).

Recall eq. (5.44), which relates the displacement vector to the scalar and vector potentials. In our axisymmetric case, this relation becomes

$$u_r = \frac{\partial \phi}{\partial r} + \frac{\partial^2 \psi_\theta}{\partial r \, \partial z},$$

$$u_z = \frac{\partial \phi}{\partial r} - \frac{1}{r}\frac{\partial}{\partial r}\left(r\frac{\partial \psi}{\partial r}\right) = \frac{\partial \phi}{\partial z} - \frac{\partial^2 \Psi_\theta}{\partial r^2} - \frac{1}{r}\frac{\partial \Psi_\theta}{\partial r}.$$

(5.68)

The wave equations for ϕ and Ψ in axisymmetric cylindrical coordinates become

$$c_L^2 \left[\frac{\partial^2 \phi}{\partial r^2} + \frac{1}{r} \frac{\partial \phi}{\partial r} + \frac{\partial^2 \phi}{\partial z^2} \right] = \frac{\partial^2 \phi}{\partial t^2},$$

$$c_T^2 \left[\frac{\partial^2 \Psi_\theta}{\partial r^2} + \frac{1}{r} \frac{\partial \Psi_\theta}{\partial r} + \frac{\partial^2 \Psi_\theta}{\partial z^2} \right] = \frac{\partial^2 \Psi}{\partial t^2}. \qquad (5.69)$$

The only nonzero components of the stress tensor are

$$\tau_{rr} = \lambda \left(\frac{u_r}{r} + \frac{\partial u_r}{\partial r} + \frac{\partial u_z}{\partial z} \right) + 2\mu \left(\frac{\partial u_r}{\partial r} \right),$$

$$\tau_{rz} = \mu \left(\frac{\partial u_r}{\partial z} + \frac{\partial u_z}{\partial r} \right), \qquad (5.70)$$

since the cylinder is axisymmetric so that the angular components of the stress tensor vanish. The boundary conditions are on a free surface, which means that on the surface $r = a$ we have

$$\tau_{rr} = 0, \qquad \tau_{rz} = 0, \qquad \text{at } r = a. \qquad (5.71)$$

To solve for the potentials we take time-harmonic solutions, Furthermore, since the wave front propagates in the z direction, we assume the potentials are harmonic in z. Therefore we can separate the r-dependent parts of the potentials and write them in the form

$$\phi = AF(r)e^{i(kz \pm \omega t)}, \qquad \omega = BG(r)e^{i(kz \pm \omega t)}, \qquad (5.72)$$

where A and B are constants and F and G are functions of r to be determined. Note that the same phase $kz \pm \omega t$ is used for the scalar and vector potentials. k is the wave number ($k = 2\pi/\lambda$) and ω is the frequency in rad/sec. Recall that the phase velocity $c = \omega/k$. Inserting eq. (5.72) into the appropriate wave equations (5.69) yields

$$\frac{d^2 F}{dr^2} + \frac{1}{r} \frac{dF}{dr} + \beta^2 F = 0,$$

$$\frac{d^2 G}{dr^2} + \frac{1}{r} \frac{dG}{dr} + \gamma^2 G = 0, \qquad (5.73)$$

CHAPTER FIVE

where

$$\beta^2 = \frac{\omega^2}{c_L^2} - k^2, \qquad \gamma^2 = \frac{\omega^2}{c_T^2} - k^2. \tag{5.74}$$

The solution of eq. (5.73) that has no singularity at $r = 0$ is Bessel's function of order zero, so that the potentials become, for progressing wave fronts,

$$\phi = AJ_0(\beta r)e^{i(kz-\omega t)}, \qquad \psi = BJ_0(\gamma r)e^{i(kz-\omega t)}. \tag{5.75}$$

Using eqs. (5.75) and (5.76), eq. (5.68) for the displacement components becomes

$$u_r = \left[A\frac{d}{dr}J_0(\beta r) - iBk\frac{d}{dr}J_0(\gamma r) \right] e^{i(kz-\omega t)},$$

$$u_z = \left[-iAkJ_0(\beta r) - \frac{B}{r}\frac{d}{dr}\left(r\frac{d}{dr}J_0(\gamma r) \right) \right] e^{i(kz-\omega t)}. \tag{5.76}$$

To satisfy the BCs at $r = a$ we insert eq. (5.76) into the stress components given by (5.70) and obtain

$$\left[2\mu\frac{d^2}{da^2}J_0(\beta a) - \frac{\omega^2\lambda}{c_L^2}J_0(\beta a) \right] A - \left[2i\mu k\frac{d^2}{da^2}J_0(\gamma a) \right] B = 0,$$

$$\left[2ik\mu\frac{d}{da}J_0(\beta a) \right] A + \left[\left(2k^2 - \frac{\omega^2}{c_T^2} \right) \frac{d}{da}J_0(\gamma a) \right] B = 0. \tag{5.77}$$

The system (5.77) is a set of two algebraic homogeneous equations for the constants A and B. Therefore, for nontrivial solutions the determinant must equal zero, yielding

$$\begin{vmatrix} 2\mu\dfrac{d^2}{da^2}J_0(\beta a) - \dfrac{\omega^2}{c_L^2}\lambda J_0(\beta a) & -2i\mu k\dfrac{d^2}{da^2}J_0(\gamma a) \\ 2ik\dfrac{d}{da}J_0(\beta a) & \left(2k^2 - \dfrac{\omega^2}{c_T^2} \right)\dfrac{d}{da}J_0(\gamma a) \end{vmatrix} = 0. \tag{5.78}$$

This is the *period equation*. It is difficult to discuss this equation in its general form, other than to perform numerical analyses for special cases. But if we have a thin cylinder, then the radius a is small; specifically, βa and γa must be small enough to neglect fourth-order terms. This is seen by expanding the Bessel function

$$J_0(x) = 1 - \tfrac{1}{4}x^2 + \tfrac{1}{64}x^4 - \cdots, \tag{5.79}$$

where x stands for βa or γa. Using eq. (5.79) in the expansion of eq. (5.78) and keeping only the quadratic terms, we get the required approximation for the period equation. From this we obtain an approximation for the longitudinal wave speed:

$$c_L = \frac{\omega}{k} \approx \sqrt{\frac{E}{\rho}\left(1 - \frac{1}{4}\sigma^2 k^2 a^2\right)}, \tag{5.80}$$

where E is the Young's modulus and σ is the Poisson's ratio of the cylinder.

Waves Propagated Over the Surface of an Elastic Body

Lord Rayleigh investigated a type of surface wave running along the planar interface between air and an isotopic elastic solid in which the amplitude of the wave damps off exponentially as it penetrates the solid. It was anticipated by Lord Rayleigh that solutions of this type might approximate the behavior of seismic waves observed during earthquakes [*Proceedings of the London Mathematical Society*, vol. 17 (1887), or *Scientific Papers*, vol. 2, p. 441.]

In studying Rayleigh's treatment of surface waves we take the following model: We consider the (x, y) plane bounded by the free surface $y = 0$. Let the upper half plane $y > 0$ be air and the lower half plane $y < 0$ be an isotropic elastic solid. We assume that monochromatic progressing waves (single frequency) are propagated in the positive x direction as a result of forces applied in the solid at some distance from the surface (for example, the forces that produce earthquakes). We make use of the scalar and vector wave equations. Since the nature of the disturbing force is not specified, there are infinitely many solutions

CHAPTER FIVE

to these wave equations. However, using Rayleigh's approach we obtain solutions that are exponentially damped.

We assume that all the dependent variables are functions of (x, y, t) and that the displacement vector $\mathbf{u} = (u, v, 0)$. The rotation vector $\boldsymbol{\Psi} = (0, 0, \Psi_z)$, where we set $\Psi_z \equiv \Psi$ and $\Psi = v_x - u_y$. The two-dimensional wave equations for the scalar potential ϕ and the vector potential $\boldsymbol{\Psi}$ are

$$c_L^2(\phi_{xx} + \phi_{yy}) = \phi_{tt},$$
$$c_T^2(\Psi_{xx} + \Psi_{yy}) = \Psi_{tt}. \quad (5.81)$$

The BC at $y = 0$ is a free surface, so that the shear stress τ_{xy} and the normal stress τ_{yy} vanish on the x axis. Hooke's law gives us

$$\tau_{xy} = 2\mu\varepsilon_{xy}, \quad \tau_{yy} = \lambda\Theta + 2\mu\varepsilon_{yy}, \quad \Theta = \varepsilon_{xx} + \varepsilon_{yy}, \quad (5.82)$$

for two-dimensional stress and strain. The displacement vector becomes

$$u = \phi_x + \Psi_y, \quad v = \phi_y - \Psi_x, \quad w = 0. \quad (5.83)$$

The strain tensor has the nonzero components

$$\varepsilon_{xx} = u_{xx} = \phi_{xx} + \Psi_{yx},$$
$$\varepsilon_{xy} = \tfrac{1}{2}(v_x + u_y) = \phi_{xy} + \tfrac{1}{2}(-\Psi_{xy} + \Psi_{yy}), \quad (5.84)$$
$$\varepsilon_{yy} = v_{yy} = \phi_{yy} - \Psi_{xy}.$$

Inserting eq. (5.84) into (5.82) yields the BCs for the components of the stress tensor at $y = 0$:

$$[\tau_{xy}]_{y=0} = \mu(2\phi_{xy} - \Psi_{xx} + \Psi_{yy}) = 0,$$
$$[\tau_{yy}]_{y=0} = \lambda\nabla^2\phi + 2\mu(\phi_{yy} - \Psi_{xy}) = 0. \quad (5.85)$$

The problem is to find time-harmonic solutions to (5.81) that exponentially decay with y, are progressing waves in the x direction, and satisfy the BCs (5.85). To this end we take the potentials in the form

$$\phi = Ae^{-ay}e^{ik(x-ct)}, \quad \Psi = Be^{-by}e^{ik(x-ct)}, \quad a > 0, \quad b > 0. \quad (5.86)$$

Note that y is positive downward. A is the amplitude of the scalar potential, B is the amplitude of the vector potential, and a and b are the *decay constants*, which are determined from the wave equations. (Clearly, the wave number k and the frequency ω are the same for each potential.) Inserting eq. (5.86) into eq. (5.81) yields

$$a^2 - k^2\left[1 - \left(\frac{c}{c_L}\right)^2\right] = 0,$$

$$b^2 - k^2\left[1 - \left(\frac{c}{c_T}\right)^2\right] = 0. \quad (5.87)$$

The wave velocity c is not equal to c_L or c_T. However, if $a = 0$ then $c = c_L$, or if $b = 0$ then $c = c_T$. (But neither a nor b can vanish.)

To satisfy the BCs we insert eq. (5.86) into eq. (5.85) and obtain

$$-2iakA + (b^2 + k^2)B = 0,$$

$$[2\mu a^2 + \lambda(a^2 - k^2)]A + 2i\mu bkB = 0. \quad (5.88)$$

Equations (5.88) are a pair of homogeneous algebraic equations for the complex constants A and B. As usual, we set the determinant equal to zero in order to have nontrivial solutions for A and B. We get

$$4\mu abk^2 - (b^2 + k^2)[2\mu a^2 + \lambda(a^2 - k^2)] = 0. \quad (5.89)$$

We eliminate a and b from eq. (5.89) by appealing to (5.87). Equation (5.89) becomes a cubic in $(c/c_T)^2$ and can be put in the form

$$s^3 - 8s^2 + 8(3 - 2r)s - 16(1 - r) = 0, \quad (5.90)$$

where

$$s = \left(\frac{c}{c_T}\right)^2, \quad r = \left(\frac{c_T}{c_L}\right)^2. \tag{5.91}$$

Using the approximation $\lambda = \mu$ (Poisson's condition), we get $r = 1/3$ and eq. (5.90) becomes

$$(s - 4)(3s^2 - 12s + 9) = 0. \tag{5.92}$$

The roots of eq. (5.92) are

$$s_{1,2,3} = 4, 2 + \frac{2}{\sqrt{3}}, 2 - \frac{2}{\sqrt{3}}. \tag{5.93}$$

It is easily seen that the only root that yields positive values for a and b is $s = 2 - 2/\sqrt{3}$. We thereby obtain the following relationship between c and c_T:

$$c = \sqrt{s}\, c_T = 0.9194 c_T. \tag{5.94}$$

For the case of an incompressible body we have $\Theta = 0$. This gives $r = 0$, so that the velocity of a Rayleigh wave becomes

$$c = 0.9553 c_T. \tag{5.95}$$

We have seen that in either case c is slightly less than the velocity of an equivoluminal wave.

We note that c is independent of frequency. This means that there is no dispersion—the wave shape is maintained. Having determined c in terms of c_T or c_L, we can then calculate the decay constants a and b, which determine the rate at which the potentials attenuate with depth. From the definition of the wave number k, both a and b are proportional to ω and $a > b$. This means that for a given frequency, irrotational waves attenuate faster than equivoluminal waves. We also see that waves of higher frequency are attenuated more rapidly than those of lower frequency.

Seismographic signatures often depict waves similar in structure to Rayleigh waves. However, seismograph records of distant earthquakes

STRESS WAVES IN ELASTIC SOLIDS

indicate dispersion, which means dependence of c on ω. This arises mainly because of the inhomogeneity of the earth, and also because of the viscoelastic properties of the earth.

PROBLEMS

1. Show that in any rigid-body rotation from **a** to **x**, $J^*J = E$, so that the Jacobian $(x, y, z)/(a, b, c)$ is the rotation matrix. Use the appropriate direction cosines in three dimensions, taking into account the orthonormal conditions.

2. Calculate the strain tensor and the compression ratio for the uniform compression or dilatation given by the transformation $\mathbf{x} = (1 + \mathbf{k})\mathbf{a}$, where **k** is a constant vector whose coordinates are (k_x, k_y, k_z).

3. Calculate the strain tensor and the compression ratio for the following transformation from **a** to **x**: $x = a + kb$, $y = rb$, $z = sc$, where k, r and s are constants.

4. Consider a square in the (x, y) plane whose vertices have coordinates $(0,0), (0, L), (L, L), (L, 0)$. Show that the shear strain ε_{xy} can be interpreted as the extension or contraction of the diagonal from $(0,0)$ to (L, L) and find the angle of shear.

5. Calculate the strain tensor for the simple tension (compression) and cross-sectional contraction (dilatation) of a solid cylinder defined by the transformation $x = a - \sigma ka$, $y = b - \sigma kb$, $z = c + kc$, where σ is Poisson's ratio and k is a constant. Relate the sign of k to tension or contraction.
Answer:

$$\varepsilon = \begin{pmatrix} -\sigma k + \frac{1}{2}\sigma^2 k^2 & 0 & 0 \\ 0 & -\sigma k + \frac{1}{2}\sigma^2 k^2 & 0 \\ 0 & 0 & k + \frac{1}{2}k^2 \end{pmatrix}.$$

6. Let the stress tensor have the components $\tau_{xx}, \tau_{xy}, \tau_{xz}, \ldots, \tau_{zz}$ and let the strain tensor have the components $\varepsilon_{xx}, \varepsilon_{xy}, \ldots, \varepsilon_{zz}$ (in Cartesian coordinates). Expand eq. (5.19) and obtain the stress components as linear functions of the strain components, making use of eq. (5.20). This is Hooke's law in Cartesian coordinates. Obtain the inverse (the strain

247

components as linear functions of the stress components) by expanding eq. (5.22) and making use of eq. (5.21).

7. Show that the equations of motion for the stress components in cylindrical coordinates can be put in the form

$$\frac{\partial \tau_{rr}}{\partial r} + \frac{1}{r}\frac{\partial \tau_{r\theta}}{\partial \theta} + \frac{\partial \tau_{rz}}{\partial z} + \frac{\tau_{rr} - \tau_{\theta\theta}}{r} = \rho \frac{\partial^2 u_r}{\partial t^2},$$

$$\frac{\partial \tau_{\theta r}}{\partial r} + \frac{1}{r}\frac{\partial \tau_{\theta\theta}}{\partial \theta} + \frac{\partial \tau_{\theta z}}{\partial z} + \frac{2}{r}\tau_{\theta r} = \rho \frac{\partial^2 u_\theta}{\partial t^2},$$

$$\frac{\partial \tau_{zr}}{\partial r} + \frac{1}{r}\frac{\partial \tau_{z\theta}}{\partial \theta} + \frac{\partial \tau_{zz}}{\partial z} + \frac{\tau_{rz}}{r} = \rho \frac{\partial^2 u_z}{\partial t^2}.$$

8. Derive eq. (5.80) from the approximation (5.79) to the period equation (5.78).

9. Derive the expressions for the components of the rotation vector in terms of the components of the displacement vector in cylindrical coordinates given by eq. (5.66).

10. Find the characteristic frequencies of the longitudinal vibrations of a one-dimensional bar of length L fixed at $x = 0$ and free at $x = L$. Hint: The free end is stress free so that $\sigma_{xx} = Eu_{xx} = 0$. Assume a solution of the wave equation for the bar of the form $u = A \cos(\omega t + \alpha) \sin kx$. Do the same for both ends free or both ends fixed.

11. Derive the equilibrium equation for a spherical shell of radius R.

12. Show that, when the displacement vector $\mathbf{u}$ is written as

$$\mathbf{u} = \nabla \phi + \text{curl } \mathbf{\Psi},$$

then eq. (5.30) is satisfied if

$$c_L^2 \nabla^2 \phi = \phi_{tt}, \qquad c_T^2 \nabla^2 \mathbf{\Psi} = \mathbf{\Psi}_{tt}.$$

Set $F = 0$ in eq. (5.30).

13. Referring to problem 12, show that a class of particular solutions of eq. (5.30) can be generated by taking

$$\phi = Ae^{-ay + i(x - \omega t)},$$

$$\mathbf{\Psi} = \mathbf{B}e^{-by + i(x - \omega t)},$$

where $w = 0$ and u and v are independent of z.

STRESS WAVES IN ELASTIC SOLIDS

14. Referring to problem 12, show that when the gradient and curl are expressed in cylindrical coordinates (r, θ, z), then for axially symmetric problems,

$$u_r = \frac{\partial \phi}{\partial r} - \frac{\partial \Psi}{\partial z}, \quad u_z = \frac{\partial \phi}{\partial z} + \frac{\partial \Psi}{\partial r} + \frac{1}{r}\Psi, \quad u_\theta = 0,$$

$$c_L^2 \nabla^2 \phi = \frac{\partial^2 \phi}{\partial t^2}, \quad c_T^2 \nabla^2 \Psi = \frac{\partial^2 \Psi}{\partial \Psi^2},$$

$$\nabla^2 = \frac{\partial^2}{\partial r^2} + \frac{1}{r}\frac{\partial}{\partial r} + \frac{\partial^2}{\partial z^2}.$$

Deduce two sets of particular solutions of these equations by taking

$$\phi = F(r)e^{i(z-\omega t)}, \quad \Psi = G(r)e^{i(z-\omega t)}.$$

CHAPTER SIX

Stress Waves in Viscoelastic Solids

Introduction

In the previous chapter we treated stress waves in elastic solids in which there is no dissipation of energy—no frictional or damping effects. The concept of an elastic solid neglects this energy dissipation and therefore neglects the attenuation of waves due to damping, which occurs in real materials. The next stage in our investigation of stress waves in continuous media is to take into account this dissipation of energy. This means that we must consider more realistic constitutive equations. The one used for an elastic solid considered stress to be a function of strain (a linear function when we use Hooke's law).

Continuous media in which the constitutive equations are more general than those in an elastic solid are called *rheological materials*. In the general case, the stress and stress-rate tensors are functions of the strain and strain-rate tensors. Higher-order stress- and strain-rate effects may also occur. For a Newtonian viscous fluid the stress tensor is a linear function of the strain-rate tensor. The function that relates the stress to the strain rate is called the viscosity coefficient, which is constant for a Newtonian fluid. The simplest type of rheological material is one that combines the properties of a Hookean elastic solid with those of a Newtonian fluid. The properties are determined by the constitutive equations. It follows that the stress tensor is a linear function of the strain and strain-rate tensor. Such a material is called *viscoelastic*.

In this chapter we shall describe the properties of a viscoelastic (VE) solid and discuss the propagation of waves in such a medium. (Note: We call this material a VE solid for convenience, although it has the properties of both an elastic solid and a viscous fluid.) A simple example is "Silly Putty," which behaves as a solid and fractures under a sharp impact, but exhibits the flow properties of a viscous fluid under gentle forces. The earth is essentially a VE medium, as shown by experiments with seismic waves.

Internal Friction

The process of stress wave propagation in a VE medium involves dissipation of energy due to frictional effects. When the wave front causes the material particles to oscillate, some of the elastic energy is converted into thermal energy so that heat is evolved (the dissipation process). The various mechanisms that describe this shifting back and forth between elastic and thermal energy are described under the general term of internal friction. Consider the case of an applied sinusoidal force at the resonant frequency of the body. For an elastic material the amplitude of the displacement vector at resonance will become infinite since there is no damping. (It is clear that the linear theory of elasticity does not hold in the neighborhood of the resonant frequency.) However, for real materials, since there is damping due to energy dissipation, the amplitude at resonance will be finite. For liquids and gases the dissipative process is due to a combination of thermal conduction, convection, and viscosity and can be attacked analytically. However, for a solid the mechanisms are more complicated and depend on the nature of the solid.

One of the most direct ways of describing the process of internal friction in an oscillating VE medium is to study the ratio $\Delta W/W$ in taking a specimen through a stress cycle. ΔW is the energy dissipated and W is the elastic energy stored in the material at maximum strain. This ratio is sometimes called the *specific loss*. It can be measured for a stress cycle without knowing anything about the nature of the forces producing internal friction. This ratio is found to depend on the speed of the cycle and on the past history or behavior of the specimen.

Other methods of measuring internal friction are more indirect. These methods assume that (1) the elastic restoring force of an elastic solid is proportional to the amplitude of oscillations; and (2) the dissipative force of a Newtonian fluid is proportional to the velocity of the fluid. As implied above, a VE solid combines these two assumptions. These assumptions clearly depend on the linearity condition, which involves small-amplitude oscillations. It follows that the ratio of successive free oscillations is constant. The natural logarithm of this ratio is called the *logarithmic decrement* δ; it is taken as a measure of the internal friction of the material. One result is that if δ is taken as the logarithm of the ratio of two successive oscillations on the same side of the equilibrium position, then it is equal to half the specific loss for low damping.

CHAPTER SIX

Another indirect method of measuring the internal friction of a solid is to study the sharpness of the resonance curve (a plot of amplitude vs. frequency) produced by an external force (forcing function) whose frequency is in the neighborhood of the resonant frequency of the body.

DISCRETE VISCOELASTIC MODELS

To understand how stress waves are propagated in VE materials it is best to start with simple discrete models, as a background. The simplest model that involves the constitutive equations of an elastic solid and a viscous fluid is a two-element one wherein the elastic property is modeled by a spring and the viscous element by a dashpot. The elements are represented as follows:

(1) The spring obeys Hooke's law, which is the constitutive equation

$$\sigma(t) = k\varepsilon(t),$$

where σ is the one-dimensional stress tensor acting on the spring and ε is the one-dimensional strain tensor. σ and ε are functions of time t. σ is the tensile or compressive force per unit cross-section area of the spring, ε is the change in length of the spring relative to its equilibrium length, and k is the spring constant (force per unit displacement of spring).

(2) The dashpot represents the viscous element, which models a Newtonian fluid whose constitutive equation is

$$\sigma = \eta \frac{d\varepsilon}{dt} = \eta\dot{\varepsilon},$$

where σ is the one-dimensional shear stress on the dashpot characterized by the viscosity coefficient η and $\dot{\varepsilon}$ is the strain rate. This constitutive equation for a Newtonian fluid tells us that the stress is proportional to the strain rate.

The coupling of (1) and (2) gives us the two-element model. There are two ways of coupling the spring and dashpot: in *series* or in *parallel*. These two types of couplings describe two basic phenomena as shown by experiment: *stress relaxation* described by series coupling (the *Maxwell model*), and *creep* described by parallel coupling (the *Voigt* or *Kelvin model*). Note that in the two-element model mass is neglected, which

STRESS WAVES IN VE SOLIDS

means we neglect the inertial effect. When we come to investigate wave phenomena we cannot neglect mass. (But, for the time being we do so.) It is clear that the above two constitutive equations are independent of the type of coupling. Also note that models involving springs and dashpots do not represent real materials. However, these models lead to a mathematical description in terms of ODEs that gives a good approximation to the VE phenomena.

Maxwell Model

A single Maxwell model is described by series-coupled elastic and viscous elements according to the self-evident scheme: *E-V*. Figure 6.1 describes this model. A spring coupled in series with a dashpot is acted on by a prescribed time-dependent strain. Let σ_E be the stress on the elastic element and let σ_V be the shear stress on the viscous element. A similar notation applies to the strain. We now define a series-coupled two-element model by the following criteria:

(1) The stress on the elastic element is equal to that on the viscous element.

(2) The sum of the strains on the elastic and viscous elements is equal to the applied strain ε.

These two defining properties of a series-coupled model are expressed mathematically by

$$\sigma_E = \sigma_V \tag{6.1}$$

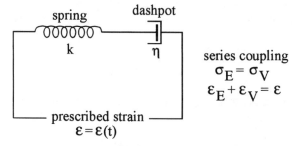

FIGURE 6.1. Maxwell model *E-V* coupling.

CHAPTER SIX

and

$$\varepsilon_E + \varepsilon_V = \varepsilon. \tag{6.2}$$

The constitutive equations may be put in the form

$$\dot{\varepsilon}_E = k^{-1}\dot{\sigma}_E, \qquad \dot{\varepsilon}_V = \eta^{-1}\sigma_V. \tag{6.3}$$

Differentiating eq. (6.2) with respect to t, using eq. (6.2) and the constitutive equations (6.3), yields

$$\dot{\sigma} + \frac{1}{\tau}\sigma = k\dot{\varepsilon}, \qquad \tau = \frac{\eta}{k}. \tag{6.4}$$

Equation (6.4) is the differential equation (ODE) governing the Maxwell model. The parameter τ is called the *relaxation time*; we will see why below. This ODE tells us that given an applied strain rate and the initial stress can solve for the stress on the Maxwell model. Clearly, the response of a Maxwell model to a constant strain is the solution to the homogeneous ODE, which is a simple exponential decay in the stress involving the relaxation time.

STRESS RELAXATION EXPERIMENT

Maxwell's equation (6.4) is a mathematical description of a stress relaxation experiment. The simplest type of such an experiment is described as follows: A specimen of VE material in the form of a thin strip is suspended vertically. The upper end is clamped to a rigid support, and a force is applied to the lower end in such a way that the extended length of the specimen is maintained constant with time. If the initial length is L_0 and the strained length is $L_1 > L_0$, then the applied strain $\varepsilon = (L_1 - L_0)/L_0 = \bar{\varepsilon}$. For $t < 0$ the applied strain is zero. At $t = 0$ $\bar{\varepsilon}$ is suddenly applied and kept constant as t increases. This is described by

$$\varepsilon = \bar{\varepsilon}H(t), \qquad H(t) = \begin{cases} 1, & t > 0, \\ 0, & t < 0. \end{cases} \tag{6.5}$$

$H(t)$ is the Heaviside or unit step function. Equation (6.4) becomes

$$\dot{\sigma} + \frac{1}{\tau}\sigma = k\bar{\varepsilon}\delta(t). \tag{6.6}$$

$\delta(t)$ is the *Dirac delta function*, which has the following properties:

$$\delta(t) = 0 \quad \text{for } t \neq 0, \qquad \delta(t) = \infty \quad \text{for } t = 0,$$

$$\int_{-\infty}^{\infty} \delta(t)\, dt = 1, \qquad \delta(t) = dH(t)/dt.$$

The solution to eq. (6.6) is

$$\sigma(t) = (\sigma_0 + k\bar{\varepsilon})e^{-t/\tau}, \qquad t > 0. \tag{6.7}$$

This solution is valid only for $t > 0$. For $t = 0$ we have $\sigma(0) = \sigma_0$. When a force is applied to the Maxwell model for times small compared to τ, the material behaves like an elastic solid where the solution is approximately $\sigma(t) = k\bar{\varepsilon}$. For times large compared to τ, the behavior is like a Newtonian fluid with viscosity coefficient $\eta = k\tau$. In any case, for an applied constant force the Maxwellian material relaxes exponentially with time.

Voigt or Kelvin Model

A single Voigt or Kelvin model is defined as a parallel-coupled spring-dashpot model as shown in fig. 6.2. By contrast to the Maxwell model (which is series coupled), this model is defined by the following properties:

(1) The strain ε_E on the elastic element is equal to the strain ε_V on the viscous element.

(2) The sum of the stresses on the elastic and viscous elements is equal to the applied stress.

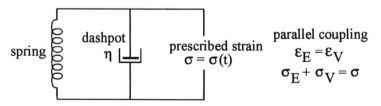

FIGURE 6.2. Single Voigt or Kelvin model $E \| V$ coupling.

CHAPTER SIX

Mathematically,

$$\varepsilon_E = \varepsilon_V,$$
$$\sigma_E + \sigma_V = \sigma,$$

where σ is the applied stress.

By using these defining properties and the constitutive equations (6.3), we obtain the ODE for the Voigt or Kelvin model:

$$\dot{\varepsilon} + \frac{1}{\tau}\varepsilon = \frac{\sigma}{\eta}, \qquad \tau = \frac{\eta}{k}, \tag{6.8}$$

where τ is called the *retardation time*. By contrast with the Maxwell model, we solve for the strain in this model for an applied stress. Note that an interchange of stress and strain allows us to go to and from a series and parallel model.

CREEP EXPERIMENT

Our experiment consists of the same strip of VE material suspended vertically with the upper end clamped to a rigid support. A known weight is hung from the lower end, which means that a constant tensile stress is suddenly applied to the lower end (a step function in time). Since we have a one-dimensional model the decrease in cross-sectional area is neglected. As before, the equilibrium length is L_0 and the stressed length is L_1, which changes with time. The problem is to determine how the tensile strain $\varepsilon = (L_1 - L_0)/L_0$ changes with time due to the applied step function in stress. The stress is given by

$$\sigma(t) = \bar{\sigma} H(t), \tag{6.9}$$

where $H(t)$ is the Heaviside step function and $\bar{\sigma}$ is the magnitude of the applied tensile stress. The ODE (6.8) becomes

$$\dot{\varepsilon} + \frac{1}{\tau}\varepsilon = \frac{\bar{\sigma} H(t)}{\eta}, \tag{6.10}$$

whose solution is

$$\varepsilon(t) = \frac{\bar{\sigma}}{k}[1 - e^{-t/\tau}H(t)] \tag{6.11}$$

with the initial condition $\varepsilon(0) = 0$. Equation (6.11) tells us the response of the Voigt model to a step stress input: the internal strain of the specimen increases exponentially with t and asymptotically approaches the strain $\varepsilon_\infty = \bar{\sigma}/k$ as $t \to \infty$. ε_∞ is the strain on the pure elastic model of modulus k. This model behaves like a Newtonian fluid for small time. Let t start from zero and change by an amount Δt. Then $e^{-\Delta t/\tau} \approx 1 - \Delta t/\tau$, so that the solution (6.11) takes the approximate form

$$\frac{\Delta \varepsilon}{\Delta t} = \frac{\sigma}{\eta}, \tag{6.12}$$

showing that the rate of change of strain is proportional to the stress. This proves that the Voigt model behaves like a Newtonian fluid for small time.

Figure 6.3 shows a family of plots of $\varepsilon(t)$ versus t for various values of τ. These are the *response curves* for a Voigt model. We see that ε tends to its asymptotic value ε_∞ at a slower rate as τ gets larger. This means that the larger the influence of the dashpot compared to the spring the longer the time it takes for the strain to approach its asymptotic value. The reason that τ is called the retardation time is that it is a measure of the time for the internal strain to "creep" to $1 - e^{-1}$ times the elastic strain ε_∞.

We can construct more complicated models by coupling Maxwell models in parallel and Voigt models in series, leading to more complicated ODEs. These models describe more accurately relaxation or

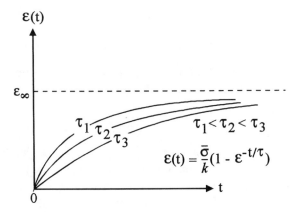

FIGURE 6.3. Plots of strain vs. time for the Voigt model.

CHAPTER SIX

creep experiments for more complicated inputs. The derivation of these ODEs is rather tedious. Therefore we take a different approach, which makes the mathematical manipulations much easier. This approach makes direct use of the constitutive equations by introducing the concepts of a *complex modulus* and a *complex compliance*.

Complex Modulus and Compliance

Many problems in VE involve oscillating stresses and strains so that σ and ε are sinusoidal functions of t. They are conveniently put in the complex form $f(t) = f_0 e^{i\omega t}$, where f stands for σ or ε, f_0 is their amplitude, and ω the frequency.

We first round out the constitutive equations by considering the effect of inertia, so that the equation of motion becomes

$$\sigma = mD^2\varepsilon, \qquad D^2 \equiv \frac{d^2}{dt^2}.$$

The complex modulus Y and the complex compliance J are defined by

$$Y = \frac{\sigma}{\varepsilon}, \qquad J = \frac{1}{Y}.$$

It is clear that Y and J are complex functions of ω. We give Y and J for an elastic element (spring), a Newtonian fluid (dashpot), and an inertial element (mass):

(1) Spring: constitutive equation $\sigma = k\varepsilon$, $Y_k = k$, $J_k = k^{-1}$
(2) Dashpot: constitutive equation $\sigma = \eta D\varepsilon$, $D \equiv d/dt$, $Y_\eta = \eta D$, $J_\eta = Y_\eta^{-1}$
(3) Mass: constitutive equation $\sigma = mD^2\varepsilon$, $Y_m = mD^2$, $J_m = Y_m^{-1}$

Note that for the spring Y and J are real, but for the dashpot and mass they are differential operators. For a single frequency ω we have

$$D = i\omega, \qquad D^2 = (i\omega)^2 = -\omega^2.$$

A Maxwell model is defined by eqs. (6.1) and (6.2). We have $\varepsilon_E = J_E \sigma$, $\varepsilon_V = J_V \sigma$, Maxwell's equation (6.4). A Voigt model is defined by setting the strains equal and summing the stresses. This gives $\sigma_E = Y_E \varepsilon$, $\sigma_V = Y_V \varepsilon$. Using the definitions of these moduli, adding them, and operating on ε gives the Voigt equation (6.8).

STRESS WAVES IN VE SOLIDS

Applying the compliances or moduli to these models suggests that the following rules can be applied to any number of coupled Maxwell or Voigt models:

(1) For series-coupled springs and dashpots we sum the compliances J_E, J_V.

(2) For parallel-coupled springs and dashpots we sum the moduli Y_E, Y_V.

Relaxation and Creep Functions for Discrete Models

Consider the ith Maxwell unit defined by the parameters k_i and τ_i. The *relaxation function* $\Psi_i(t)$ is defined as the stress response of the ith Maxwell unit to a unit step input in strain. Therefore eq. (6.7) becomes

$$\psi_i(t) = k_i e^{-t/\tau}. \tag{6.13}$$

We now couple N Maxwell units in parallel, using the condition that the stresses add and the strain on each unit is the same. We obtain (for zero IC)

$$\sigma(t) = \sum_i \sigma_i(t) = \bar{\varepsilon} \sum_i k_i e^{-t/\tau_i}.$$

Setting $\bar{\varepsilon} = 1$ we get

$$\psi(t) = \sum_i \psi_i(t) = \sum_i k_i e^{-t/\tau_i}. \tag{6.14}$$

This tells us that the relaxation function of N Maxwell units coupled in parallel is equal to the sum of the relaxation functions of each unit. This *discrete relaxation spectrum* is characterized by the set of N τ_is.

In a similar manner, the *creep function* $\phi_i(t)$ is defined as the strain response of the ith Voigt unit to a unit step function in stress. Equation (6.11) becomes

$$\phi_i(t) = \frac{1}{k_i} H(t)[1 - e^{-t/\tau_i}].$$

259

CHAPTER SIX

The retardation function for a model defined as N Voigt units coupled in series is

$$\phi(t) = H(t) \sum_i \frac{1}{k_i}[1 - e^{-t/\tau_i}]. \tag{6.15}$$

The *retardation spectrum* is given by the set of N retardation times τ_i.

Thus far we have treated discrete models. However, when we consider wave propagation in VE materials we must deal with continuous media. Therefore the above discussion must be considered as background material which introduces the reader to the nature of VE materials. A continuous VE medium may be considered as the limiting case of a very large number of VE units coupled in series or parallel. As the number of units becomes infinite in a finite region, the models tend to a continuous distribution of matter, leading to continuous relaxation and retardation spectra. These will be discussed below.

Continuous Maxwell Model

Consider a model composed of an infinite number of Maxwell units coupled in parallel in a finite region. Each unit is then characterized by the parameters k and τ, which form a continuous spectrum. We define the *density distribution function* $f(k, \tau)$ as the number of Maxwell units whose parameters k and τ lie between (k, τ) and $(k + dk, \tau + d\tau)$. If we think of the upper half plane in (k, τ) space, then $f(k, \tau) \, dk \, d\tau$ is the number of Maxwell units lying in the area $dk \, d\tau$ in (k, τ) space. Since f is a continuous function of (k, τ), there are an infinite number of maxwell units in the differential area $dk \, d\tau$. The term "number of units" is therefore not to be taken literally. The density distribution of f means that f is a measure of the density of the units in the area $dk \, d\tau$. Extending this concept to a global distribution throughout k, τ space, we have, in the above sense, the number N of Maxwell units in the space (k, τ), which is given by

$$N = \iint_0^\infty f(k, \tau) \, dk \, d\tau.$$

In order to make use of this distribution function, we first must determine the stress response of a single Maxwell unit to a prescribed strain input. To this end we solve the Maxwell ODE given by eq. (6.4) and obtain

$$\sigma(t; k, \tau) = \sigma_0 e^{-t/\tau} + k \int_0^t e^{-(t-\bar{t})/\tau} \dot{\varepsilon}(\bar{t}) \, d\bar{t}. \tag{6.16}$$

This may be obtained by the method of variation of parameters. We now integrate the integral in eq. (6.16) by parts and obtain

$$\sigma(t; k, \tau) = \sigma_0 e^{-t/\tau} + k\varepsilon(t) - \frac{k}{\tau} \int_0^t \varepsilon(\bar{t}) e^{-(t-\bar{t})/\tau} \, d\bar{t}, \tag{6.17}$$

where we set $\varepsilon(0) = 0$. Note that eq. (6.17) depends on a single pair of parameters (k, τ). Equation (6.17) may be interpreted as follows: The first term on the right-hand side is the stress response if there were no applied strain. The second term represents the instantaneous or purely elastic response and is not affected by the dashpot. The third term represents the "memory" the model has by virtue of its being viscoelastic (as shown below). Note that if τ becomes infinite for a finite value of k, then $\sigma(t) = k\varepsilon(t)$ (purely elastic response).

Memory Function

The expression $(1/\tau)e^{-(t-\bar{t})/\tau}$ is called the *memory function*. When it is multiplied by $k\varepsilon(\bar{t})$ and the result integrated from 0 to t, it has the effect of allowing $k\varepsilon(\bar{t})$ to "remember" all the past effects of the applied strain from zero time up to the present time t. We may consider the memory function as an operator that operates on $k\varepsilon$ at time $\bar{t}$ and by integrating over all time from 0 to t allows $k\varepsilon$ to remember the past. For a given k and t, the larger is τ the less the model remembers, since it decays exponentially faster. Recall that for infinite τ the integrand decays to zero in zero time, giving a purely elastic response (no memory). On the other hand, if $\tau = 0$ then there is "infinite memory" in the sense that there is no exponential decay. The above analysis is for a single relaxation time τ. We now use these results for a continuous Maxwell model.

Continuous Distribution Function

In order to investigate the continuous distribution of Maxwell units we must make use of the density distribution function $f(k, \tau)$. In order to

CHAPTER SIX

determine the effect of the whole spectrum of (k, τ) on the stress we multiply both sides of eq. (6.17) by $f(k, \tau)$ and integrate over all positive k and τ. We get

$$\sigma(t) = \iint_0^\infty f(k,\tau)\sigma(t; k, \tau)\, dk\, d\tau$$

$$= \varepsilon(t)\iint_0^\infty kf(k,\tau)\, dk\, d\tau$$

$$- \int_0^\infty \int_0^\infty \int_0^t \frac{k}{\tau} f(k,\tau)\varepsilon(\bar{t})e^{-(t-\bar{t})/\tau}\, dk\, d\tau\, \overline{dt}. \qquad (6.18)$$

We set

$$F(\tau) = \int_0^\infty kf(k,\tau)\, dk, \qquad K = \int_0^\infty F(\tau)\, d\tau. \qquad (6.19)$$

$F(\tau)$ is a continuous distribution function that expresses how τ is continuously distributed over the model, and K is a positive constant representing the integral of this distribution function over all positive τ. Equation (6.19) then becomes

$$\sigma(t) = K\varepsilon(t) - \int_0^\infty \int_0^t \frac{1}{\tau} F(\tau)\varepsilon(\bar{t})e^{-(t-\bar{t})/\tau}\, d\tau\, \overline{dt}. \qquad (6.20)$$

In practice $F(\tau)$ depends on the material and is obtained experimentally. (This is beyond the scope of this book.)

Example. Step input. Let the applied strain be a step input of the form $\varepsilon(t) = H(t)\bar{\varepsilon}$. Then eq. (6.20) becomes (after a little manipulation)

$$\sigma(t) = \bar{\varepsilon}\int_0^\infty F(\tau)e^{-t/\tau}\, d\tau.$$

We can interpret K by setting $t = 0$. We get

$$\sigma(0) = \bar{\varepsilon}\int_0^\infty F(\tau)\, d\tau = \bar{\varepsilon}K.$$

This tells us that K is the spring constant for the continuous Maxwell model.

STRESS WAVES IN VE SOLIDS

Relaxation Function for a Continuum

The relaxation function for a discrete spectrum has already been discussed. We now extend the concept of the relaxation function to a continuum. The continuous relaxation function $\Psi(t)$ is defined as the normalized response of a continuous Maxwell model to a unit step function in strain. Mathematically, we have

$$\Psi(t) = \frac{K - \sigma(t)}{K} = \frac{\int_0^\infty F(t)[1 - e^{-t/\tau}]\,d\tau}{\int_0^\infty F(\tau)\,d\tau}. \tag{6.21}$$

The normalization property tells us that $\Psi(0) = 0$ and $\Psi(\infty) = 1$.

CONTINUOUS VOIGT MODEL

Each Voigt unit is characterized by the parameter k and the retardation time τ. For the continuum (k, τ) range over the upper half plane in (k, τ) space. Let $g(k, \tau)$ be the density distribution function, so that $g(k, \tau)\,dk\,d\tau$ is the "number" of Voigt units in the differential area $dk\,d\tau$. The strain response of a single Voigt unit to a step stress input of magnitude $\bar{\sigma}$ is given by eq. (6.11). For a continuous Voigt model we multiply the response by $g(k, \tau)\,dk\,d\tau$ and integrate over all positive k and τ, obtaining

$$\varepsilon(t) = H(t) \int_0^\infty \frac{1}{k}[1 - e^{-t/\tau}g(k, \tau)]\,dk\,d\tau. \tag{6.22}$$

We set

$$G(\tau) = \int_0^\infty \frac{1}{k} g(k, \tau)\,dk, \qquad \frac{1}{K} = J = \int_0^\infty G(\tau)\,d\tau. \tag{6.23}$$

Then (6.22) becomes

$$\varepsilon(t) = H(t)\bar{\sigma}\left(J - \int_0^\infty G(\tau)e^{-t/\tau}\right)d\tau. \tag{6.24}$$

This is the required strain response of a continuous Voigt model to a step function in stress. The distribution function $G(\tau)$ describes the

CHAPTER SIX

continuous distribution of retardation times integrated over τ. J is called the *equivalence compliance* of the model. The method essentially depends on obtaining the density distribution function $g(k, \tau)$ which is best done by experimental methods.

Creep Function

The normalized creep function $\phi(t)$ for the continuous Voigt model is defined as the strain response to a unit step function in stress. It is given by

$$\phi(t) = \frac{H(t) \int_0^\infty G(\tau)(1 - e^{-t/\tau}) \, d\tau}{J}. \qquad (6.25)$$

THREE-DIMENSIONAL VE CONSTITUTIVE EQUATIONS

In general, rheological materials have constitutive equations that relate stress, stress rates, strain, strain rates, and higher-order rate terms. For discrete models the constitutive equations can be put in the form

$$P\sigma = Q\varepsilon,$$

where P and Q are polynomials in the operator D. These polynomials define the type of material under consideration. For example, for a single Maxwell model, we have $P = D + 1/\tau, Q = kD$; for a single Voigt model, we have $P = 1/\eta, Q = D + 1/\tau$. These can easily be verified from eqs. (6.4) and (6.8).

In this section we generalize to a three-dimensional continuous medium. σ and ε are replaced by τ_{ij} and ε_{ij}, the components of the stress and strain tensors, respectively. The constitutive equations for such a rheological material become

$$P\tau_{ij} = Q\varepsilon_{ij}, \qquad D \equiv \frac{\partial}{\partial t}, \qquad (6.26)$$

where P and Q are polynomials in the partial derivative operator D. There are six equations. Recall from chapter 5 that the stress-strain relationship for a Hookean solid is given by eqs. (5.19), which we repeat for convenience:

$$\tau_{ij} = \lambda \delta_{ij} \Theta + 2\mu \varepsilon_{ij}, \qquad \Theta = \varepsilon_{ii} \equiv \varepsilon_{11} + \varepsilon_{22} + \varepsilon_{33}. \qquad (6.27)$$

These are the six constitutive equations for an elastic solid. But we are interested in rheological or at least VE materials. We therefore must generalize these to the corresponding constitutive equations for rheological materials. We appeal to the constitutive equations given by (6.26) and attempt to determine the proper operators P and Q that could be inserted into eq. (6.17) for a given rheological model. In eq. (6.27) the parameters describing the elastic material are the Lamé constants λ and μ. We want to convert them into partial derivative operators [compatible with eq. (6.26)]. To this end we set

$$\lambda^* = \frac{p}{q}\lambda, \qquad \mu^* = \frac{p}{q}\mu, \qquad (6.28)$$

where p and q are polynomials in the partial derivative operator D such that (λ^*, μ^*) reduce to (λ, μ) for an elastic material. The six rheological constitutive equations then become

$$\tau_{ij} = \delta_{ij}\lambda^* \Theta + 2\mu^* e_{ij}. \qquad (6.29)$$

For a Maxwell model these operators become

$$\lambda^* = \frac{\lambda \tau D}{1 + \tau D}, \qquad \mu^* = \frac{\mu \tau D}{1 + \tau D}. \qquad (6.30)$$

For a Voigt model they become

$$\lambda^* = \lambda(1 + \tau D), \qquad \mu^* = \mu(1 + \tau D). \qquad (6.31)$$

Equations of Motion for a VE Material

Recall that the Navier equations are the equations of motion in terms of the displacement vector for an elastic solid. To extend these equations to a VE medium we merely replace the Lamé constants by the operators λ^*, μ^*. The generalization of the Navier equations for a VE

CHAPTER SIX

medium then becomes

$$\mu^* \nabla^2 u_i + (\lambda^* + \mu^*)\frac{\partial \Theta}{\partial x_i} = \rho(-F_i + u_{i,tt}), \qquad i = 1,2,3. \quad (6.32)$$

The equation of motion for the dilatation becomes

$$c_L^{*2} \nabla^2 \Theta = -\operatorname{div} \mathbf{F} + (\operatorname{div} u)_{tt} = -\operatorname{div} \mathbf{F} + \Theta_{tt}, \quad (6.33)$$

where the operator $c^{*2} = c^{*2}(D)$ is the square of the generalized longitudinal wave speed and is given by

$$c_L^{*2} = \frac{\lambda^* + 2\mu^*}{\rho}. \quad (6.34)$$

The equation of motion for the rotation vector becomes

$$c_T^{*2} \nabla^2 \boldsymbol{\psi} = (-\operatorname{curl} F + \boldsymbol{\psi}_{tt}), \quad (6.35)$$

where c_T^{*2} is the square of the generalized transverse wave speed and is given by

$$c_T^{*2} = \frac{\mu^*}{\rho}. \quad (6.36)$$

The important point is that for wave propagation in a VE material (or any rheological material) the longitudinal and transverse wave speeds are generalized to operators that are quotients of polynomials in the D operator. This clearly arises from the fact that the constitutive equations for a rheological continuum can be cast in the form of time-varying partial differential equations relating stress, strain, and their rates and possibly higher-order terms, as shown by eq. (6.26).

ONE-DIMENSIONAL WAVE PROPAGATION IN VE MEDIA

For a one-dimensional medium the displacement vector $\mathbf{u} = (u,0,0)$ and the dilatation $\Theta = u_x$, so that the generalized Navier equations (5.34) reduce to a single equation:

$$c_L^{*2} u_{xx} = u_{tt}. \quad (6.37)$$

STRESS WAVES IN VE SOLIDS

Equation (6.37) is the generalized wave equation for the displacement of a one-dimensional VE bar. The structure of the operator c_L^{*2} determines the type of VE material the bar is composed of. Note that eq. (6.37) is second order in the operator $\partial/\partial x$, but the order of $D \equiv \partial/\partial t$ depends on the generalized wave speed (which determines the type of VE material). The solution of this generalized wave equation gives the propagation of irrotational or longitudinal waves in the VE bar. By using Maxwell and Voigt models as examples of VE materials, we shall demonstrate that wave propagation in VE media involves the phenomena of attenuation and dispersion.

Maxwell Model

We investigate wave propagation in a thin VE bar that behaves like a Maxwell material. Using eqs. (6.30) and (6.34) we get

$$c_L^{*2} = \frac{c_L^2 \tau D}{1 + \tau D}, \tag{6.38}$$

which is the generalized wave speed for the Maxwell solid. Observe that as the relaxation time becomes infinite the material becomes elastic since $c_L^* \to c_L$. Note that this is compatible with the asymptotic solution of Maxwell's equation for large relaxation times. Inserting eq. (6.38) into (6.37) yields

$$c_L^2 u_{xxt} = u_{ttt} + \frac{1}{\tau} u_{tt}. \tag{6.39}$$

Equation (6.39) appears to be third order in the operator D. However, by choosing the proper initial condition on u we can reduce it to a second-order PDE in D. To see this, we integrate eq. (6.39) with respect to t and obtain

$$c_L^2 u_{xx} = u_{tt} + \frac{1}{\tau} u_t + F(x),$$

where $F(x)$ is arbitrary. The initial condition on u is $u(x,0) = f(x)$. Inserting this expression in the above equation gives $c_L^2 (d^2 f(x)/dt^2) = F(x)$. We choose the IC such that $F(x) = 0$. This is a mild restriction on the IC. However, if we do not want this restriction we need only add the proper function of x to the solution, which does not change its

CHAPTER SIX

character. For simplicity, we set $F(x) = 0$ and obtain

$$c_L^2 u_{xx} = u_{tt} + \frac{1}{\tau} u_t. \tag{6.40}$$

Equation (6.40) is sometimes called the *damped wave equation*. The damping term $(1/\tau)u_t$ represents the effect of the dashpot in causing viscous damping. Experiment bears out the fact that this viscous damping is proportional to the particle velocity u_t. Again, note that this PDE approaches the undamped wave equation for an elastic material as τ becomes infinite.

We now seek a time-harmonic solution of eq. (6.40) in an infinite medium (we neglect the effect of boundary conditions but merely demand that the solution be bounded at infinity). To this end we set

$$u = \bar{u} e^{i(\omega t - \zeta x)}, \tag{6.41}$$

where $\bar{u}$ is the amplitude, ω is the real radial frequency, and ζ is called the *complex wave number*. Inserting eq. (6.41) into (6.40) yields the following quadratic equation for ζ:

$$c_L^2 \zeta^2 - \omega^2 + i \frac{\omega}{\tau} = 0. \tag{6.42}$$

Since ζ is complex we set

$$\zeta = k + i\alpha, \tag{6.43}$$

where k is the real part and α is the imaginary part of ζ. Inserting eq. (6.43) into (6.42) and equating real and imaginary parts yields

$$k^2 = \frac{\theta^2}{2c_L^2 \tau^2} \left[1 + \sqrt{1 + \frac{1}{\theta^2}} \right], \quad \theta = \omega\tau, \tag{6.44}$$

$$\alpha^2 = \frac{\theta^2}{2c_L^2 \tau^2} \left[-1 + \sqrt{1 + \frac{1}{\theta^2}} \right]. \tag{6.45}$$

Note that θ is dimensionless. This is the valid parameter to use when discussing the frequency dependence of the complex wave number.

Inserting eq. (6.43) into (6.41) yields

$$u = \bar{u} e^{i(\omega t - kx)} e^{\alpha t}. \tag{6.46}$$

Equation (6.46) represents an attenuated progressing wave of wave number k and attenuation or decay factor α, which must be negative for stability, so that we must take the negative root for α in eq. (6.45).

The exponent in eq. (6.41) can be put in the form: $i(\omega t - \zeta x) = i\zeta(\bar{c}t - x)$, so that we have

$$\bar{c} = \frac{\omega}{\zeta} = \frac{\omega}{k + i\alpha} = \frac{\omega k}{k^2 + \alpha^2} - i \frac{\alpha}{k^2 + \alpha^2}. \tag{6.47}$$

$\bar{c}$ is called the *generalized wave speed*. Both its real and imaginary parts depend on ω, k, and α. If the real part of the complex wave number $k = 0$, then the generalized wave speed is purely imaginary.

For high-frequency waves and/or large τ, we have $\theta \gg 1$ so that the complex wave number is approximated by

$$\zeta \approx k \approx \frac{\theta}{c_L \tau}, \qquad \alpha = 0.$$

This tells us that for very large θ the complex wave number reduces to the wave number for an elastic solid (no damping, $\tau \to \infty$). In this case the period of the stress wave is short compared with the relaxation time. Recall that in this case $k = 2\pi/\lambda$, where λ is the wavelength, and $c_L = \omega/k$.

The decay constant α increases as $\omega^{1/2}$ for $\theta \ll 1$ and approaches zero as τ becomes very large.

Voigt Model

We now investigate wave propagation in a one-dimensional bar whose VE is characterized by a Voigt model. Using eqs. (6.31) and (6.34) the square of the generalized longitudinal wave speed for the Voigt model becomes the operator

$$c_L^{*2} = c_L^2(1 + \tau D). \tag{6.48}$$

CHAPTER SIX

Inserting eq. (6.48) into the generalized wave equation (6.37) gives the PDE for the Voigt model, which is

$$c_L^2 \tau u_{xxt} + c_L^2 u_{xx} = u_{tt}, \tag{6.49}$$

which is a third-order PDE. As for the Maxwell model we seek time-harmonic solutions given by eq. (6.41), where ω is the same but ζ will be different for the Voigt model. To obtain this complex wave number, we insert eq. (6.41) into (6.49) and get the following quadratic equation in ζ:

$$-c_L^2 (i\omega\tau\zeta^2 + \zeta^2) + \omega^2 = 0. \tag{6.50}$$

Using eq. (6.43) and equating real and imaginary parts yields

$$k^2 = \frac{\theta^2}{2c_L^2 \tau^2 (1+\theta^2)} \left(1 + \sqrt{1+\theta^2}\right),$$

$$\alpha^2 = \frac{\theta^2}{2c_L^2 \tau^2 (1+\theta^2)} \left(-1 + \sqrt{1+\theta^2}\right). \tag{6.51}$$

The reader should compare these expressions with the corresponding ones for the Maxwell model given by eq. (6.45).

Radially Symmetric Waves for a VE Bar

In chapter 5 the propagation of radially symmetric waves in a Hookean elastic bar was analyzed. Recall that scalar and vector potentials were used since there was no energy dissipation in the medium. For a VE medium we have energy dissipation and therefore it would appear that we cannot use this method. However, we notice that eq. (5.69) give the scalar and vector wave equations, which involve the longitudinal and transverse wave speeds. For VE materials these wave speeds become the generalized wave speeds that are the operators given by eqs. (6.34) and (6.36). These operators are the Lamé constants λ^*, μ^*, which define the type of VE material. Therefore, what were the potentials ϕ, ψ for an elastic material are now the "generalized potentials" that allow us to treat dissipative materials when using the generalized wave speeds. Recall from chapter 5 that we used cylindrical coordinates. The wave equations for a VE material are generalizations of eq. (5.69) in the above sense. They are given by

$$c_L^{*2}\left[\frac{\partial^2\phi}{\partial r^2}+\frac{1}{r}\frac{\partial\phi}{\partial r}+\frac{\partial^2\phi}{\partial z^2}\right]=\frac{\partial^2\phi}{\partial t^2},$$
$$c_T^{*2}\left[\frac{\partial^2\psi}{\partial r^2}+\frac{1}{r}\frac{\partial\psi}{\partial r}+\frac{\partial^2\psi}{\partial z^2}\right]=\frac{\partial^2\psi}{\partial t^2},$$
(6.52)

where $\Psi = \Psi_\theta$ is the angular component of the rotation vector Ψ given by the second equation (5.66).

Electromechanical Analogy

We go back to the discrete model formulation from a different point of view: its analogy with electrical circuit theory. To this end we investigate the relationship between the mechanical series and parallel coupling of masses, springs, and dashpots and corresponding couplings of simple electrical circuits composed of coils, capacitors, and resistors.

In order to find these relationships we must give analogies between the mechanical variables such as strain, strain rate, and stress and the electrical variables such as charge, current, and voltage. We shall also relate the mechanical parameters of mass, spring constant, and viscosity coefficient to the electrical parameters of inductance, capacitance, and resistance.

The electrical analogy of strain ε is the charge Q on an electron. Since the current I in an electrical circuit is given by $I = DQ$ ($D = d/dt$), the mechanical analogue of current is the strain rate $D\varepsilon$. The mechanical analogue of voltage is stress. The electrical analogue of mass m is the inductance L (whose unit is the henry). If a current I is passed through a coil a voltage V is induced across the coil. The equation relating voltage to current is

$$V = L\frac{dI}{dt}.$$

The positive constant L is the inductance. This equation tells us that the voltage induced across a coil is proportional to the time rate of change of the current. It is clear why L is analogous to m, since the

CHAPTER SIX

stress is related to the inertial force by Newton's equation of motion

$$\sigma_x = \rho u_{tt},$$

where σ_x is analogous to ∇V, the density ρ is related to the mass, and the displacement u is analogous to the strain.

For an elastic mechanical element, the spring constant or Young's modulus is analogous to the capacitance C (whose unit is the farad). C is defined as the ratio of charge Q on a capacitor to the voltage V across the capacitor. It follows that $V = Q/C$. If F is the force on a spring and u is its displacement, the equation of motion is $F = -ku$. Relating F to voltage and u to charge we get the analogy that $k = 1/C$.

Now consider the dashpot, which is a viscous element. The equation for the dashpot is $\sigma = \eta D\varepsilon$, where η is the viscosity coefficient. The electrical analogy involves the voltage across a resistor whose resistance is R. It follows that η is analogous to R. Ohms law tells us that $V = RI$. On physical grounds this makes sense, since a circuit with resistance dissipates energy, as does the dashpot. In general, for an R, L, C circuit we replace R by the electrical impedance Z so that $Z = V/I$. For a circuit having only a capacitor, Z is called the *capacitative reactance*. For a circuit having only a coil, Z is called the *inductive reactance*. For an R, L, C circuit for an oscillating input voltage or current, Z turns out to be a complex variable in ω, as shown below. The mechanical impedance $Z = \sigma/D\varepsilon$.

Table 6.1 summarizes these analogies.

We point out that the phenomenological aspects of these mechanical and electrical models are entirely different. For instance, there is no

TABLE 6.1
Electrical-Mechanical Analogues

Electrical	Mechanical
Charge Q	Strain ε
Current I	Strain rate $D\varepsilon$
Voltage V	Stress σ
Capacitance C	Compliance $J = \varepsilon/\sigma$
Inductance L	Mass
Resistance R	Viscosity coefficient η
Impedance Z	Impedance $Z = \sigma/D\varepsilon$

STRESS WAVES IN VE SOLIDS

mechanical counterpart to charge, current, or voltage. However, these analogies are important in allowing engineers to treat complicated discrete mechanical systems by their electrical analogues and thereby relate mechanical properties to corresponding voltages and currents using alternating currents (ac). We also point out that these mechanical and electrical relationships are independent of the type of coupling (series or parallel). For a mechanical system, these relations among the stress, strain, and strain rate, as shown above, are the constitutive equations defining the material. A similar statement may be made for the electrical system where the electrical "constitutive equations" relate charge, current, and voltage. We put the constitutive equations for an electrical circuit (which relate voltage and current for each of the three elements) in the form

$$V_R = RI_R, \quad V_L = LDI_L, \quad DV_c = \frac{I_C}{C}, \quad I = DQ. \quad (6.53)$$

The notation is obvious since the subscripts on I and V refer to the element.

We now treat series- and parallel-coupled electrical circuits.

Series R-L-C Circuit

We consider an electrical circuit consisting of three elements: a resistor of resistance R, a capacitor of capacitance C, and a coil of inductance L. We couple these elements in series and represent the resulting model symbolically by R-L-C. This circuit is shown in fig. 6.4. The series three-element circuit is characterized by the two following properties:

(1) The current through each element is the same. This gives

$$I_R = I_L = I_C = I.$$

(2) The sum of the voltages across each element is equal to the applied voltage V. Thus

$$V_R + V_L + V_C = V.$$

(Recall that, for a Voigt model the strains on each element are the same, while the sum of the stresses on the elements is equal to the applied stress. Also recall that a Voigt model is parallel coupled.)

CHAPTER SIX

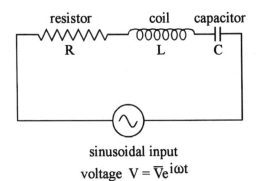

FIGURE 6.4. Series *R-L-C* circuit.

We derive the OD for the current I in the series *R-L-C* circuit by appealing to eq. (6.53). (Recall that I is the same in each element.) We get

$$\left(D^2 + \frac{R}{L}D + \frac{1}{LC}\right)I = \frac{1}{L}DV. \tag{6.54}$$

Equation (6.54) represents a damped harmonic oscillator with $V(t)$ as the applied forcing function. A more interesting case arises when V is a sinusoidal function of t rather than being a constant. We assume that V is monochromatic in the sense that we have a single frequency ω. The ODE can be rewritten by recognizing that the operator $D = i\omega$ because of the sinusoidal input. Equation (6.54) then becomes

$$\left(-\omega^2 + \frac{R}{L}i\omega + \frac{1}{LC}\right)I = \frac{1}{L}i\omega V. \tag{6.55}$$

From this equation we can calculate the impedance Z. We write Z in the form

$$Z = R + iX,$$
$$X = X_L - X_C, \quad X_L = L\omega, \quad X_C = \frac{1}{C\omega}. \tag{6.56}$$

X_L is the inductive reactance, X_C is the capacitative reactance, and X is the total reactance. Since $Z = V/I = R + iX$, it follows that Z is a complex variable in ω. For a pure resistive circuit $L = 1/C = 0$, so that

$Z = R$ (which is the real part of Z and is independent of ω). For very high frequency, $X \approx X_L$, so that the capacitance can be neglected. For very low frequencies, $X \approx X_C$, so that the coil can be neglected. We now interpret Z as a vector in the complex plane where the positive real axis is R, the positive imaginary axis is X_L, and the negative imaginary axis is X_C. Figure 6.5 shows a plot of Z in the complex plane for the case $X > 0$, which means that $X_L > X_C$. This means that we have an R-L-C circuit where the effect of the coil overbalances the effect of the capacitor; the reactive part of the circuit is inductive. The angle δ made by Z and R is called the phase angle. It is positive if Z is in the first quadrant (as shown in the figure), zero if $Z = R$, and negative if Z is in the fourth quadrant where $X_C > X_L$.

We now put Z in the form

$$Z = |Z|e^{i\delta}. \tag{6.57}$$

We know that $Z = V/I$, so that eq. (6.56) gives Z as a complex variable where the modulus (magnitude of Z) and argument (phase angle) are given by

$$|Z| = \sqrt{R^2 + \left(L\omega - \frac{1}{C\omega}\right)^2},$$
$$\tan \delta = \frac{L\omega - \frac{1}{C\omega}}{R}. \tag{6.58}$$

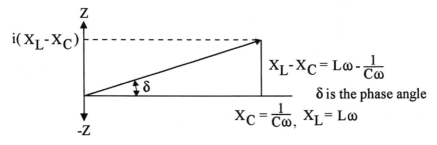

FIGURE 6.5. Impedance Z shown in the complex plane.

CHAPTER SIX

Equation (6.58) tells us that $|Z|$ is a minimum when $Z = R$, and the ω that minimizes $|Z|$ is given by

$$\omega = \omega_r = \frac{1}{\sqrt{LC}}. \tag{6.59}$$

We call ω_r the *resonant frequency*. It is the frequency that maximizes I, since we have

$$I = \frac{V}{Z} e^{-i\delta}. \tag{6.60}$$

Note that for the resonant frequency $\delta = 0$. We can use eq. (6.60) to plot I versus ω for a unit V. This gives a frequency response curve that peaks at $\omega = \omega_r$ for $I_{\max} = R$.

We are now in a position to give the analogue of an R-L-C (series) circuit to a mechanical three-element model composed of a mass defined by m, a dashpot defined by η, and a spring defined by k. We call this an M-V-E (mass-viscous-elastic) system. From eq. (6.56) we see that the impedance of an R-L-C circuit is the sum of the impedances of the resistor, capacitor, and coil. (It is the vector sum since R is real and the reactances are imaginary.) The corresponding impedance of a mechanical system is given by the ratio of resulting stress to strain rate. It is the sum of the impedances due to the mass, spring, and dashpot. This is given by

$$Y = \frac{\sigma}{D\varepsilon} = Y_M + Y_D + Y_E, \tag{6.61}$$

where Y is the modulus of the system, Y_M the modulus of the inertial element (mass), Y_V the modulus of the viscous element (dashpot), and Y_E the modulus of the elastic element (spring). The moduli of these elements are given by

$$Y_M = mD^2, \quad Y_V = \eta D, \quad Y_E = k, \tag{6.62}$$

so that the modulus of the system is

$$Y = mD^2 + \eta D + k = -m\omega^2 + i\eta\omega + k, \tag{6.63}$$

STRESS WAVES IN VE SOLIDS

which is a differential operator operating on $D\varepsilon = i\omega\varepsilon$ to produce σ. Since for a series circuit the impedances add, and for the corresponding mechanical system the compliances add, and since the compliance is analogous to the electrical impedance, it follows that the *series electrical system is analogous to a parallel mechanical system consisting of a mass, spring, and dashpot*. It is easily seen that the ODE for this mechanical system is

$$(mD^2 + \eta D + k)\varepsilon = (-m\omega^2 + i\eta\omega + k)\varepsilon = \sigma(t). \quad (6.64)$$

We may put eq. (6.64) into the form

$$(-\omega^2 + i\tau\omega_0^2\omega + \omega_0^2)\varepsilon = \sigma, \quad \text{where } \tau = \frac{\eta}{k}, \quad \omega_0 = \sqrt{\frac{k}{m}}. \quad (6.65)$$

ω_0 is the natural frequency that would occur if eq. (6.65) were reduced to a simple harmonic oscillator, $(D^2 + \omega_0^2)\varepsilon = 0$. For the case $m = 0$, eq. (6.65) reduces to the ODE given by eq. (6.8) for a Voigt model. The ODE is reduced from a second-order to a first-order equation so that one of the two initial conditions on ε is lost. A plot of ε versus t for a unit σ from eq. (6.64) gives a frequency response curve analogous to the I versus ω curve for the R-L-C circuit. Note that for $m = 0$ the response curve is given by fig 6.3, which shows no maximum. This tells us that, if the inertial term is missing (Voigt model), there is no maximum in the response curve.

Parallel R|L|C Circuit

We now treat a parallel $R|L|C$ circuit and its mechanical analogue in the same manner as above. Figure 6.6 shows the parallel coupling of the

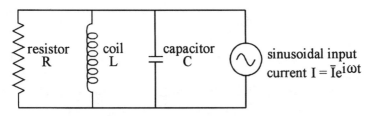

FIGURE 6.6. Parallel $R|L|C$ circuit with sinusoidal current input.

CHAPTER SIX

three-element electrical circuit. The following properties characterize this circuit:

(1) The voltage across each element is the same,

$$V_R = V_L = V_C.$$

(2) The sum of the currents through each element equals the applied current I, which we assume to be sinusoidal of frequency ω:

$$I_R + I_L + I_C = \bar{I}e^{i\omega t}.$$

Using the fact that $V_R = V_L = V_C = V$ and adding the currents in the electrical constitutive equations (6.53) yields the following ODE for the parallel electrical circuit:

$$\left(D^2 + \frac{D}{RC} + \frac{1}{LC}\right)V = \frac{1}{C}D\bar{I}e^{i\omega t}$$

or

$$\left(-\omega^2 + \frac{i\omega}{RC} + \frac{1}{LC}\right)V = \frac{1}{C}i\omega\bar{I}e^{i\omega t}. \tag{6.66}$$

Note that if we compare this ODE with that for the series circuit given by (6.54), we see that for the series circuit the quadratic operator in D operates on I to produce DV, while this operator operates on V to produce DI for the parallel circuit.

Let Z_R, Z_L, Z_C be the impedances of the corresponding elements and Z the impedance due to the applied current I (the impedance of the circuit). Since the voltages are the same across each element and the current in each element adds up to I, we obtain the following result:

$$\frac{1}{Z_R} + \frac{1}{Z_L} + \frac{1}{Z_C} = \frac{1}{Z}. \tag{6.67}$$

This is the principle of a parallel $R|L|C$ circuit: the sum of the reciprocal of the impedances of each element equals the reciprocal of the impedance of the circuit. In accordance with the standard notation in electrical circuits, the reciprocal of the impedance is called the *admittance A*. This principle becomes: the sum of the admittances of

STRESS WAVES IN VE SOLIDS

each element equals the admittance of the circuit, which is

$$A = A_R + A_L + A_C = \frac{1}{R} + \frac{1}{LD} + CD = \frac{1}{R} + \frac{1}{Li\omega} + Ci\omega. \quad (6.68)$$

Note that A (like Z) is a complex variable in ω. It is easily seen by summing admittances that the same ODE (6.66) is produced.

The analogous mechanical model to the parallel $R|L|C$ circuit is seen by defining the admittance of a mechanical system in the same manner as that of an electrical system. Since the mechanical admittance A is the reciprocal of the impedance, we have

$$A_M = \frac{1}{mD} = \frac{1}{mi\omega}, \quad A_V = \frac{1}{\eta D} = \frac{1}{\eta i\omega}, \quad A_E = \frac{1}{k},$$

$$A = \frac{D\varepsilon}{\sigma} = A_M + A_V + A_E.$$

Recall that the complex compliance J of a mechanical system is defined as the ratio of strain to stress, and the sum of the compliances of the elements of a series-coupled mechanical system is equal to the compliance of the system. A Maxwell system with a mass added in series is defined as a series M-V-E system. We have $J_M + J_V + J_E = J = \varepsilon/\sigma$. The conclusion is that *a parallel $R|L|C$ circuit is analogous to a series M-V-E mechanical system*:

$$J_M = \frac{1}{mD^2} = \frac{1}{-m\omega^2}, \quad J_V = \frac{1}{\eta D} = \frac{1}{\eta i\omega}, \quad J_E = \frac{1}{k}, \quad (6.69)$$

$$J = J_M + J_V + J_E.$$

Since $J = \varepsilon/\sigma$ we obtain the ODE defining the M-V-E system, as

$$\left(D^2 + \frac{k}{\eta}D + \frac{k}{m}\right)\sigma = kD^2\varepsilon$$

or

$$\left(-\omega^2 + i\frac{\omega}{\tau} + \omega_0^2\right)\sigma = -\omega^2 k\varepsilon, \quad \tau = \frac{\eta}{k}, \quad \omega_0^2 = \frac{k}{m}. \quad (6.70)$$

It is easily seen that as the mass becomes infinite this ODE reduces to the one defining a Maxwell model.

279

CHAPTER SIX

PROBLEMS

1. A spring of spring constant k_1 is coupled in series with a Voigt model of spring constant k_2 and dashpot of viscosity coefficient η_2. Find the ODE relating stress to strain on the system.

2. The spring in problem 1 (same notation) is coupled in parallel to a Maxwell model. Draw this four-element model. Find the ODE relating stress to strain on the system.

3. A Maxwell model of spring constant k_1 and viscosity coefficient η_1 is coupled in series with a Voigt model of spring constant k_2 and viscosity coefficient η_2. Draw the resulting system. Find the ODE relating stress to strain on the system.

4. A series R_1, L_1, C_1 circuit is coupled in series with a parallel R_2, L_2, C_2 circuit. Find the ODE relating current to voltage on this system for an applied voltage. Find the impedance of this system.

5. A mechanical *M-V-E* system and an *R-L-C* system (both in series and in parallel) obey the following generic ODE:

$$D^2 u + 2\gamma\, Du + \omega_0^2 u = \tilde{f} e^{i\omega t},$$

where γ is related to the damping coefficient and we have a sinusoidal forcing function.

(a) Relate $u(t)$ to the strain or stress for the mechanical systems and to current or voltage for the electrical systems.

(b) Find the solution to the homogeneous ODE (zero forcing function) for the initial conditions $u(0) = u_0$, $Du(0) = v_0$.

(c) Find the solution to the nonhomogeneous equation by setting $u(t)$ equal to a sinusoidal solution with the same frequency ω. Find the resonant frequency and the maximum value of u at resonance. What happens when there is no damping?

6. The generalized Young's modulus for a VE material is

$$E^* = \frac{\mu^*(3\lambda^* + 2\mu^*)}{\lambda^* + \mu^*}.$$

Find the generalized Young's modulus for a Maxwell cylinder and for a Voigt cylinder.

7. Using the results of problem 3 find time-harmonic solutions for the one-dimensional VE wave equation of a bar of infinite length for a Maxwell and a Voigt model, where the Young's modulus is the appropriate time-differential operator, and compare to the results given in the text.

CHAPTER SEVEN

Wave Propagation in Thermoelastic Media

Introduction

In the previous chapters we have treated wave propagation in fluids, elastic solids, and viscoelastic media under essentially isothermal conditions. This means that temperature effects on the material were not considered. In this chapter we shall discuss propagating waves in an elastic solid where the effects of temperature on the medium are taken into account, so that we no longer have an isothermal condition. Such a material is called a *thermoelastic medium*.

The process of heat transfer within the material and between the material and its environment is governed by Fourier's law of heat conduction. In chapter 4 mention was made of Fourier's PDE, which governs unsteady transfer in a viscous fluid. There we showed how the viscosity term $\nu \nabla^2 v$ in the Navier-Stokes equation under certain conditions is actually Fourier's equation, representing the dissipation of energy in a viscous fluid. In order to treat propagating waves in a thermoelastic medium we must generalize the energy equation to take into account Fourier's law of heat conduction. We must also generalize the stress-strain relations for an elastic medium to take into account the effect of temperature. This stress-strain-temperature relation is given below.

Duhamel-Neumann Law

An elastic body under deformation, and hence strain, suffers local temperature gradients, thus producing a *thermal stress field*. To get a handle on the complicated relations that involve stress, strain, and temperature effects leading to the *Duhamel-Neumann law*, we start with a simple model that can explain thermal stresses. Consider an elastic body in equilibrium occupying a volume element $dV_0 = dx_0 \, dy_0 \, dz_0$ (a parallelepiped) where the sides are parallel to the coordinate axes. We neglect rotational motion since this does not contribute to the strain field (we treat a linear medium with small deformations). The material is assumed to be homogeneous as well as isotropic. Therefore, if the thermal gradient produced by the strain field is small, then the change in each side of the volume element is proportional to the temperature T, so that we have

$$\frac{dx - dx_0}{dx_0} = \frac{dy - dy_0}{dy} = \frac{dz - dz_0}{dz_0} = \alpha T, \quad \alpha > 0, \quad (7.1)$$

where α is called the *coefficient of linear thermal expansion*. Since the material is homogeneous the same value of α is valid for each side of the expanding volume element. The unit of α is reciprocal temperature.

The *coefficient of cubical expansion* is α^3 and is given by

$$\alpha^3 = \frac{dV}{T dV_0}.$$

Since we shall deal with the strain fields due to temperature gradients produced by free thermal expansion, and since the components of the strain tensor involve relative changes in line elements, we shall be concerned with the coefficient of linear expansion rather than the coefficient of cubical expansion.

We first treat a one-dimensional model: a specimen consisting of an elastic wire whose equilibrium length is L_0. Suppose heat is supplied to the specimen so that its length is increased by a small amount, to length L. The resulting strain on the wire has two causes: (1) The elastic tensile strain e the wire would have under an isothermal condition. This is related to the tensile stress by Hooke's law. (2) The tensile strain due purely to a small increase in temperature, which is equal to αT. Let t be the resultant tensile stress. Then we have

$$t = Ee - E\alpha T. \quad (7.2)$$

The resultant tensile stress t and the elastic stress Ee are negative (by convention). Equation (7.2) tells us that the resultant tensile stress on the wire is equal to the sum of the isothermal tensile stress and the purely thermal stress. This model will act as a guide to the investigation of the three-dimensional stress field.

CHAPTER SEVEN

We now treat the three-dimensional thermal stress field. Let ε'_{ij} be the (ij)th component of the part of the strain tensor due to the free thermal expansion. The assumption is that the magnitude of the strain field induced by the free thermal expansion of a volume element is small enough so that only the normal components of the strain tensor are affected by the temperature T. It follows that

$$\varepsilon'_{ij} = \alpha T \delta_{ij}, \tag{7.3}$$

where δ_{ij} is the Kronecker delta. Recall from chapter 5 that the dilatation is $\Theta = \varepsilon_{ii}$ (summed over i). This is the sum of the normal components of the strain tensor. For a homogeneous medium for the above conditions the relation between the dilatation and the temperature is

$$\Theta = 3\alpha T. \tag{7.4}$$

The components of the shear strain and shear stress are not affected by T. Let $\Phi = \tau_{ii}$, where Φ is the sum of the normal components (the trace) of the stress tensor τ. The relation between Θ and Φ is given by eq. (4.21). It is

$$\Phi = (3\lambda + 2\mu)\Theta,$$

where λ and μ are the Lamé constants. Using eq. (7.4) we get

$$\Phi = 3\alpha(3\lambda + 2\mu)T. \tag{7.5}$$

Equation (7.5) tells us that the trace of part of the stress tensor due to thermal expansion is proportional to the temperature. Recall that for the isothermal case the relation between the stress and the strain tensors is given by Hooke's law, eq. (4.19), that is,

$$\tau_{ij} = \lambda \delta_{ij} \Theta + 2\mu \varepsilon_{ij}.$$

To account for the stress Φ due to thermal expansion we must subtract eq. (7.5) from the above equation. We obtain

$$\tau_{ij} = \lambda \Theta \delta_{ij} + 2\mu E_{ij} - 3\alpha(3\lambda + 2\mu)T\delta_{ij}. \tag{7.6}$$

Equation (7.6) gives the (ij)th component of the resultant stress tensor as a function of the isothermal strain and the thermal strain. The reason for subtracting off the Φ term is that the stresses and strains are negative in tension. Note that the Kronecker delta tells us that only the normal components of the stress tensor τ involve the stress due to thermal expansion. Equation (7.6) is called the Duhamel-Neumann law. This law expresses the constitutive equations for a linear thermoelastic medium. It was deduced by J. M. C. Duhamel in 1838, who proceeded on the hypothesis that an elastic body can be modeled as a system of material points undergoing molecular interactions. It was also deduced by Franz Neumann, who used a continuum point of view.

Equations of Motion

Using the Duhamel-Neumann law we can obtain the equations of motion for a three-dimensional thermoelastic medium in terms of the displacement vector **u**. This is done by constructing the equations of motion that relate the divergence of the stress tensor to the inertial force plus the resultant of the external forces. This is given in tensor form by

$$\tau_{ij,j} = \rho(u_{i,tt} - F_i). \tag{7.7}$$

Note that in the tensor formulation the subscript j after the comma in the (ij)th component of the stress tensor means differentiation with respect to the coordinate x_j and summed over j. F_i is the ith component of the external force. Equation (7.7) involves the components of the stress tensor. To obtain the equations of motion in terms of the displacement components we insert eq. (7.6) into (7.7). We get

$$\mu \nabla^2 u_i + (\lambda + \mu)\Theta_{,i} = \rho(-F_i + \beta T_{,i} + u_{i,tt}), \tag{7.8}$$

where

$$\beta = \frac{(3\lambda + \mu)\alpha}{\rho}. \tag{7.9}$$

The quantity $T_{,i}$ is the ith component of grad T. For the isothermal case we set $\beta = 0$, which reduces to the Navier equations for an elastic medium [eq. (4.30)]. The system of three PDEs for **u** given by eq. (7.8) is the generalization of the Navier equations for a thermoelastic material. These equations involve both the displacement field **u** and the tempera-

CHAPTER SEVEN

ture field T. They are to be solved for an appropriate set of initial and boundary conditions which prescribe the displacement and/or tractions and temperature on the surface. The surface tractions can be expressed in terms of the displacement derivatives by using the Duhamel-Neumann law.

There are two cases to be considered: *uncoupled* and *coupled*. For the uncoupled case we solve for the unsteady temperature field independently and use the resulting temperature distribution in the generalized Navier equations to solve for the displacement field. For the coupled case we must solve simultaneously for the displacement and the temperature fields.

The Uncoupled Case

The assumption is that we can obtain the temperature field by solving the unsteady Fourier heat conduction equation, which is given by

$$\kappa \nabla^2 T = -Q,$$

where κ is the thermal diffusivity and Q is the heat source. This is a parabolic PDE and requires one initial condition and two boundary conditions for its solution. The assumption that we can solve for the temperature field independently of the displacement field is an approximation to the more realistic case of the coupled system.

The Coupled Case

The energy equation based on thermodynamic considerations (which will not be considered here) tells us that, instead of solving Fourier's equation, we must solve a more complicated combined heat and displacement equation of the form

$$\kappa \nabla^2 T = -Q + \eta \Theta_t, \qquad \eta = \frac{\beta T_0}{K}, \qquad (7.10)$$

where K is the thermal conductivity, T_0 is a reference temperature, and η is called the *coupling coefficient*. We shall call eq. (7.10) the *coupled heat equation*. In a sense, eq. (7.10) is a generalization of Fourier's heat equation that also involves the displacement field. Clearly, if $\eta = 0$ it reduces to Fourier's equation. For $\eta > 0$ the system is coupled and we must simultaneously solve the three generalized Navier equations and the coupled heat equation.

Suppose the external force $\mathbf{F}$ is derivable from a scalar potential Φ such that $\mathbf{F} = \mathbf{grad}\,\Phi$. Suppose we are also looking for irrotational waves. Then there exists another scalar potential ϕ such that $u = \mathrm{grad}\,\phi$. Then the generalized Navier equations (7.8) can be put in the form

$$(\lambda + 2\mu)\phi_{,jji} = (\Phi + \beta T)_{,i} + \rho\phi_{,tti}. \tag{7.11}$$

Integration of these three equations with respect to x_i yields

$$(\lambda + 2\mu)\phi_{,jj} = (\Phi + \beta T) + \rho\phi_{,tt} + f(t), \tag{7.12}$$

where $f(t)$ is an arbitrary function.

Plane Harmonic Waves

We shall begin our investigation of stress wave propagation in a thermoelastic medium by a discussion of plane harmonic waves. Recall that a plane wave is characterized by the property that the wave front is a planar surface normal to the direction of wave propagation. Also recall that in a longitudinal wave the material particles vibrate in the direction of wave propagation, while in a transverse wave the particles vibrate normal to that direction, in the plane of the wave front. We first consider one-dimensional wave propagation in the x direction so that $\mathbf{u} = (u, 0, 0)$. Let σ be the stress and ε the strain. The Duhamel-Neumann equation (7.6) reduces to the following for the one-dimensional case:

$$\sigma = (\lambda + 2\mu)\varepsilon - \alpha(3\lambda + 2\mu)T. \tag{7.13}$$

Neglecting the external force F, the equations of motion (7.8) reduce to

$$c_L^2 u_{xx} - u_{tt} = \beta T_x, \quad c_L^2 = \frac{\lambda + 2\mu}{\rho}, \quad \beta = \frac{(3\lambda + 2\mu)\alpha}{\rho}, \tag{7.14}$$

where c_L is the longitudinal wave speed and β is proportional to the coupling coefficient η where $\eta = T_0/K$. The coupled energy equation

(7.10) becomes

$$\kappa T_{xx} - T_t = \eta \varepsilon_t. \tag{7.15}$$

To obtain time-harmonic solutions for $u(x,t)$ and $T(x,t)$ we assume solutions of the form

$$u = u_0 e^{i(\zeta x - \omega t)}, \quad T = T_0 e^{i(\zeta x - \omega t)}, \tag{7.16}$$

where u_0, T_0 are the amplitudes of the displacement and temperature, ω is the frequency, and ζ is the complex wave number. (Note that we assumed the same frequency and wave number for u and T.) Inserting eq. (7.16) into (7.15) yields a pair of homogeneous algebraic equations for u_0, T_0:

$$\eta \zeta \omega u_0 + (\kappa \zeta^2 + i\omega) T_0 = 0,$$
$$(c_L^2 \zeta^2 - \omega^2) u_0 + i\beta \zeta T_0 = 0. \tag{7.17}$$

In order to insure nontrivial solutions for u_0, T_0 the determinant of the system (7.17) must vanish. This gives the following quadratic in ζ^2 as a function of ω:

$$(c_L^2 \zeta^2 - \omega^2)(\kappa \zeta^2 + i\omega) - i\eta \beta \omega \zeta^2 = 0. \tag{7.18}$$

It is convenient to introduce dimensionless variables $\bar{\zeta}, \bar{\omega}, e$ defined by

$$\bar{\zeta} = \frac{\kappa}{c_L}, \quad \bar{\omega} = \frac{\kappa}{c_L^2} \omega, \quad e = \frac{\eta \beta}{c_L^2}. \tag{7.19}$$

Equation (7.18) becomes (omitting the overbars)

$$(\zeta^2 - \omega^2)(\zeta^2 + i\omega) - ie\omega \zeta^2 = 0. \tag{7.20}$$

The uncoupled case is obtained if we set $e = 0$ in eq. (7.20). We get

$$(\zeta^2 - \omega^2)(\zeta^2 + i\omega) = 0. \tag{7.21}$$

Setting the first factor in eq. (7.21) equal to zero gives $\zeta^2 = \omega^2$. In dimensional variables we get

$$\zeta^2 = \frac{\omega^2}{c_L^2}, \qquad (7.22)$$

which is the expression relating the wave number to the frequency and wave speed for the case of longitudinal wave propagation in a linearly elastic medium. (Recall that $\zeta = 2\pi/\lambda$ where λ is the wavelength.) Inserting eq. (7.22) into the solution for $u(x,t)$ in eq. (7.16) yields

$$u(x,t) = Ae^{i(\omega/c_L)(x-c_L t)} + Be^{i(\omega/c_L)(x+c_L t)}. \qquad (7.23)$$

Setting the second factor in eq. (7.21) equal to zero gives $\zeta^2 = -i\omega$. In dimensional form the two roots of the wave number become

$$\zeta = \pm\sqrt{\frac{\omega}{2\kappa}}(1-i). \qquad (7.24)$$

The correct root that insures exponential damping for $x > 0$ is the one with the minus sign. (For $x < 0$ we take the plus sign.) The solution for $T(x,t)$ from eq. (7.16) becomes

$$T(x,t) = T_0 e^{-\sqrt{\omega/2\kappa}\,x} e^{i\sqrt{\omega/2\kappa}(x-ct)}, \qquad c = \sqrt{2\kappa\omega}. \qquad (7.25)$$

Equation (7.25) tells us that the temperature is a *thermal wave*, which is a damped progressing wave whose wave speed c and damping coefficient both vary as the square root of the frequency. Note that, since c is a function of frequency, this thermal wave is dispersive, which means that the waveform (which depends on the wave speed) changes with frequency and is thus dispersed.

The coupled case occurs when $e \neq 0$. Then eq. (7.20) is a quadratic in ζ^2. We, of course, can solve this quadratic directly; but we shall be content here to use a perturbation method. This means that we let the coupling coefficient e be very small and therefore calculate the effect of a small perturbation of the coupling on the solution. To this end we expand ζ in a power series in e in the form $\zeta = \sum_0^n e^i \zeta_i$, where ζ_i is the

CHAPTER SEVEN

ith-order approximation to the root ζ in eq. (7.20). This is called an nth-order perturbation. For our purpose we only need the first-order perturbation. We therefore set

$$\zeta = \zeta_0 + e\zeta_1, \qquad (7.26)$$

where ζ_0 is the zeroth-order (uncoupled) approximation, and ζ_1 is the first-order approximation to the root. We insert eq. (7.26) into eq. (7.20) and obtain

$$(\zeta_0^2 - \omega^2 + 2e\zeta_0\zeta_1)(\zeta_0^2 + i\omega + 2e\zeta_0\zeta_1) - ie\omega\zeta_0^2 = 0.$$

The uncoupled roots are obtained by setting the coefficient of e^0 equal to zero, yielding

$$(\zeta_0^2 - \omega^2)(\zeta_0^2 + i\omega) = 0,$$

which is eq. (7.21).

The first-order coupling expression is obtained by setting the coefficient of e equal to zero. We obtain

$$2\zeta_1(\zeta_0^2 - \omega^2) + 2\zeta_1(\zeta_0^2 + i\omega) - i\omega\zeta_0 = 0. \qquad (7.27)$$

Equation (7.27) yields two cases.

Case 1: Pure Elastic Wave Propagation

$\zeta_0^2 - \omega^2 = 0$ for the uncoupled case. From eq. (7.27) we obtain

$$\zeta_1 = \pm\frac{1}{2}\frac{\omega}{\omega^2+1} \pm i\frac{1}{2}\frac{\omega^2}{\omega^2+1} = \zeta_{11} \pm i\zeta_{12}, \qquad (7.28)$$

where ζ_{11}, ζ_{12} are the real and imaginary parts of ζ_1. It follows that $\zeta = \zeta_0 + e(\zeta_{11} + i\zeta_{12})$. The exponent in the plane wave solution for u and T, given by eq. (7.16), is $i(\zeta x - \omega t) = i[(\zeta_0 + e\zeta_{11})x - \omega t] - e\zeta_{12}x$. Therefore $\zeta_{12} > 0$ for stability.

We now revert to dimensional coordinates and introduce a characteristic time τ and a characteristic wave number k defined by

$$\tau = \frac{\kappa}{c_L^2}, \qquad k = \frac{c_L}{\kappa}. \qquad (7.29)$$

WAVES IN THERMOELASTIC MEDIA

The diffusivity κ has dimensions of (length)^2/time. (This is easily seen by studying the dimensionality of Fourier's heat conduction equation.) In dimensional form the complex wave number becomes

$$\zeta = \pm \frac{\omega}{c_L} + \frac{e}{2}\left(\frac{k}{\tau^2\omega^2 + 1} + i\frac{k\tau\omega}{\tau^2\omega^2 + 1}\right). \qquad (7.30)$$

Inserting this expression into eq. (7.16) yields the time-harmonic solutions for u and T for the case of pure elastic wave propagation. Since T is of the same form we only display u:

$$u(x,t) = A \exp\left(-\frac{1}{2}\frac{ek\tau\omega x}{\tau^2\omega^2 + 1}\right) \times \exp i\left[\omega t - \left(\frac{\omega}{c_L} + \frac{1}{2}\frac{ek}{\tau^2\omega^2 + 1}\right)x\right]$$

$$+ B \exp\left(-\frac{1}{2}\frac{ek\tau\omega x}{\tau^2\omega^2 + 1}\right)$$

$$\times \exp i\left[\omega t + \left(\frac{\omega}{c_L} + \frac{1}{2}\frac{ek}{\tau^2\omega^2 + 1}\right)x\right]. \qquad (7.31)$$

It is clear that the complex constants A and B depend on the initial and boundary conditions in a given problem, and that we must take the real part of the solution. The first term of the right-hand side represents a progressing wave and the second a regressing wave. Note that the damping coefficient is frequency dependent. Upon examining the wave number we see that the frequency appears in the dimensionless form $\omega\tau$. The solution for $u(x,t)$ and $T(x,t)$ can be written entirely in terms of the dimensionless frequency by setting $\omega = \bar{\omega}/\tau$. The smallness or largeness of $\bar{\omega}$ rather than of ω is seen to be important in investigating the asymptotic behavior of the solutions with respect to frequency. For small $\bar{\omega}$, the damping varies as $\bar{\omega}$, while for large $\bar{\omega}$, it varies as $1/\bar{\omega}$. Since the damping coefficient is a differentiable function of frequency, there must exist a value of $\bar{\omega}$ that maximizes the damping. This value is $\bar{\omega}_M = 1$ or $\omega_M = c_L^2/\kappa$, and the maximum value of the damping coefficient is $-k/2$.

It is clear that the progressing and regressing wave fronts are dispersive, since the wave number is frequency dependent. It is convenient to define the dimensionless group velocity by $\bar{c} = c_g/c_L$. Since $c_g = d\omega/d\zeta$; upon using eq. (7.30) we get for the dimensionless group

291

CHAPTER SEVEN

velocity, to within order e,

$$\bar{c}_g = 1 - \frac{e\bar{\omega}}{(1+\bar{\omega}^2)^2}. \tag{7.32}$$

Notice that the group velocity is different from the longitudinal wave velocity only when the first-order perturbation in e is taken into account. It also follows that $\bar{c}_g(0) = 1$. For small $\bar{\omega}$ $c_g \approx 1 - e\bar{\omega}$, and for large $\bar{\omega}$ $\bar{c}_g \approx 1 - e/\bar{\omega}^3$. The minimum value of c_g is $\bar{c}_{g\,min} = 1 - e/4$. As $\bar{\omega} \to \infty$, $\bar{c}_g \to 1$. Figure 7.1 shows a plot of $\bar{c}_g$ vs. $\bar{\omega}$.

Let c be the phase velocity, or velocity of wave propagation. (Do not confuse c with c_L, which is independent of frequency.) Let the dimensionless phase velocity be defined by $\bar{c} = c/c_L$. We determine $\bar{c}$ to within e by setting $\omega t - \zeta x = 0$ and using the real part of ζ. We get

$$\bar{c} = 1 + \frac{e}{2\bar{\omega}(1+\bar{\omega}^2)}. \tag{7.33}$$

Case 2: Pure Thermal Wave Propagation

In this case, $\zeta_0^2 + i\omega = 0$ and eq. (7.27) becomes

$$2\zeta_1(\zeta_0^2 - \omega^2) - i\omega\zeta_0 = 0,$$

where $\zeta_0 = \pm\sqrt{i\omega} = \pm(1-i)\sqrt{\omega/2}$. The complex wave number becomes

$$\zeta = \pm\sqrt{\frac{\omega}{2}}(1-i) + \frac{e}{2}\sqrt{\frac{\omega}{2}}\frac{1+\omega}{1+\omega^2}(1-i). \tag{7.34}$$

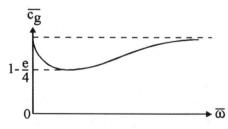

FIGURE 7.1. Plot of dimensionless group velocity vs. dimensionless frequency.

Inserting eq. (7.34) into the solution for $u(x,t)$ and $T(x,t)$ in eq. (7.16) gives the following expression for the exponent in order to insure a stable solution:

$$i(\zeta x - \omega t) = i\left(-\sqrt{\frac{\omega}{2}}x - \omega t + \frac{e}{2}\sqrt{\frac{\omega}{2}}\frac{1+\omega}{1+\omega^2}x\right)$$
$$-\sqrt{\frac{\omega}{2}}\left(1 + \frac{e}{2}\frac{1+\omega}{1+\omega^2}\right)x. \qquad (7.35)$$

The damping coefficient is

$$-\sqrt{\frac{\omega}{2}}\left(1 + \frac{e}{2}\frac{1+\omega}{1+\omega^2}\right)x.$$

THREE-DIMENSIONAL THERMAL WAVES; GENERALIZED NAVIER EQUATION

We now investigate the propagation of thermoelastic waves in a three-dimensional medium. We need the vector form of the generalized Navier equations (7.8). It is given by

$$\mu \nabla^2 \mathbf{u} + (\lambda + \mu)\nabla\Theta = \rho(-\mathbf{F} + \beta\,\nabla T + \mathbf{u}_{tt}), \qquad (7.36)$$

where the dilatation is given by $\Theta = \nabla \cdot \mathbf{u}$. We use the vector identity

$$\nabla^2 \mathbf{u} = \operatorname{grad} \nabla \cdot \mathbf{u} - \operatorname{curl} \mathbf{\Psi},$$

where the rotation vector $\mathbf{\Psi}$ is given by

$$\mathbf{\Psi} = \operatorname{curl} \mathbf{u}.$$

Then eq. (7.36) becomes

$$-\mu \operatorname{curl} \mathbf{\Psi} + (\lambda + 2\mu)\nabla\Theta = \rho(-\mathbf{F} + \beta\,\nabla T + \mathbf{u}_{tt}). \qquad (7.37)$$

Equation (7.37) is the vector form of the generalized Navier equations for a thermoelastic material in terms of the curl of the rotation vector and the gradients of the dilatation and the temperature. It reduces to the linearly isotropic, isothermal elastic case for $\beta = 0$. The term

CHAPTER SEVEN

$\beta \operatorname{grad} T$ may be considered an additional force derivable from a potential. If the external force F is also derivable from a potential then it may be combined with the $\beta \operatorname{grad} T$ term.

We can eliminate the curl Ψ term by taking the divergence of each term of eq. (7.37) and using the fact that div curl $\Psi = 0$. The generalized Navier equation becomes

$$(\lambda + 2\mu) \nabla^2 \Theta = \rho(-\operatorname{div} \mathbf{F} + \beta \nabla^2 T + \Theta_{tt}). \quad (7.38)$$

This is a second-order linear PDE for the dilatation in terms of the external force F and the temperature gradient. Irrotational waves are produced by this PDE with the longitudinal wave velocity $c_L = \sqrt{(2\lambda + \mu)/\rho}$.

We can now eliminate Θ and T from eq. (7.37) by taking the curl of each term and recognizing that curl grad $f = 0$, where f is a scalar. We obtain

$$\mu \nabla^2 \Psi = \rho(-\operatorname{curl} \mathbf{F} + \Psi_{tt}). \quad (7.39)$$

Equation (7.39) is a second-order linear PDE for the rotation vector, which obeys the same PDE as the isothermally elastic case. The fact that the rotation vector is independent of the temperature is not surprising, because within the framework of the linear theory the temperature affects only the volume or dilation and is thus independent of rotation.

We now introduce the wave operators $\Box_1$ and $\Box_2$ defined by

$$\Box_i = c_i^2 \nabla^2 - D^2, \quad i = 1, 2, \quad (7.40)$$

where

$$c_1^2 = c_L^2 = \frac{\lambda + 2\mu}{\rho}, \quad c_2^2 = c_T^2 = \frac{\mu}{\rho}, \quad D^2 = \frac{\partial^2}{\partial t^2}. \quad (7.41)$$

Then eq. (7.38) becomes

$$\Box_1 \nabla \cdot \mathbf{u} = \Box_1 \Theta = -\nabla \cdot \mathbf{F} - \beta \nabla^2 T \quad (7.42)$$

and eq. (7.39) becomes

$$\Box_2 \operatorname{curl} \mathbf{u} = -\operatorname{curl} \mathbf{F}. \quad (7.43)$$

Since Θ depends on T, we now combine the generalized dilatational equation (7.42) with the coupled energy equation (7.10). To this end we define the heat operator L by

$$L = \kappa \nabla^2 - D, \quad D = \frac{\partial}{\partial t}. \tag{7.44}$$

Then eq. (7.10) takes the form

$$LT = -Q + \eta D\Theta \tag{7.45}$$

and eq. (7.38) becomes

$$\Box\Theta = \frac{\beta}{\kappa}(L + D)T, \tag{7.46}$$

where, for simplicity, we neglect Q and F. We now eliminate Θ from eq. (7.46) by operating on eq. (7.45) with $\Box$, using the result in eq. (7.46) operated on by D. We get

$$\left[\Box L - \frac{\eta\beta}{\kappa}D(L + D)\right]T = 0. \tag{7.47}$$

We now operate on eq. (7.46) by L and use eq. (7.45), thus eliminating T from eq. (7.46). We obtain

$$\left[\Box L - \frac{\eta\beta}{\kappa}D(L + D)\right]\Theta = 0. \tag{7.48}$$

Setting

$$\Box L - \frac{\eta\beta}{\kappa}D(L + D) = H. \tag{7.49}$$

We note that the same operator H operates on T and Θ, yielding two fourth-order PDEs of the form

$$Hf = 0, \tag{7.50}$$

where f stands for T or Θ.

CHAPTER SEVEN

We now consider time-harmonic solutions for f in one dimension of the form

$$f = e^{i(\zeta x - \omega t)}. \tag{7.51}$$

Using the operator H given by eq. (7.49) and using eqs. (7.40) and (7.44), we get the following equation for the wave number ζ:

$$(c^2\zeta^2 - \omega^2)(\kappa\zeta^2 - i\omega) - i\beta\eta\omega\zeta^2 = 0,$$

which is the same as eq. (7.18). This result is not surprising since eq. (7.18) was obtained by using the time-harmonic solutions for the two second-order PDEs for T and Θ and requiring that their amplitudes be finite (by setting the determinant equal to zero). The same result was obtained above by the fact that the same fourth-order operator operated on both T and Θ.

CHAPTER EIGHT

Water Waves

Introduction

A treatise on wave propagation phenomena would not be complete without a chapter on water waves. The study of water waves is essentially the part of hydrodynamics that deals with an inviscid, incompressible, irrotational fluid. As a first approximation the compressibility of water is small enough to be neglected. As shown below, these properties of water waves allow for simplification in the equations of motion. Moreover, water waves are essentially two-dimensional, giving an additional simplification. Wave motion in water and other fluids involves a free surface—a surface on which no external forces act. Therefore we say that water waves are surface waves. The water particles are subject to the action of gravity, which is an external force that cannot be neglected in the equations of motion.

Surface wave problems have interested many illustrious mathematicians and physicists, such as Euler, the Bernoulli brothers, Lagrange, Cauchy, Poisson, and Poincaré, and also the British school of Airy, Stokes, Kelvin, Rayleigh, Lamb, and the American school of K. Friedrichs, F. John, J. Keller, J. Stoker, and so on. The literature is vast, and the more modern investigations concentrate on nonlinear problems and stochastic processes that occur in turbulence. In this chapter we can explore only the basic foundations of the theory, leaving a more extensive study to an investigation of the literature.

Irrotational, Incompressible, Inviscid Flow; Velocity Potential and Equipotential Surfaces

Since water waves are essentially inviscid, irrotational, and incompressible, we start by reviewing some basic properties of such flow fields. The inviscid nature of the flow means that we can neglect the $\nu \nabla^2 v$ term in

CHAPTER EIGHT

the Navier-Stokes equations. The *continuity equation* or conservation of mass can be written as

$$\nabla \cdot (\rho \mathbf{v}) = -\rho_t, \qquad (8.1)$$

where $\nabla \cdot$ is the divergence operator, ρ is the particle density, and $\mathbf{v}$ the particle velocity. For an incompressible flow the density is constant so that $\rho_t = 0$. The conservation of mass becomes

$$\nabla \cdot \mathbf{v} = 0. \qquad (8.2)$$

For an irrotational fluid we have

$$\mathbf{curl}\,\mathbf{v} = 0. \qquad (8.3)$$

Written out, we have

$$\mathbf{curl}\,\mathbf{v} = \mathbf{i}(w_y - v_z) + \mathbf{j}(u_z - w_x) + \mathbf{k}(v_x - u_y),$$

where $(\mathbf{i}, \mathbf{j}, \mathbf{k})$ are the unit vectors in the (x, y, z) directions and $\mathbf{v} = (u, v, w)$. It is shown in any textbook on vector analysis that an irrotational flow field means that there exists a potential function ϕ such that

$$\nabla \phi = \mathbf{v}.$$

Written out, we have

$$u = \phi_x, \qquad v = \phi_y, \qquad w = \phi_z.$$

ϕ is call the *velocity potential*. In this sense, an irrotational flow field is associated with potential flow, where the particle velocity is derived from the velocity potential. If a velocity potential exists at one instant then, provided the density of the fluid is constant or a function of the pressure only, the velocity potential exists for all time. This statement was proved first by Lagrange in 1791 and then rigorously by Cauchy in 1827 [21, p. 17].

We now introduce the concept of streamlines, which are defined as a family of curves in space which is everywhere in the direction of the velocity $\mathbf{v}$ field. This means that the tangent to every point on a streamline is in the direction $\mathbf{v}$ at that point. Let (dx, dy, dz) be an element of the tangent plane. Then the streamlines are determined by solving the following system of ODEs:

$$\frac{dx}{u} = \frac{dy}{v} = \frac{dz}{z}. \tag{8.4}$$

It is clear that for steady flow the streamlines are independent of time and coincide with the particle trajectories. For unsteady flow the streamlines are time dependent, so that they coincide with the particle velocity but not with the particle path.

Setting the velocity potential equal to a constant generates an *equipotential surface*. Varying the constant generates a family of such surfaces. We now show that these surfaces are orthogonal to the streamlines. Let ϕ_i be a constant that generates the ith surface. On each surface we have

$$d\phi = \phi_x\, dx + \phi_y\, dy + \phi_z\, dz = u\, dx + v\, dy + w\, dz = \mathbf{v} \cdot d\mathbf{t} = 0, \tag{8.5}$$

where $\mathbf{t}$ is a unit vector in the tangent plane whose direction cosines are proportional to dx, dy, dz, so that $\mathbf{t}$ lies on an equipotential surface. We see that $\mathbf{v}$ is normal to this surface, thus completing the proof that the streamlines cut each equipotential surface orthogonally. For irrotational incompressible flow the combination of the continuity equation and the condition of irrotationality leads to the fact that, for steady flow, the velocity potential is a harmonic function, thus satisfying Laplace's equation

$$\nabla^2 \phi = 0. \tag{8.6}$$

As mentioned previously, this is an elliptic PDE. Since the pressure is derivable from a potential, p also satisfies Laplace's equation. Each component of an external force that is derivable from a potential satisfies Laplace's equation. In addition, for water waves p is a function of ρ, so that for an incompressible fluid ρ and p are constant.

Euler's Equations

We now write down Euler's equations for a fluid under the action of a pressure p (force per unit area) and an external force F per unit mass. This expresses the conservation of linear momentum for a fluid con-

CHAPTER EIGHT

tinuum. The vector form of Euler's equations is

$$\frac{d\mathbf{v}}{dt} = \mathbf{v}_t + (\mathbf{v} \cdot \nabla)\mathbf{v} = -\frac{1}{\rho}\nabla p + \mathbf{F}, \tag{8.7}$$

Here $\mathbf{F} = \mathbf{i}X + \mathbf{j}Y + \mathbf{k}Z$, where (X, Y, Z) are the (x, y, z) components of the external force per unit mass. Recall that the convective part of the acceleration $(\mathbf{v} \cdot \text{grad})\mathbf{v}$ is nonlinear. Written out in extended form, Euler's equations become

$$u_t + uu_x + vu_y + wu_z = -\frac{1}{\rho}p_x + X,$$

$$v_t + uv_x + vv_y + wu_z = -\frac{1}{\rho}p_y + Y,$$

$$w_t + uw_x + vw_y + vw_z = -\frac{1}{\rho}p_z + Z.$$

For the case of water waves we have $\mathbf{F} = (0, 0, -g)$, which means that the external force is due to gravity and is derivable from a potential. g is the acceleration of gravity and acts in the $-z$ direction.

Two-Dimensional Fluid Flow

As a first approximation water waves exhibit two-dimensional irrotational, incompressible, inviscid flow. Let the flow be in the xy plane. This means that the flow takes place in a series of planes parallel to xy and is the same in each of these planes. The particle velocity becomes $\mathbf{v} = (u, v, 0)$, where u and v are functions of (x, y, t), and $w = 0$.

We introduced the concept of streamlines above. If we draw a streamline through each point of a closed curve we obtain a *stream tube*. If we reduce the area of the curve to an infinitesimal we get a *stream filament*, which is essentially a streamline. In the steady motion of a fluid, a stream tube behaves like an actual tube through which the fluid flows. There can be no flow across the walls of the tube for, by definition, the velocity is tangent to the walls in the direction of flow. This means we have conservation of mass within the stream tube. Now consider a stream tube in steady flow. A section AB of the tube is

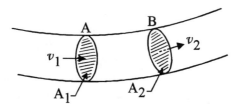

FIGURE 8.1. Part of a stream filament AB illustrating conservation of mass.

shown in fig. (8.1) where the fluid flows with a velocity v_1 into the section at A across area A_1 and flows out of the section at B across area A_2. It is clear that the velocity is normal to each area. The volume of fluid flowing across each area per unit time is given by the velocity times the area. Each area is small enough so that there is no velocity change across it. Since the flow is incompressible the same volume of fluid flows in at A as flows out at B. This leads to a simple form of the continuity equation:

$$A_1 v_1 = A_2 v_2.$$

The continuity equation tells us that the product of the flow speed and the cross-sectional area is constant along a stream tube of a fluid in steady flow. A stream tube must be either circular or end on a boundary; if it ends in space in an open curve, not on a boundary, it ends where the cross-sectional area of the tube is zero so that the speed is infinite. The same holds true for a streamline, which is the limit as the cross-sectional area becomes infinitesimal.

Stream Function

Consider the unsteady two-dimensional flow of a fluid. Figure 8.2 shows curves B and C going from a fixed point A to a variable point P. Let the flow go from left to right. Let R be the region bounded by the closed curve formed by B and C. The flux of fluid flowing across the curves B or C is defined as the rate at which fluid flows across a curve. The continuity equation tells us that the fluid flux into R through the curve B is equal to the flux across C out of R. Point A is fixed and P has coordinates (x, y). This fluid flux depends solely on (x, y) and the time t. We denote this flux by the *stream function* Ψ where $\Psi = \Psi(x, y, t)$. The existence of a stream function is merely a consequence

CHAPTER EIGHT

of the assertion of continuity and incompressibility. Thus a stream function exists for a viscous incompressible fluid. The stream function is constant along a streamline: if the point P moves about in such a manner that the value of the stream function does not change, it will trace out a curve such that no fluid crosses it—the curve is hence a streamline. Each streamline is associated with a particular stream function.

In fig. 8.2, if P is displaced vertically upward by an infinitesimal amount δy to the point Q, then the flow from left to right across PQ tells us that the increment of the stream function is $\delta \Psi = u\, \delta y$. If P is displaced horizontally to the right by an amount δx, then the decrement of the stream function is $\delta \Psi = -v\, \delta x$. In the limit we obtain the following relations between the velocity and the stream function:

$$u = \Psi_y, \qquad v = -\Psi_x. \tag{8.8}$$

COMPLEX VARIABLE TREATMENT

An elegant way of relating the velocity potential to the stream function is through the application of the theory of complex variables. We have introduced the concept of complex variables in chapter 2, p. 00. If $z = x + iy$ is a complex variable and $f(z)$ is an analytic function of z such that $f = u(x, y) + iv(x, y)$, then the Cauchy-Riemann equations

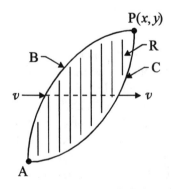

FIGURE 8.2. Flux of fluid across curves B and C through region R. Point A is fixed. Point P is variable in the (x, y) plane.

exist. They are

$$u_x = v_y, \quad u_y = -v_x. \tag{8.9}$$

Both u and v are harmonic functions in that they satisfy Laplace's equations.

In the notation of complex variables, the velocity **v** can be written as the complex variable

$$\mathbf{v} = u + iv = \phi_x + i\phi_y. \tag{8.10}$$

We see that the real part of **v** is the x component of the velocity u while the imaginary part is the y component v. We next define the analytic function $f(z)$ as

$$f(z) = \phi(x, y) + i\Psi(x, y). \tag{8.11}$$

This definition gives the velocity potential as the real part of $f(z)$ and the stream function as the imaginary part of $f(z)$. $f(z)$ is called the complex potential. Using the Cauchy-Riemann equations we get

$$\phi_x = \Psi_y, \quad \phi_y = -\Psi_x, \quad u = \Psi_y, \quad v = -\Psi_x. \tag{8.12}$$

The Cauchy-Riemann equations tells us that curves of constant $\phi(x, y)$ are orthogonal to curves of constant $\Psi(x, y)$. This tells us that, in two dimensions, curves of constant stream function cut curves of constant velocity potential orthogonally. We also see that $\phi(x, y)$ and $\Psi(x, y)$ are both harmonic functions.

Example. Suppose $f(z) = z^2$, which is an analytic function of z. It is easily seen that

$$f(z) = x^2 - y^2 + 2ixy,$$

which means that $\phi = x^2 - y^2$, $\Psi = 2xy$. Curves of constant ϕ and constant Ψ form two families of orthogonal hyperbolas.

Another property of the complex potential $f(z)$ is that

$$\frac{\overline{df}}{dz} = u + iv = \mathbf{v}, \tag{8.13}$$

where the bar over df/dz means the complex conjugate of df/dz. (The complex conjugate of f is obtained by replacing i by $-i$).

Another way of looking at the stream function is to revisit fig. 8.2. Take the line integral of $d\Psi$ along any curve from the fixed point A to the field point $P(x, y)$. We have

$$\Psi(x, y) = \int_A^P [\Psi_x \, dx + \Psi_y \, dy] = \int_A^P [-\phi_y \, dx + \phi_x \, dy]$$

$$= \int_A^P [-v \, dx + u \, dy]. \quad (8.14)$$

The integrand $-v \, dx + u \, dy = \mathbf{v} \cdot \mathbf{n} \, ds$, where ds is the element of arc length along the curve and n is the normal to the curve. We obtain

$$\Psi(x, y) = \int_A^P \mathbf{v} \cdot \mathbf{n} \, ds. \quad (8.15)$$

f(z) as a Mapping Function

In order to adequately discuss the two-dimensional irrotational, incompressible, inviscid flow over objects in the complex plane we need to study how the analytic function $f(z)$ is used to map a curve to and from the z plane and the w plane, where $w = f(z) = u + iv$. The coordinates of the z plane are (x, iy) and those of the w plane are (u, iv). The Jacobian of the mapping function $f(z)$ that maps a domain from the z to a corresponding domain in the w plane is given by $|df(z)/dz|^2$. This fact is left for a problem.

(1) A trivial case is $w = z$. This means that any curve in the z plane is mapped into the same curve in the w plane and vice versa. Applying this mapping to two-dimensional flow, we let the free-stream velocity U be in the positive x direction. If $w = Uz$ then $w = U\phi + iU\Psi$, which means that the motion is a uniform stream of velocity U parallel to the positive x direction. The stream function is a family of straight lines parallel to the x axis, and the velocity function is a family orthogonal to the first family.

(2) We now consider $w = 1/z$. Using polar coordinates we have

$$w = \rho e^{i\phi}, \quad z = r e^{i\theta}.$$

This yields $\rho = 1/r$, $\phi = -\theta$. Thus a unit circle in the z plane maps into a unit circle in the w plane in the sense that a counterclockwise rotation in the z plane gives a clockwise rotation in the w plane. A point outside the circle in the z plane is mapped into its inverse conjugate in the w plane and vice versa. This is seen in fig. 8.3, where the point P in the z plane is mapped into its image P' in the w plane. (The transformation from θ to $-\theta$ yields the conjugate.) In fluid flow, $w = U/z$, where U is the free-stream velocity.

(3) For $w = z + c^2/4z$, we shall show that a circle of radius r is mapped into an ellipse whose foci are $\pm c$. Setting $z = re^{i\theta}$, $w = u + iv$, we obtain

$$u = \left(r + \frac{c^2}{4r}\right)\cos\theta, \quad v = \left(r - \frac{c^2}{4r}\right)\sin\theta.$$

This gives

$$\frac{u^2}{\left[r + \dfrac{c^2}{4r}\right]^2} + \frac{v^2}{\left[r - \dfrac{c^2}{4r}\right]^2} = 1,$$

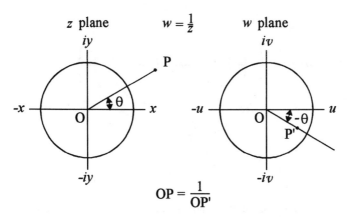

FIGURE 8.3. Inverse mapping $w = 1/z$ from z to w plane.

CHAPTER EIGHT

which is an ellipse in the w plane such that

$$a = r + \frac{c^2}{4r}, \qquad b = r - \frac{c^2}{4r},$$

where a is the semimajor axis and b the semiminor axis. It is easily shown that the foci are $\pm c$, where $c^2 = a^2 - b^2$. The circles r and $1/r$ clearly yield the same ellipse. For the case $r = 1$ the ellipse degenerates into the line $v = 0$, from $u = -c$ to $u = c$. This transformation maps both the interior and exterior of a circle of radius r onto the full w plane, which has a rectilinear slit from $w = -c$ to $w = c$. This transformation is important in certain aerodynamic applications, because this mapping maps the outside of certain circles onto the outside of certain airfoils. Figure 8.4 shows a family of ellipses and a family of confocal hyperbolas orthogonal to these ellipses.

These hyperbolas have the equations

$$\frac{u^2}{\cos^2 \theta} - \frac{v^2}{\sin^2 \theta} = 1.$$

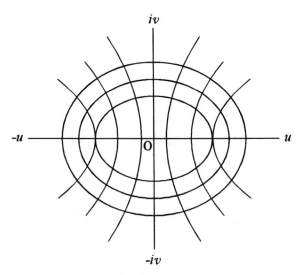

FIGURE 8.4. Family of ellipses and orthogonal confocal hyperbolas in the w plane.

We shall show below the importance of this transformation in the flow around a cylinder. The transformation given by this example is called the *Joukowski transformation*. N. Joukowski was an eminent aerodynamicist who contributed significantly to aerodynamic flow problems.

D'Alembert's Paradox

Before we treat two-dimensional potential flow around a cylinder in detail, we shall prove that the drag and lift forces on a body in potential flow vanish. This is called *D'Alembert's paradox*. Consider a long straight cylindrical tube in which an inviscid fluid flows with a constant free-stream velocity U. If we place a body B in the tube, the flow in the neighborhood of B will be distorted, but at a great distance upstream or downstream of the body the flow will be essentially undisturbed. This is shown in fig. 8.5. To hold the body at rest we require, in general, a drag force and a couple or overturning moment due to the force. Let the force F be this resultant force on the body. Consider two cross-sections A_1 and A_2 at a great distance from B. The fluid between these cross-sections exhibits streamlined flow. The resultant of all the thrusts on the body is

$$F = -\rho A_1 U^2 + \rho A_2 U^2,$$

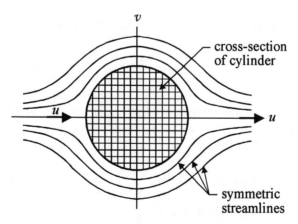

FIGURE 8.5. Two-dimensional symmetrical streamlines over the cross-section of a cylinder, illustrating D'Alembert's paradox.

CHAPTER EIGHT

which vanishes since $A_1 = A_2$. Bernoulli's theorem tells us that

$$\frac{p}{\rho_0} + \frac{1}{2}U^2 + W = \text{const},$$

that is, the potential energy due to the pressure forces plus the kinetic energy plus the work W done by the resultant drag and lift forces on the body is a constant at every point on each streamline. The pressure is the same on each cross-sectional area. Thus $F = 0$, which proves D'Alembert's paradox.

Flow around a Cylinder

Consider a circular cylinder of unit radius normal to the z plane placed in a body of water whose free-stream velocity U is in the positive x direction. This is a constant velocity far enough away from the cylinder so as not to be affected by it. We wish to determine the incompressible, irrotational, inviscid flow over the projection of the cylinder in the z plane. We take this body as a circular disk of unit radius whose center is the origin in the z plane. Recall that the complex potential is $f(z) = \phi + i\Psi$. A slight modification of the transformation given by the example 3 above is used for the complex potential. We have

$$f(z) = \phi + i\Omega = U\left(z + \frac{1}{z}\right). \tag{8.16}$$

The velocity potential and the stream function become

$$\phi = U\left(r + \frac{1}{r}\right)\cos\theta, \qquad \Psi = \left(r - \frac{1}{r}\right)\sin\theta. \tag{8.17}$$

The stream function is zero on the unit circle and the part of the x axis outside the circle. Clearly, we have $r > 1$ for flow around the disk. The components of the velocity **v** become

$$u = U[1 - r^{-4}(\cos^2\theta - \sin^2\theta)], \qquad v = -2Ur^{-4}\cos\theta\sin\theta. \tag{8.18}$$

The velocity on the boundary is

$$u = U(1 - \cos 2\theta), \qquad v = -U\sin 2\theta. \tag{8.19}$$

WATER WAVES

The streamlines are symmetric with respect to the y axis, for changing x does not change the stream function. Also, the streamlines above the x axis are reflections of those below the axis. Thus we have "fore and aft" and "up and down" symmetry. The stagnation points are those points on the boundary for which $\mathbf{v} = \mathbf{0}$. Those are the values for which $\theta = 0$ and $\theta = \pi$. The speed is greatest at the points where $\theta = \pm \pi/2$. The pattern of the streamlines around the disk is shown in fig. 8.5. The pressure distribution can be obtained from Bernoulli's equation in the form

$$\frac{p}{\rho_0} + \frac{1}{2}\mathbf{v} \cdot \mathbf{v} = \text{const.} \qquad (8.20)$$

We showed above that there is no resultant force holding the cylinder in place (D'Alembert's paradox).

We note that the flow in the $w = u + iv$ domain yields streamlines over an elliptical disk, since the mapping function $f(z) = U(z + 1/z)$ maps the unit circle into an ellipse.

Vortex Motion

Thus far we have been concerned with irrotational motion, but water waves in general exhibit rotational motion. Therefore the next concept we shall investigate is the notion of circulation in a fluid, which produces vortex or circular motion. The subject of vortex motion was first investigated by H. Helmholtz (1821–94), a German mathematical physicist who was also an anatomist. Other and simpler proofs of his theorems on vortex motion were given by Lord Kelvin.

Theorem of Kelvin

We derive an important theorem due to Lord Kelvin. Consider a closed curve C which moves with the fluid. The circulation Γ around C is defined by the line integral

$$\Gamma = \oint \mathbf{v} \cdot d\mathbf{s} = \oint [u\,dx + v\,dy + w\,dz], \qquad (8.21)$$

CHAPTER EIGHT

where ds is the element of arc length tangent to C and the integral is taken around C.

We now investigate how Γ behaves with time. To avoid confusion we denote differentiation with respect to the coordinates by δ, retaining d for differentiation with respect to time. ds can be written as the difference $\delta \mathbf{r}$ between the radius vectors $\mathbf{r}$ at the ends of the element. The circulation can then be written as $\Gamma = \oint \mathbf{v} \cdot \delta \mathbf{r}$. Since the contour C is moving with time, we note that in differentiating the integral with respect to time we must taken into account this changing contour. We get

$$\frac{d}{dt}\oint \mathbf{v} \cdot \delta \mathbf{r} = \oint \frac{d\mathbf{v}}{dt} \delta \mathbf{r} + \oint \mathbf{v} \cdot d\frac{\delta \mathbf{r}}{\delta t}.$$

Since $\mathbf{v} = d\mathbf{r}/dt$ we have

$$\mathbf{v} \cdot d\frac{\delta \mathbf{r}}{dt} = \mathbf{v} \cdot \delta \frac{d\mathbf{r}}{dt} = \mathbf{v} \cdot \delta \mathbf{v} = \delta \frac{1}{2}(v^2).$$

For the case of a gravitational field, we can write $\mathbf{F} = (0, 0, -g)$, where g is vertically downward. Euler's equation of motion given by eq. (8.7) becomes

$$\frac{d\mathbf{v}}{dt} = -\frac{1}{\rho_0}\nabla p + (0, 0, -g).$$

Upon using Stokes's theorem we get

$$\oint \frac{d\mathbf{v}}{dt} \cdot \delta \mathbf{r} = \iint_S \text{curl}\left(\frac{d\mathbf{v}}{dt}\right) \cdot d\mathbf{S} = -\iint_S \text{curl}\left(\frac{1}{\rho_0}\text{grad } p - \mathbf{F}\right) = 0.$$

The surface integral is taken around the closed surface having C as its boundary. We have used the fact that curl(grad p) = 0. It is clear that curl $\mathbf{F} = 0$ since $\mathbf{F} = (0, 0, -g)$. Gathering our results, we have shown that $(d/dt)\Gamma = 0$, which means that

$$\oint \mathbf{v} \cdot d\mathbf{s} = \Gamma = \text{const}$$

for the case where all the external forces acting on the fluid are derivable from a potential. This is *Kelvin's theorem* or the law of *conservation of circulation*.

SMALL-AMPLITUDE GRAVITY WAVES

Before investigating in detail the properties of water waves, it is appropriate here to show what the condition of linearity in the equations of motion means in terms of wave propagation. As we know, the acceleration term in Euler's equation of motion is given by the left-hand side of eq. (8.7), which is $dv/dt = v_t + (\mathbf{v} \cdot \mathbf{\nabla})v$. The convective term is nonlinear. To determine under what conditions we can neglect this term, we shall study a traveling surface water wave in the xy plane. Let $\eta = \eta(x, t)$ be the vertical displacement of a wave from its equilibrium configuration at $\eta = 0$ on the free surface. If we consider a sinusoidal progressing wave of amplitude A, wavelength λ, and wave speed c, we may set

$$\eta = A \cos\left[\frac{2\pi}{\lambda}(x - ct)\right]. \qquad (8.22)$$

This sinusoidal wave exhibits linearity since it satisfies the one-dimensional wave equation for η, which is a linear PDE. We want to get an estimate of the ratio of the convective term to the linear term: $|vv_x/v_t|$. From eq. (8.22) we calculate the magnitudes of v, v_x, and v_t. The ratio then becomes

$$\left|\frac{2\pi A}{\lambda}\right|.$$

This ratio tells us that we must have $A \ll \lambda$ in order for the nonlinear term to be neglected. This means not only is the wave of small amplitude—which it must be in order to satisfy the wave equation—but also the wavelength must be long.

WATER WAVES IN A STRAIGHT CANAL

We now investigate the case of water waves traveling along a straight canal with a horizontal bed and parallel vertical sides. Let the axis of x be parallel to the length of the canal and let y be vertically upward

CHAPTER EIGHT

(origin at the floor). This motion is two dimensional in the *xy* plane. Let the ordinate of the equilibrium surface of the water be denoted by h, which is the height of the canal of breadth b. Let the free surface of the water wave have the ordinate $h + \eta$, where $\eta = \eta(x, t)$. The geometry is shown in fig. 8.6. In this model we assume that the vertical acceleration of the water particles is small enough to be neglected. Let ξ be the horizontal displacement of a fluid particle from its equilibrium configuration. Then $\xi = \xi(x, t)$ and the particle velocity becomes $\mathbf{v} = (u, 0, 0)$, where $u = \xi_t$. Since the vertical component of the particle acceleration is zero, the y component of Euler's equations of motion reduces to $p_y = 0$, so that $\mathbf{p} = p(\eta(x), t)$. Now p is the static pressure due to the depth below the surface. The definition of a free surface means that $p = 0$ on that surface, so that $p = 0$ when $y = \eta$. Hence the static pressure at any point (x, y) in the fluid is $p = g\rho(\eta - y)$. The maximum static pressure $p_0 = g\rho h$, since this is the maximum depth of water. g is in the downward direction. We therefore have

$$p - p_0 = -g\rho(y - \eta(x)) + g\rho h. \qquad (8.23)$$

This yields the following expression for grad p:

$$p_x = g\rho\eta_x. \qquad (8.24)$$

This is independent of y, so that the horizontal acceleration is the same for all fluid particles in a plane normal to x.

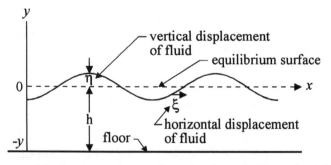

FIGURE 8.6. Geometry of a two-dimensional canal of finite depth h.

Euler's equation of motion becomes

$$u_t + u u_x = -\frac{1}{\rho} p_x. \tag{8.25}$$

In the case of small-amplitude flow this equation is linearized and thus becomes

$$u_t = -\frac{1}{\rho} p_x = -g \eta_x. \tag{8.26}$$

Since $u = \xi_t$, eq. (8.26) becomes

$$\xi_{tt} = -g \eta_x. \tag{8.27}$$

The equation of continuity may be found by considering a slab of fluid of height h, breadth b, and thickness δx. This is shown in fig. 8.7. The flux of fluid flowing per unit cross-sectional area into the slab across the plane normal to x is ξbh. The flux of fluid flowing out of the slab per unit cross-sectional area across the plane normal to $x + \delta x$ is $hb(\xi + \xi_x \delta x)$. The divergence of this flux is the net outflow horizontally through the slab, which is $hb\xi_x \delta x$. Since there are no sources or sinks of fluid in the slab, this outflow of fluid must be balanced by the inflow of fluid in the direction of the inward normal to the area $b \delta x$. This vertical flux of fluid per unit cross-sectional area is $\eta b \delta x$. The balance of fluxes is

$$-(\xi hb)_x \delta x = \eta b \, \delta x.$$

The continuity equation may be put in the form

$$\eta = -h \xi_x. \tag{8.28}$$

Eliminating η between eqs. 8.27 and 8.28 yields

$$\xi_{tt} = g h \xi_{xx}. \tag{8.29}$$

This is the one-dimensional wave equation for the horizontal displacement ξ. We note that the square of the wave speed is $c^2 = gh$. We may also eliminate ξ and obtain the wave equation for η with the same wave speed. The wave equation was treated in detail in chapter 3. We recall

CHAPTER EIGHT

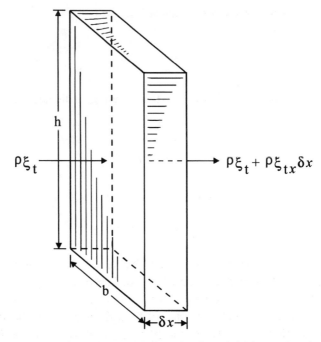

FIGURE 8.7. Slab of fluid of height h, breadth b, and thickness δx, showing flux of fluid normal to the parallel sides at x and $x + \delta x$ and illustrating divergence.

that the general solution for ξ or η is $F(x - ct) + G(x + ct)$, where F represents a progressing wave and G a regressing wave. A thorough discussion of the Cauchy problem was also given there.

Lagrangian Representation

Thus far we have been using the Eulerian method of representing the spatial coordinates. We now discuss Lagrange's method, which, we recall, uses the particle coordinates as the independent spatial variables. We still consider a rectangular canal of depth h and breadth b. Let ξ be the displacement at time t of a particle from its equilibrium position x, where x represents a particle coordinate. We shall follow the time history of each water particle. Again, let η be the elevation of the free surface from its equilibrium configuration. The equation of motion per unit breadth of a stratum of water whose undisturbed thickness is δx is given by

$$\rho h \, \delta x \, \xi_{tt} = -p_x \, \delta x (h + \eta),$$

where $h + \eta$ represents the area per unit breadth of the stratum and $p_x \delta x$ represents the pressure difference for any two particles at x and $x + \delta x$. We still assume that the pressure on any particle depends only on the depth of the fluid below the free surface (note that there is zero pressure on the free surface) so that

$$p_x = g\rho\eta_x.$$

Then Euler's equation of motion becomes

$$\xi_{tt} = -g\left(1 + \frac{\eta}{h}\right)\eta_x. \tag{8.30}$$

The equation of continuity is obtained by equating the flux of fluid consisting of the same particles in the disturbed and undisturbed states. We obtain

$$(\delta x + \xi_x \, \delta x)(h + \eta) = h \, \delta x,$$

which becomes

$$1 + \frac{\eta}{h} = (1 + \xi_x)^{-1}. \tag{8.31}$$

We eliminate η from eqs. (8.30) and (8.31) and obtain

$$\xi_{tt} = gh\frac{\xi_{xx}}{(1 + \xi_x)^3}. \tag{8.32}$$

A similar equation can be obtained for η. Equation (8.32) is a one-dimensional nonlinear wave equation for ξ because of the denominator in the right-hand side of the equation. This is the general equation in terms of Lagrange's coordinates for a long wave in a canal of depth h and vertical sides. (Note that x represents a particle coordinate.)

To linearize the wave equation given by eq. (8.32) we linearize the continuity equation (8.31) and obtain

$$\eta = -h\xi_x, \tag{8.33}$$

so that eq. (8.32) becomes the linear wave equation

$$\xi_{tt} = gh\xi_{xx}. \tag{8.34}$$

CHAPTER EIGHT

The same wave equation can be deduced for η. Note the important point that for the linear theory, it makes no difference whether we use the Eulerian or Lagrangian representation. This observation was also made in chapter 4.

KINEMATICS OF THE FREE SURFACE

For simplicity we use the Eulerian treatment. Since the surface is composed of water particles, it is clear that the kinematic condition on the free surface is that the surface moves with the fluid. This means that the velocity of a water particle on the surface is equal to the velocity of the surface. Mathematically, $dy/dt = d\eta/dt$, which gives

$$\frac{dy}{dt} = y_t + \left(\frac{dx}{dt}\right) y_x = \frac{d\eta}{dt} = \eta_t + \left(\frac{d\eta}{dx}\right) \eta_x.$$

In linearizing this equation we omit the nonlinear term which is the product of two small numbers, and obtain

$$\frac{dy}{dt} = \eta_t.$$

This is the linearized BC that occurs on the free surface.

Recall that the velocity potential ϕ exists for our flow such that $\mathbf{v} = \text{grad } \phi$. Since the only external force is that due to gravity, Euler's equation of motion can be written as

$$\nabla\left(\phi_t + \frac{p}{\rho} + gy\right) = 0. \tag{8.35}$$

Integration yields

$$\phi_t + \frac{p}{\rho} + gy = f(t), \tag{8.36}$$

where $f(t)$ is an arbitrary function of time. Equation (8.36) is an energy equation; it is the linearized form of Bernoulli's equation for unsteady potential flow. It is easily seen that we can neglect $f(t)$ by incorporating it into the potential or adjusting one of the ICs.

Vertical Acceleration

We now investigate the two-dimensional propagation of water waves in the xy plane (x horizontal and y vertical downward) where the vertical acceleration is not neglected. This is a more accurate interpretation of water waves than the preceding theory. We shall see that the amplitude of the wave diminishes as we go downward from the free surface.

The velocity potential $\phi = \phi(x, y, t)$ satisfies the two-dimensional Laplace equation

$$\phi_{xx} + \phi_{yy} = 0. \tag{8.37}$$

For convenience, we make a slight change in the coordinate system by letting the origin be on the free surface where y acts vertically downward from that surface. It is clear that the horizontal component of the acceleration is $u_{tt} = \phi_{xtt}$. Integrating the linear Euler's equation (8.26) with respect to x yields

$$\frac{p}{\rho} = -\phi_t - gy + f(t), \tag{8.38}$$

where $f(t)$ is arbitrary. Now, the pressure gradient is zero on the free surface. If η is the elevation (or depression) of the surface at time t with respect to $(x, 0)$, we have

$$\eta = -\frac{1}{g}\phi_t, \tag{8.39}$$

which holds on $y = \eta = 0$, for the linear theory. The function $f(t)$ and the additive constant are to be absorbed into ϕ_t.

We first treat the case of infinite depth so that $h = \infty$. The fluid then occupies the region $y \le 0$ in the xy plane. We look for a progressing wave solution for $\phi(x, y, t)$ traveling in the positive x direction. To this end, we separate variables as follows: We set

$$\phi(x, y, t) = f(y)\sin(kx - \omega t), \tag{8.40}$$

where $f(y)$ is a function of y to be determined from Laplace's equation. ω is the radial frequency and k is the wave number such that $k = 2\pi/\lambda$, where λ is the wavelength. We note that $f(y)\cos(kx - \omega t)$ is also a

CHAPTER EIGHT

solution, and the general solution is a linear combination of these two linearly independent solutions. Inserting eq. (8.40) into Laplace's equation (8.37) yields

$$\frac{d^2 f}{dy^2} - k^2 f = 0, \qquad (8.41)$$

which has solutions e^{ky}, e^{-ky}. We must omit the latter since ϕ must not increase exponentially as we go down from the free surface. The velocity potential becomes

$$\phi = Ae^{ky} \sin(kx - \omega t), \qquad y \leq 0. \qquad (8.42)$$

The velocity **v** has components

$$u = \phi_x = kAe^{ky} \cos(kx - \omega t), \qquad v = \phi_y = kAe^{ky} \sin(kx - \omega t). \qquad (8.43)$$

This clearly tells us that the particle velocity decays exponentially as we go into the fluid, so that the floor (at infinite depth) has zero velocity.

On the free surface the linearized BC is

$$\eta_t = y_t, \qquad y = 0. \qquad (8.44)$$

Also, on the free surface $p = 0$ so that eq. (8.39) holds. Differentiating this equation with respect to t and setting $\eta_t = \phi_y$ yields

$$\phi_{tt} = -g\phi_y, \qquad y = 0. \qquad (8.45)$$

We may now get ω in terms of k by inserting eq. (8.42) into (8.45). We obtain

$$\omega^2 = gk. \qquad (8.46)$$

We determine the particle paths by integrating eq. (8.43) with respect to t, obtaining

$$x - x_0 = \left(\frac{k}{\omega}\right) Ae^{ky} \sin(kx - \omega t),$$

$$y - y_0 = \left(\frac{k}{\omega}\right) Ae^{ky} \cos(kx - \omega t).$$

From this equation we obtain the circle

$$(x - x_0)^2 + (y - y_0)^2 = \left(\frac{k}{\omega}\right)^2 A^2 e^{2ky}. \tag{8.47}$$

This tells us that the particle paths are concentric circles whose origin is (x_0, y_0) and whose radii decrease exponentially with increasing depth.

It is clear that the velocity potential satisfies the one-dimensional wave equation where the wave speed c is given by

$$c = \frac{\omega}{k} = \sqrt{\frac{g}{k}} = \sqrt{\frac{g\lambda}{2\pi}}. \tag{8.48}$$

We may consider regressing waves, in which case we can replace eq. (8.42) by $\phi = Be^{ky} \sin(kx + \omega t)$. To satisfy a general set of ICs and BCs we combine the progressing and regressing waves and perform a linear superposition of these waves by a Fourier analysis for a discrete frequency spectrum, or a Fourier transform analysis if the frequency spectrum is continuous.

Standing Waves

The simplest type of standing wave is formed when two simple harmonic waves of equal amplitude on the free surface travel in opposite directions. The surface elevations are given by

$$\eta_1 = \tfrac{1}{2}a \sin(kx - \omega t), \qquad \eta_2 = \tfrac{1}{2}a \sin(kx + \omega t).$$

η_1 represents a progressing wave and η_2 a regressing wave. The result of superposition of these two waves is the free surface

$$\eta = a \sin kx \cos \omega t.$$

The free surface is a standing, amplitude-modulated, harmonic wave. Thus we see that a standing wave is *stationary* in the sense that it is not propagated. The points for which $kx = n\pi$ ($n = -2, -1, 0, 1, 2, \ldots$) are at rest on the free surface and are called the *nodes*. The points for which $kx = (2n + 1)\pi/2$ are points of maximum displacement and are called *loops*. When $\cos \omega t = \pm 1$ the free surface has the form $\eta =$

CHAPTER EIGHT

$\pm a \sin kx$, which represents the maximum departure from the equilibrium configuration. When $\cos \omega t = 0$ the free surface is the equilibrium configuration. It is clear that this type of standing wave has the following properties: (1) The shape of the free surface is sinusoidal, of small amplitude, and motionless. (2) Every point on the free surface exhibits simple harmonic motion.

We may treat standing waves in the setting of an infinitely deep canal by studying the velocity potential. We put it in the following form:

$$\phi(x, y, t) = e^{i\omega t}\overline{\phi}(x, y) \tag{8.49}$$

where $\overline{\phi}(x, y)$ satisfies

$$\overline{\phi}_{xx} + \overline{\phi}_{yy} \equiv \nabla^2 \overline{\phi} = 0, \qquad y \leq 0, \quad -\infty < x < \infty,$$

BC at free surface: $\overline{\phi}_y - \left(\dfrac{\omega^2}{g}\right)\overline{\phi} = 0, \qquad y = 0,$ (8.50)

BC on floor: $\overline{\phi}_y = 0, \qquad y = -\infty.$

The free-surface elevation becomes

$$\eta(x, 0, t) = -\left(\frac{i\omega}{g}\right) e^{i\omega t}\overline{\phi}(x, 0). \tag{8.51}$$

The functions

$$\phi(x, y, t) = e^{ky}\begin{Bmatrix} \cos kx \\ \sin kx \end{Bmatrix} e^{i\omega t} \tag{8.52}$$

are standing waves. They are sinusoidal functions of t whose amplitudes are modulated as follows: they are sinusoidal functions of x that decay exponentially in y. The wave number k satisfies the free-surface condition and thus becomes

$$k = \frac{\omega^2}{g}, \tag{8.53}$$

so that the wavelength λ is

$$\lambda = \frac{2\pi}{k} = \frac{2\pi g}{\omega^2}.$$

This tells us that the wavelength varies inversely as the square of the frequency, and the wave number varies directly as the square of the frequency.

Since the oscillations decay exponentially, the BC on the floor is $\bar{\phi} = 0$. the free surface η becomes

$$\eta(x, t) = -\left(\frac{i\omega}{g}\right)\left\{\begin{array}{c}\cos mx \\ \sin mx\end{array}\right\} e^{i\omega t}. \tag{8.54}$$

The standing-wave solutions given by eq. (8.52) can be formed by the interaction of a progressing and a regressing wave of the form

$$\phi(x, y, t) = e^{ky} \sin(kx - \omega t) + e^{ky} \sin(kx + \omega t). \tag{8.55}$$

We can obviously also use cosine waves.

Two-Dimensional Waves of Finite Depth

We now consider water waves in a canal of finite depth h. The solutions for $\bar{\phi}$ are the harmonic functions

$$\bar{\phi}(x, y) = \cosh k(y + h)\left\{\begin{array}{c}\cos kx \\ \sin kx\end{array}\right\}. \tag{8.56}$$

Recall that the BC on the floor is

$$\bar{\phi}_y = 0, \quad y = -h.$$

This is satisfied by eq. (8.56). Recall also that the BC at the free surface is

$$\bar{\phi}_y - \left(\frac{\omega^2}{g}\right)\bar{\phi} = 0, \quad y = 0.$$

CHAPTER EIGHT

Substituting eq. (8.56) into this BC yields the following expression for the wave number:

$$k = \left(\frac{\omega^2}{g}\right) \coth kh. \tag{8.57}$$

It is clear that this expression reduces to $k = \omega^2/g$ for the case of infinite depth.

BOUNDARY CONDITIONS

We now present a more general discussion of the boundary conditions that occur on the ocean floor and on the free surface by considering the three-dimensional case. The free surface of the fluid is a surface S that separates the water medium from air. Any water particle on S remains on S. (We neglect spray.) We now make a simple change in the coordinate system for the three-dimensional case. We have a right-handed coordinate system, where the coordinate z is negative downward from the origin on the equilibrium surface, x is in the horizontal direction, and y is directed laterally so that the xy plane is normal to z. Let ζ be the elevation of the free surface from its equilibrium configuration. This coordinate system is shown in fig. 8.8. Let $f(x, y, z, t) = 0$ be the equation of S (either free or fixed). Let the particle velocity $\mathbf{v} = (u, v, w)$, where $u = \dot{x}$, $v = \dot{y}$, $w = \dot{z}$. Calculating $\dot{f}$, we have

$$\begin{aligned} \dot{f} &= uf_x + vf_y + wf_z + f_t \\ &= (u, v, w) \cdot (f_x, f_y, f_z) + f_t \\ &= \mathbf{v} \cdot \nabla f + f_t = 0, \end{aligned} \tag{8.58}$$

which holds on S. It is clear that grad f is normal to S so that the velocity potential ϕ satisfies the condition

$$\frac{\partial \phi}{\partial n} = v_n = \frac{f_t}{\sqrt{(f_x)^2 + (f_y)^2 + (f_z)^2}}, \tag{8.59}$$

where v_n is the normal component of the particle velocity on the surface S and n is the unit vector normal to S.

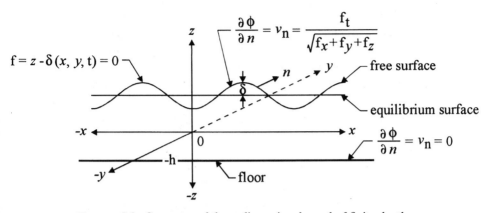

FIGURE 8.8. Geometry of three-dimensional canal of finite depth.

For the case of a fixed surface, the ocean floor or walls or floor of a water tank, we have the condition

$$\frac{\partial \phi}{\partial n} = v_n = 0 \qquad \text{on a fixed surface.} \qquad (8.60)$$

For the case of a free surface S, a BC is that the pressure gradient on S is zero. If we measure the pressure vertically downward from the free surface we may set $p = 0$ on S. In general, we assume that the elevation of the free surface is given by

$$z = \zeta(x, y, t) \quad \text{or} \quad f = z - \zeta(x, y, t) = 0.$$

This equation holds for any particle on S. Hence, we obtain

$$u\zeta_x + v\zeta_y - w + \zeta_t = \phi_x \zeta_x + \phi_y \zeta_y - \phi_z + \zeta_t = 0$$

$$\text{on free surface:} \quad z = \zeta(x, y, t). \qquad (8.61)$$

As a consequence of Bernoulli's law (by setting $p = 0$ on the free surface) we obtain

$$g\zeta + \phi_t + \tfrac{1}{2}\left(\phi_x^2 + \phi_y^2 + \phi_z^2\right) = 0 \qquad \text{on a free surface.} \qquad (8.62)$$

CHAPTER EIGHT

We see that the velocity potential must satisfy the two nonlinear BCs given by eqs. (8.61) and (8.62) on the free surface. The elevation ζ must also satisfy these BCs. This is in sharp contrast to a fixed surface, where the velocity potential must satisfy only one BC.

With respect to the free surface, the problem is nonlinear in the sense that the particle velocity must be solved on an unknown free surface since the equation of the surface must be found from the BCs.

FORMULATION OF A TYPICAL SURFACE WAVE PROBLEM

We now give a mathematical description of a typical three-dimensional surface wave problem with a floor of variable depth, such as breakers rolling on a sloping beach. We merely formulate this problem without solving it. The purpose here is to bring into focus some of the difficulties involved in such a nonlinear problem. Our physical model is an ocean whose floor is level in deep water and slopes upward to a beach. The coordinate system has its origin on the equilibrium surface at the point where the floor begins to slope upward. The z axis points downward, the x axis is horizontal, and the y axis is lateral in the horizontal plane, forming a right-handed coordinate system. Let $\xi = \xi(y, t)$ be the abscissa of the interface between the beach and the water, which is not known a priori. Let $h(x, y, 0)$ be a prescribed fixed function that defines the slope:

$$-h(x, y) \leq z \leq 0, \qquad -\infty < y < \infty, \qquad -\infty < x < \infty.$$

The geometry of this model is shown in fig. 8.9. The formulation of this problem is given by

$$\nabla^2 \phi = 0 \quad \text{for } -\infty < x < \infty, \quad \xi \leq y < \infty, \quad 0 > z > h(x, y).$$

The problem is clearly nonlinear since neither ξ nor the elevation $\zeta(x, y, t)$ is known in advance but must be determined as part of the solution. The BC at the floor is

$$\frac{\partial \phi}{\partial n} = 0 \quad \text{for } z = -h(x, y).$$

The kinematic BC is given by eq. (8.62) with the added term $p(x, y, t)/\rho$, which is zero everywhere except where there is a disturbance. The ICs are the conditions of rest in the equilibrium state.

WATER WAVES

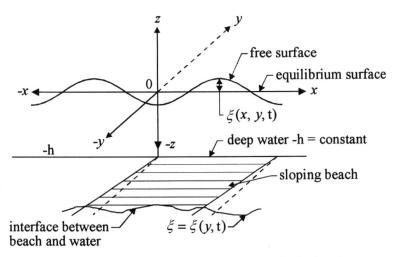

FIGURE 8.9. Geometry of surface waves on a sloping beach.

As indicated, the domain in which the velocity potential is to be determined is not known, since the free surface and the position of the water-beach interface are not known. Another difficulty is the instability of the wave as it approaches the beach. The leading edge of the wave curls over and breaks, causing the instability. Therefore, the mathematical formulation of a wave traveling toward the shore would necessitate the existence in the solution of a singularity that shows the instability.

EXAMPLE OF INSTABILITY

We now give an example of the onset of instability that occurs in nonlinear elasticity. This example is not so unrealistic since, although the details are different, the principle is the same as for water waves breaking on a beach. The nonlinear one-dimensional wave equation for the displacement $u(x, t)$ of a particle of the bar from its equilibrium position due to an axial stress running through the bar is

$$c^2 u_{xx} - u_{tt} = 0,$$

where the wave speed c is a known function of the uniaxial strain $\varepsilon(x, t) = u_x$ through the stress-strain curve (σ vs. ε), where σ is the axial stress. It turns out that $c^2 = (1/\rho_0)(d\sigma/d\varepsilon)$. In the nonlinear range of the stress-strain curve, the slope $d\sigma/d\varepsilon$ changes with strain. Suppose the slope increases with increasing strain. Then if we plot the large-amplitude strain profile (ε vs. x in the nonlinear range), the leading edge of the profile steepens since the higher-amplitude portion of ε travels with higher speed. Thus the leading edge steepens to a vertical slope where instability sets in. Figure 8.10a shows a plot of $d\sigma/d\varepsilon$ versus ε and fig. 8.10b shows profiles of ε versus x for several values of time. We have therefore exhibited the phenomenon of steepening of the leading edge of the $\varepsilon(x)$ profile, thus causing an instability. In this case of stress propagation through a bar, the instability leads to a shock wave, since we cannot have an instability that ruptures the bar, for such an instability would show three values of stress for a given

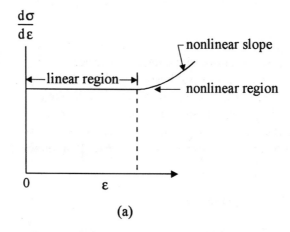

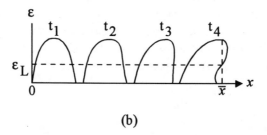

FIGURE 8.10. a. Plot of slope $d\sigma/d\varepsilon$ vs. ε. b. Profiles of ε vs. x for various times, showing the onset of instability at $t = t_3$. t_4 shows three values of ε at x. ε_L is the limit of linear strain.

value of strain. We easily see the analogy with wave profile breaking on the beach. However, in this case a shock wave does not form but the wave ruptures into spray.

Approximation Theories

As we have seen, realistic theories of water waves are, in general, nonlinear and thus cannot be attacked by classical methods. Numerical methods have been developed, but these methods only handle special cases and therefore do not allow for generalization. We shall resort to analyzing two approximation theories that will shed some more light on the theory of water waves. J. J. Stoker, in his classic work on water waves [35, Chap. 2], gives these approximation theories: (1) the theory of small-amplitude water waves and (2) shallow-water wave theory. It is of interest to review Stoker's treatment of these important approximation theories, which takes advantage of perturbation methods.

Small-Amplitude Waves

Up to now we have been treating small-amplitude water waves by using a linear theory. In this approach we use a method of perturbation to handle nonlinearities. We assume that the velocity **v** and the elevation ζ above the equilibrium surface are small quantities.

We introduce the parameter ε (the perturbation parameter) and expand the velocity potential and the elevation in powers of this parameter. We have

$$\phi = \sum_j \varepsilon^j \phi_j \tag{8.63}$$

and

$$\zeta = \sum_j \varepsilon^j \zeta_j, \tag{8.64}$$

where the unknown coefficients $\phi_j(x, y, z, t)$ and $\zeta_j(x, y, z, t)$ are on the free surface in equilibrium. (Note that we consider a three-dimensional model.) We are not interested in a rigorous treatment, which involves questions of convergence, but shall assume convergence from the physics of the situation.

CHAPTER EIGHT

From the fact that ϕ satisfies the three-dimensional Laplace equation, it follows that the corresponding expansion coefficients also satisfy Laplace's equation; thus

$$\nabla^2 \phi_j = 0. \tag{8.65}$$

The BC on a fixed surface (floor or walls) is given by $v_n = \partial \phi / \partial n = 0$, where the operator $\partial / \partial n$ means differentiating along the outward-drawn normal to the surface. It follows that

$$\frac{\partial \phi_j}{\partial n} = 0, \quad j = 1, 2, \ldots, \text{ on a fixed surface}. \tag{8.66}$$

We next consider the free surface S given by $z = \zeta(x, y, z, t)$. We have two BCs on this surface: (1) The pressure is constant (atmospheric pressure). (2) The water particles remain on S.

(1) $p = p_0$ on S and Bernoulli's law holds:

$$\phi_t + \left(\phi_x^2 + \phi_y^2 + \phi_z^2\right) + g\zeta = 0 \quad \text{on } S: \ z = \zeta(x, y, z, t). \tag{8.67}$$

We insert the expansions given by eqs. (8.63) and (8.64) into eq. (8.67). After performing the required calculations and equating like powers of ε, we obtain the following expressions for the expansion coefficients:

$$\zeta_0 = 0,$$

$$g\zeta_1 + \phi_{1,t} = 0, \tag{8.68}$$

$$g\zeta_2 + \phi_{2,t} = -\tfrac{1}{2}\left[(\phi_{1,x})^2 + (\phi_{1,y})^2 + (\phi_{1,z})^2\right] \quad \text{on } z = 0.$$

The zeroth power of ε means that the water is in equilibrium. Therefore $\phi_0(x, y, z, t) = 0$. We use the notation $\phi_{1,t} \equiv \partial \phi_1 / \partial t$, etc. Equation (8.68) gives expressions for the first two expansion coefficients. The third equation of (8.68) tells us that ϕ_2 and ζ_2 are given in terms of ϕ_1 and ζ_1. The nth equation of the system (8.68) for the expansion coefficients is

$$g\zeta_n + \phi_{n,t} = F_{n-1}, \tag{8.69}$$

where F_{n-1} is a known function of $\phi_{k,x}$, $\phi_{k,y}$, $\phi_{k,z}$, and ζ_k for $k \leq n - 1$. All the calculations are performed for $z = 0$. Equation

(8.69) is called a *recursive system*, meaning that, if the first-order through the $(n-1)$st-order set of expansion coefficients are given then we can calculate the nth expansion coefficients from this equation. The calculations are started by solving for the lowest-order expansion coefficients. We then solve for the higher-orders recursively.

(2) The water particles remain on S. This leads to the kinematic BC

$$\zeta_x \phi_x + \zeta_y \phi_y - \phi_z + \zeta_t = 0 \quad \text{on } S: \quad z = \zeta(x, y, t) = 0. \quad (8.70)$$

Inserting the expansions given by eqs. (8.63) and (8.64) into eq. (8.70) and equating like powers of ε yields the following system for the first two expansion coefficients:

$$z = \zeta_0 = \zeta_{0,t} = \zeta_{0,x} = \zeta_{0,y} = 0,$$
$$-\phi_{1,z} + \zeta_{1,t} = 0, \quad (8.71)$$
$$-\phi_{2,z} + \zeta_{2,t} = -\phi_{1,x}\zeta_{1,x} - \phi_{1,y}\zeta_{1,y} \quad \text{on } S: \quad z = \zeta_0(x, y) = 0.$$

The nth equation is of the form

$$\zeta_{n,t} - \phi_{n,z} = G_{n-1}, \quad (8.72)$$

G_{n-1} is a known function of the expansion coefficients ϕ_k and ζ_k, where $k \leq n-1$. This system is to be satisfied for $z = 0$.

The first-order theory is obtained by keeping only terms of order ε. For convenience we set $\varepsilon\phi_1 = \phi$ and $\varepsilon\zeta_1 = \zeta$. The free-surface BCs become

$$g\zeta + \phi_t = 0,$$
$$\zeta_t - \phi_z \quad \text{for } z = 0. \quad (8.73)$$

By eliminating ζ from these equations we obtain

$$\phi_{tt} + g\phi_z = 0 \quad \text{for } z = 0. \quad (8.74)$$

It is easily seen that the same equation holds on the boundary for ζ. These are the linearized BCs that hold on the free surface. The solution for ϕ is obtained by solving Laplace's equation in three-dimensions, $\nabla^2\phi = 0$, using the BC given by eq. (8.74) and the BC on the fixed surface ($\partial\phi/\partial n = 0$).

CHAPTER EIGHT

We see that a significant simplification results from this first-order theory. The BCs are linearized, and also the domain of the solution is fixed so that the velocity potential can be solved as a classical problem in potential theory. Once we have ϕ then we can solve for **v** and ζ.

Shallow-Water Theory

We revisit the propagation of water waves in a straight canal of finite depth h from the point of view of the more exact nonlinear theory. The definition of shallow-water theory is that we assume the approximation that the depth is small compared to a characteristic length, which we take here as the wavelength of the surface wave. In other words, the wavelength is large compared to the depth. Hence the term "long waves" [21], for example. In this theory we do not assume that the displacement and slope of the water surface are small so that, as mentioned, the theory is nonlinear. There are many applications of this approximation theory: the theory of tides, the breaking of waves on beaches, hydraulics concerning roll waves and flood waves in rivers, etc.

We shall consider a two-dimensional model and take as our coordinate system the x axis horizontally and the z axis vertically downward with the origin on the equilibrium surface $z = 0$. The vertical elevation of the free surface is $\zeta = \zeta(x, z, t)$. To review, the continuity equation is

$$u_x + w_z = 0. \tag{8.75}$$

The dynamic BC on the free surface is that the pressure is atmospheric: $p = p_0$ on $z = \zeta$. The kinematic BC on the free surface is

$$\frac{d\zeta}{dt} = u\zeta_x + \zeta_t = w \quad \text{on } z = \zeta(x, t). \tag{8.76}$$

The BC on a fixed surface (the floor) is

$$\frac{dh}{dt} = uh_x = w \quad \text{on } z = h. \tag{8.77}$$

We now integrate the continuity equation with respect to z and obtain

$$w = -\int_h^\zeta u_x \, dz.$$

WATER WAVES

We use this expression for w in the BC on the free surface, eq. (8.76), to obtain

$$\frac{\partial}{\partial x}\int_h^\zeta u\,dz = u\zeta_x + uh_x + \int_h^\zeta u_x\,dz = -\zeta_t. \tag{8.78}$$

No approximations were used up to now. We now invoke an approximation in shallow-water theory which says that the z component of the particle acceleration has no effect on the pressure in the fluid. This means that the z component of Euler's equations of motion is

$$0 = -p_z - \rho_0 g.$$

Integrating this equation with respect to z yields

$$p - p_0 = \rho_0 g(\zeta - z), \qquad -h \le z \le \zeta. \tag{8.79}$$

From this equation we deduce that p_x is independent of z. This means that the x component of the particle acceleration is independent of z and hence $\mathbf{u} = u(x,t)$. The x component of Euler's equation of motion is

$$u_t + uu_x = -\rho_0 p_x = -g\zeta_x. \tag{8.80}$$

Since u is independent of z, we have $\int_{-h}^\zeta u\,dz = u(\zeta + h)$. Using this expression, eq. (8.78) becomes

$$[u(\zeta + h)]_x = -\zeta_t. \tag{8.81}$$

Equations (8.80) and (8.81) are the equations of shallow-water theory for v and ζ. They are nonlinear, coupled, and not easily solved. An approach to solving the nonlinear problem is a perturbation method given in the next section.

We may linearize eqs. (8.80) and (8.81) by neglecting terms of second order (the products of small quantities). The linearized equations are

$$\begin{aligned} u_t &= -g\zeta_x, \\ (hu)_x &= -\zeta_t. \end{aligned} \tag{8.82}$$

CHAPTER EIGHT

We can eliminate ζ from these equations and obtain

$$g(hu)_{xx} = u_{tt}. \tag{8.83}$$

This is a linear second-order PDE for u with the variable coefficient $h(x)$. The wave speed is $c = \pm\sqrt{gh(x)}$, which tells us that the wave speed varies as the square root of the depth. If h is constant then the wave speed is constant.

Perturbation Method of Solving the Shallow-Water Problem

We now treat surface waves in shallow water by a perturbation method of K. O. Friedrichs as discussed in [35]. The model is three dimensional with a coordinate system whose origin is on the equilibrium surface $z = 0$; z is directed downward, x horizontal, and y lateral. For the convenience of the reader we restate the mathematical model. The continuity equation for incompressible flow is

$$\nabla \cdot \mathbf{v} = 0. \tag{8.84}$$

The irrotationality condition is

$$\operatorname{curl} \mathbf{v} = 0. \tag{8.85}$$

Euler's equations of motion are

$$v_t + (\mathbf{v} \cdot \nabla)\mathbf{v} = -\frac{1}{\rho_0}\operatorname{grad} p + \mathbf{F}, \qquad \mathbf{F} = (0, 0, -g). \tag{8.86}$$

The BCs at the free surface are

$$\frac{d\zeta}{dt} = \zeta_t + u\zeta_x + v\zeta_y = w \qquad \text{at } z = \zeta,$$
$$p = p_0 \qquad \text{at } z = \zeta. \tag{8.87}$$

The BC at the fixed surface on the ocean floor is

$$uh_x + vh_y + w = 0 \qquad \text{at } z = -h. \tag{8.88}$$

In order to compare the relative magnitudes of various physical quantities, we make the model dimensionless. Let d be a constant representing the average depth h and let k be a characteristic length such as the wavelength. One of the requirements of shallow-water theory is that the ratio k/d is assumed to be very large (long waves). Stoker introduced the following dimensionless variables given by the overbars:

$$\bar{x} = \frac{x}{k}, \quad \bar{y} = \frac{y}{k}, \quad \bar{z} = \frac{z}{d}, \quad \tau = \frac{t}{k}\sqrt{gd},$$

$$\bar{u} = \frac{u}{\sqrt{gd}}, \quad \bar{v} = \frac{v}{\sqrt{gd}}, \quad \bar{w} = \frac{w}{\sqrt{gd}}, \quad (8.89)$$

$$\bar{p} = \frac{p}{\rho_0 gd}, \quad \bar{\zeta} = \frac{\zeta}{d}, \quad \bar{h} = \frac{h}{d}.$$

The ratio k/d is a *stretching parameter*, which stretches the dimensionless coordinates $\bar{x}$ and $\bar{y}$ compared to $\bar{z}$. Note that k/d is very large. To see this, we note that

$$\frac{x}{z} = \left(\frac{k}{d}\right)\frac{\bar{x}}{\bar{z}}, \quad \frac{y}{z} = \left(\frac{k}{d}\right)\frac{\bar{x}}{\bar{z}}.$$

The same ratios hold for the components of the particle velocity, and it is easily seen that these ratios hold for the components of the particle acceleration. This means that the z component of the acceleration is very small compared with the x and y components, and hence can be ignored. We used this approximation of neglecting the vertical acceleration when we studied the two-dimensional shallow-water model of finite depth.

To develop a perturbation theory, Stoker uses the perturbation parameter defined by

$$\varepsilon = \left(\frac{d}{k}\right)^2. \quad (8.90)$$

This process of stretching the horizontal coordinates compared to the vertical coordinate combined with the perturbation method with respect to ε is the characteristic feature of this approach to shallow-water theory.

We now transform the mathematical model given by eqs. (8.84) through (8.88) into dimensionless variables by using eq. (8.89). We omit

CHAPTER EIGHT

the overbars for convenience.

$$\varepsilon(u_x + v_y) + w_z = 0; \tag{8.91}$$

$$w_y - v_z = 0, \quad u_z - w_x = 0, \quad v_x - u_y = 0; \tag{8.92}$$

$$\varepsilon[u_t + uu_x + vu_y + p_x] + wu_z = 0,$$
$$\varepsilon[v_t + uv_x + vv_y + p_y] + wv_z = 0, \tag{8.93}$$
$$\varepsilon[w_t + uw_x + vw_y + p_z + 1] + ww_z = 0;$$

$$\varepsilon[\zeta_t + u\zeta_x + v\zeta_y] = w \quad \text{at } z = \zeta,$$
$$p = 1 \quad \text{at } z = \zeta; \tag{8.94}$$

$$\varepsilon[uh_x + vh_y] + w = 0 \quad \text{at } z = -h. \tag{8.95}$$

(Obviously we set the dimensionless density equal to unity.)

We now construct a perturbation method by expanding the dependent variables in a power series in ε:

$$v = \sum_j \varepsilon^j v_j,$$
$$\zeta = \sum_j \varepsilon^j \zeta_j, \tag{8.96}$$
$$p = \sum_j \varepsilon^j p_j,$$

where the expansion coefficients are functions of the dimensionless coordinates and the ζ_js are evaluated at $z = 0$. Inserting these power series into eqs. (8.91)–(8.95) and equating like powers of ε yields the following relations among the expansion coefficients of zero order:

$$\begin{aligned} w_{0,z} &= 0, \\ w_0 u_{0,z} &= 0, \\ w_0 v_{0,z} &= 0, \\ w_0 w_{0,z} &= 0, \\ w_{0,y} = v_{0,z}, \quad u_{0,z} &= w_{0,x}, \quad v_{0,x} = u_{0,y}, \\ w_0 &= 0 \quad \text{at } z = \zeta_0, \\ p_0 &= 1 \quad \text{at } z = \zeta_0, \\ w_0 &= 0 \quad \text{at } z = -h. \end{aligned} \tag{8.97}$$

These equations yield the following facts:

$$u_0 = u_0(x,y,t), \quad v_0 = v_0(x,y,t),$$
$$w_0 = 0, \quad p_0(x,y,\zeta_0,t) = 0. \tag{8.98}$$

This shows, among other things, that the dimensionless pressure is atmospheric at the free surface and is taken as zero.

The zeroth-order theory yields the following approximation to the original hydrodynamic system and BCs:

$$u_{0,x} + v_{0,y} = 0; \tag{8.99}$$

$$w_{0,y} - v_{0,z} = 0, \quad u_{0,z} - w_{0,x} = 0, \quad v_{0,x} - u_{0,y} = 0; \tag{8.100}$$

$$u_{0,t} + u_0 u_{0,x} + v_0 u_{0,y} = -p_{0,x},$$
$$v_{0,t} + u_0 v_{0,x} + v_0 v_{0,y} = -p_{0,y}, \tag{8.101}$$
$$w_{0,t} + u_0 w_{0,x} + v_0 w_{0,y} = -p_{0,z} - 1,$$

$$\zeta_{0,t} + u_0 \zeta_{0,x} + v_0 \zeta_{0,y} = 0 \quad \text{at } z = \zeta_0,$$
$$p_0 = 1 \quad \text{at } z = \zeta_0; \tag{8.102}$$

$$u_0 h_x + v_0 h_y = 0 \quad \text{at } z = -h. \tag{8.103}$$

Note that the zeroth-order approximation is nonlinear since it involves the convective term in the equations of motion and the gradient of ζ in the BC for the free surface. We can, of course, omit these nonlinear terms, in which case we obtain the linear version of the shallow-water theory already discussed.

We now obtain the first-order theory by equating the coefficients of the first power of ε:

$$w_{1,z} = -u_{0,x} - v_{0,y}; \tag{8.104}$$

$$w_1 u_{1,z} = -[u_{0,t} + u_0 u_{0,x} + v_0 u_{0,y} + p_{0,x}],$$
$$w_1 v_{1,z} = -[v_{0,t} + u_0 v_{0,x} + v_0 v_{0,y} + p_{0,y}], \tag{8.105}$$
$$w_1 w_{1,z} = -[w_{0,t} + u_0 w_{0,x} + v_0 w_{0,y} + p_{0,z} + 1];$$

$$w_1 = \zeta_{0,t} + u_0 \zeta_{0,x} + v_0 \zeta_{0,y} \quad \text{at } z = \zeta_0; \tag{8.106}$$

$$w_1 = u_0 h_x + v_0 h_y \quad \text{at } z = -h. \tag{8.107}$$

CHAPTER EIGHT

Equation (8.105) can be integrated at once (since u_0 and v_0 are independent of z) to yield

$$w_1 = -(u_{0,x} + v_{0,y})z - [u_0 h_x + v_0 h_y]_{z=-h}. \tag{8.108}$$

We use eq. (8.107) to get the result.

In a similar fashion we obtain

$$p_0(x, y, z, t) = \zeta_0(x, y, t) - z. \tag{8.109}$$

This shows that the hydrostatic pressure (in dimensionless coordinates) is zero on the free surface and increases linearly with depth. In the previous treatment of shallow-water theory, this expression for the pressure arose from the a priori assumption that the vertical component of the acceleration can be neglected. In this perturbation method this expression is derived.

We continue this perturbation process to obtain higher-order approximations. Equating the coefficients of ε^n we obtain the following relations between the nth and the $(n-1)$st expansion coefficients.

The continuity equation becomes

$$[w_t]_n = -[u_x + v_y]_{n-1}, \tag{8.110}$$

where the notation $[\ ]_k$ means that the brackets contain the kth expansion coefficients, where $k = n, n-1$, respectively.

Euler's equations of motion become

$$[wu_z]_n = -[u_t + uu_x + vu_y + wu_z + p_x]_{n-1},$$
$$[wv_z]_n = -[v_t + uv_x + vv_y + wv_z + p_y]_{n-1}, \tag{8.110}$$
$$[ww_z]_n = -[w_t + uw_x + vw_y + ww_z + p_z]_{n-1}.$$

The nth expansion coefficients all satisfy the same irrotationality equations.

The BCs become

$$[w]_n = -[\zeta_t + u\zeta_x + v\zeta_y]_{n-1}, \quad [p]_n = 1 \quad \text{at } z = \zeta_0, \tag{8.111}$$

$$[w]_n = -[uh_x + vh_y]_{n-1} \quad \text{at } z = -h, \tag{8.112}$$

WATER WAVES

where $[\;]_{n-1} = 0$ for $n - 1 < 0$. In this scheme we can calculate recursively the nth expansion coefficients, knowing the $(n - 1)$st and given the zeroth-order expansion coefficients. In general, all perturbation methods enjoy this iterative property.

TIDAL WAVES

H. Lamb [21, pp. 250–362] has a very extensive treatment of tidal waves. He begins with a general discussion of the theory of small oscillations, forced oscillations, treats waves in canals (which we covered to some extent), equations of a dynamical system referred to rotating axes, Laplace's kinetic theory of tides on a rotating globe, kinetic stability of the ocean, etc. It is not our desire to treat this subject as extensively as Lamb. Rather, we shall present a somewhat simplified treatment, emphasizing the physics of tides.

Equilibrium Theory

We shall start by a preliminary investigation of tidal action, where we neglect the rotation of the earth. Figure 8.11 shows a polar section in the (x, y) plane of the earth as a great circle of radius a whose center O is a distance R from the center C of the moon. A point P on the earth's surface is a distance r from the center of the moon. The center

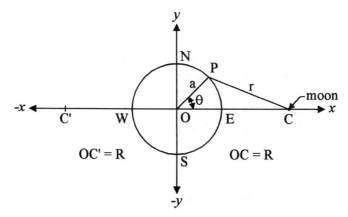

FIGURE 8.11. Polar section of earth as a great circle. The moon, whose center is at C, a distance R from the center of the earth.

CHAPTER EIGHT

O of the (x, y) coordinate system (c.s.) of the earth is a distance ρ from an inertial c.s. (say, fixed in the pole star). Let $\ddot{\rho}$ be the acceleration of the center of the earth, M the mass of the earth, m the mass of moon, F the gravitational force the moon exerts on the earth, $\mathbf{i}$ the unit vector in the x direction, and γ the gravitational constant, where $\gamma = 6.664 \times 10^{-8}$ g^{-1} cm^3 sec^{-2}. Newton's law of motion gives

$$\mathbf{F} = \frac{\mathbf{i}\gamma M m}{R^2} = M\ddot{\boldsymbol{\rho}}.$$

The potential V due to the force $\mathbf{F}$ is the potential energy per unit mass and is given by $-\nabla V = -\mathbf{i}(\partial V/\partial x)$ since $\mathbf{F}$ is in the $\mathbf{i}$ direction. It follows that

$$\ddot{\boldsymbol{\rho}} = \frac{\mathbf{i}\gamma m}{R^2} = -\nabla V, \quad \text{where } V = -\frac{\gamma m}{R^2}x. \tag{8.113}$$

Let V_m be the potential due to the moon's attraction to point P. Then

$$V_m = -\frac{\gamma m}{r} = -\frac{\gamma m}{R\sqrt{1 - 2\dfrac{a}{R}\cos\theta + \dfrac{a^2}{R^2}}}$$

$$= -\frac{\gamma m}{R}\left[1 - \frac{1}{2}\left(-\frac{2a}{R}\cos\theta + \frac{a^2}{R^2}\right) + \frac{3}{8}\left(\frac{4a^2}{R^2}\cos^2\theta \cdots\right)^2 + \cdots\right].$$
(8.114)

The point P on the earth's surface is subjected to the force of gravity as well as the attraction of the moon. If V_g is the potential due to this force, then the total force on P per unit mass due to gravity and the attraction f the moon is $-\nabla(V_m + V_g)$. Let the total acceleration of P be $\ddot{\boldsymbol{\rho}}_P$. Then Newton's law yields

$$\ddot{\boldsymbol{\rho}}_P = -\nabla(V_m + V_g). \tag{8.115}$$

The acceleration relative to the nonrotating earth is

$$\ddot{\boldsymbol{\rho}}_{P/E} = \ddot{\boldsymbol{\rho}}_P - \ddot{\boldsymbol{\rho}} - \nabla(V_m + V_g - V). \tag{8.116}$$

WATER WAVES

It follows that a particle at P has the same motion relative to the earth as if the earth were fixed and the particle were acted upon by a potential given by

$$V + V_g - V_m = -\frac{\gamma m}{R} + \frac{\gamma m}{R^2} a \cos\theta + g\zeta$$

$$= -\frac{\gamma m}{R}\left[1 - \frac{1}{2}\frac{a^2}{R^2}(1 - 3\cos^2\theta) \cdots \right] + g\zeta, \quad (8.117)$$

where $g\zeta$ is the potential due to gravity and ζ is the elevation of the water surface above mean sea level (the equilibrium surface). By setting the potential given by eq. (8.117) equal to a constant C and setting the mass M of the earth equal to $a^2 g/\gamma$, we obtain the equation of the equipotential surface. Solving for the elevation yields

$$\zeta = \frac{1}{2}\frac{m}{M}\left(\frac{a}{R}\right)^3 a(3\cos^2\theta - 1) + \frac{C}{g}.$$

The average value of ζ over the surface of the earth must vanish. This gives

$$\int_0^\pi \zeta 2\pi a^2 \sin\theta \, d\theta = 0.$$

Substituting the above value of ζ and integrating yields $C = 0$, and ζ becomes

$$\zeta = \frac{1}{2}\frac{m}{M}\left(\frac{a}{R}\right)^3 a(3\cos^2\theta - 1). \quad (8.118)$$

Equation (8.118) tells us that high tide is at $\theta = 0$ and π, which are the points E and W in fig. 8.11. While this equilibrium theory accounts for two tides daily, it gives the wrong phase to the tides. Moreover, no tidal wave motion is exhibited. A closer approximation to the observations is given by the dynamical theory given below.

Dynamical Theory

In this theory, we account for the motion of the moon with respect to the earth, thus allowing for tidal waves. Our model is a canal of depth h

CHAPTER EIGHT

and smooth sides encircling the earth at the equator. The centrifugal force due to the rotation of the earth is taken into account by the measured value of g. The Coriolis force is negligible. Therefore we consider the earth to be without rotation, with the moon rotating around it. Figure 8.12 shows a cross-section of the earth whose latitude is a great circle of radius a. Let O be the center of the earth, C the center of the moon, P a point on the water surface at a fixed longitude, and OC the line of action of the moon making an angle θ with the line OP. Let Ψ be the angle between OP and the x axis (the longitude) and ωt the angle between the x axis and OC. ω is the constant angular velocity of the moon's rotation around the earth relative to the meridian fixed on the earth. Relative to an inertial system (fixed on the pole star) ω is the excess of the eastward angular velocity of the earth over that of the moon. From the figure $\theta = \omega t + \Psi$.

Our plan is to calculate the tangential force on the water particle at P due to the moon's gravitational attraction and use this as the external force in the equation of motion. We then calculate the height of the tide.

The tangential force F_T on P due to the moon is calculated from the potential V_T, which is approximated by

$$V_T = -\frac{\gamma m}{R}\left[1 - \frac{1}{2}\frac{a^2}{R^2}(1 - 3\cos^2\theta)\right], \qquad (8.119)$$

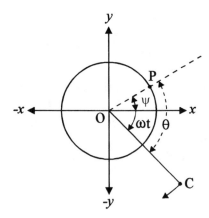

FIGURE 8.12. Longitudinal section of earth whose surface is a great circle. The moon at C rotates around the earth with constant angular velocity ω.

where we neglect the potential $g\zeta$ since it is in the radial direction. To obtain the force due to this potential, we calculate the gradient of this potential in the tangential direction. We get

$$-\frac{\partial V_T}{a\,\partial \Psi} = -\frac{\partial V_T}{a\,\partial \theta} = -\frac{3}{2}\frac{\gamma m a}{R^3}\sin 2(\omega t + \Psi) = F_T. \quad (8.120)$$

We add this force as the external forcing function to the wave equation for the horizontal displacement ζ and obtain

$$\frac{\partial^2 \xi}{a^2\,\partial t^2} = gh\frac{\partial^2 \xi}{a^2\,\partial \theta^2} - f \sin 2(\omega t + \Psi) \quad \text{where } f = \frac{3}{2}\frac{\gamma m a}{R^3}. \quad (8.121)$$

We seek a particular integral of this nonhomogeneous wave equation that has the same frequency and phase as the forcing function. We finally obtain

$$\xi = -\frac{1}{4}\frac{fa^2}{c^2 - \omega^2 a^2}\sin 2(\omega t + \Psi), \qquad c^2 = gh. \quad (8.122)$$

This expression for the horizontal displacement shows infinite amplitude at the resonance frequency, which, of course, is unrealistic. We must take into account the damping action of the water. However, as long as we stay away from resonance eq. (8.122) is a valid approximation. We may obtain the vertical displacement η by integrating the continuity equation with respect to y. We get

$$\int \xi_x \, dy = y\xi_x = -\int \eta_y \, dy = -\eta,$$
$$\eta = -y\frac{\partial \xi}{a\,\partial \theta}. \quad (8.123)$$

Inserting (8.122) into (8.123), we get for the vertical displacement

$$\eta = \frac{1}{2}\frac{afy}{c^2 - \omega^2 a^2}\cos 2(\omega t + \Psi). \quad (8.124)$$

These equations for ξ and η tell us that the frequency 2ω of the tides is double that of the revolution of the moon. Therefore there are two tides a day, in agreement with the equilibrium theory. If $gh > \omega^2 a^2$ the

CHAPTER EIGHT

tides are direct, which means that high tide is in phase with the moon. However, if $gh < \omega^2 a^2$ then the tides are inverted, meaning that low tide is in phase with the moon.

Problems

1. If $f(z) = \phi + i\psi$ and $f(z)$ is real when $y = a$, show that $\psi = 0$ when $y = a$.

2. Show that the transformation $w = z^2$ maps the upper half of the z plane on the whole of the w plane.

3. Sketch the streamlines represented by $w = \phi + i\psi = Az^2$ and show that the speed is everywhere proportional to the distance from the origin. Sketch the family of lines of constant potential.

4. Prove that the transformation $w = -i(z-1)/(z+1)$ maps the region within the unit circle in the z plane, indented at the points $z = 1, z = -1$ by a very small radius, on the region in the w plane within a semicircle of very large radius indented at the origin.

5. Show that the Jacobian of the mapping function that maps the z domain to the corresponding w domain is given by $|df(z)/d(z)|^2$.

6. Prove that $w = A\cos(2\pi/\lambda)(z + ih - ct)$ is the complex potential for the propagation of simple harmonic surface-water waves of small elevation in a canal of depth h, the origin being at the equilibrium surface. Express A in terms of the amplitude a of the surface oscillations. Find the stream function and the potential function. Show that the wave speed is $c^2 = (g\lambda/2\pi)\tanh(2\pi h/\lambda)$. Prove that each particle describes an ellipse about its equilibrium position. Obtain corresponding results when the water is infinitely deep.

7. Two-dimensional long waves are traveling parallel to the x axis in water of variable depth h. Prove that the elevation ζ of the free surface from the equilibrium configuration satisfies

$$(gh\zeta_x)_x = \zeta_{tt}.$$

8. Apply problem 6 to the following problem: Let the depth of a canal be h_1 for $x < 0$ and h_2 for $x > 0$. A progressing surface wave of the

form $\zeta = a \sin k(x - c_1 t)$, where $c_1^2 = gh_1$, travels along the portion of depth h_1. Obtain the amplitudes of the reflected and transmitted waves (which involve progressing and regressing waves).

9. Show that the potential given by eq. (8.117) is the same as if the earth were at rest in an inertial system and two equidistant moons of mass $(1/2)m$ were located at C and C' in fig. 8.11. (C' is the image of C about the origin.)

10. The ratio of the mass of the earth to that of the moon is 80 and the ratio of the distance of the moon from the center of the earth to the radius of the earth is 60. For the equilibrium theory of the tides find the elevation at high tide and the depression at low tide from mean sea level. *Answer*: 1.22 ft.

11. The ratio of the masses of the sun and earth is 332,000 and the ratio of the distance of the sun from the earth to the radius of the earth is 23,200. What is the ratio of the height of the solar tide to that of the lunar tide? *Answer*: 0.46.

12. Find the amplitude of the horizontal and vertical displacements due to the lunar tides in an equatorial canal of 2 miles depth. *Answer*: 165 ft., 0.16 ft.

CHAPTER NINE

Variational Methods in Wave Propagation

Introduction; Fermat's Principle

Up to now, we have investigated wave propagation phenomena in various media by use of the field equations, which are hyperbolic PDEs that were derived from the fundamental conservation laws. This is essentially a *differential principle*, in the sense that the dynamic variables (displacement, velocity, stress, strain, etc.) were derived from the field equations (equations of motion and energy equation) in terms of PDEs, so that point-by-point solutions of these field equations can be obtained when considering appropriate IV and BV problems. By contrast, the approach used in this chapter is based on an *integral principle* or a *global approach*, which means that we consider the entire motion of a dynamical system over a finite period of time.

The global approach uses the calculus of variations from which the equations of motion are derived. They are expressed in two different ways: (1) Lagrange's equations of motion: second-order PDEs given in terms of generalized coordinates and velocities; and (2) Hamilton's equations: first-order PDEs given in terms of generalized coordinates and momenta. We stress that these two types of equations of motion, though they are given in terms of PDEs, arise from an integral principle, which is the variational principle in the calculus of variations.

The fundamental problem of the calculus of variations is to optimize an integral over a fixed finite time of a function (the integrand) that contains the features of a dynamical system. The integrand will include a variable that describes a set of paths over which the integral is taken; the actual path is the one that optimizes the functional. The integral is called a *functional*. In dynamics, "optimizing the functional" usually means finding its *minimum* value. The actual path of a dynamical system (over a given time interval) is the one that minimizes the functional.

An example in optics is the *principle of least time* or *Fermat's principle*. The functional is the time integral over a given time interval.

VARIATIONAL METHODS

The integrand is a function of the velocity of light and a set of possible light paths in the x, y plane given by $y = y(x)$. The calculus of variations will allow us to find the minimum value of the functional. Of the set of possible paths, the actual path that a light beam takes is the one that minimizes the functional. Obviously, if the refractive index is constant then the path is a straight line, but for a variable index of refraction the problem is not trivial.

Here is another example: For a simple harmonic oscillator the equation of motion is $m\ddot{x} = -kx$, where m is the mass, k is the spring constant, and x is the displacement of the mass from equilibrium. Let the velocity of the mass be $v = dx/dt = v\, dv/dx$. The equation of motion can then be written as $mv\, dv/dx = -kx$ so that time is eliminated. Integrating with respect to x gives $\frac{1}{2}mv^2 + \frac{1}{2}kx^2 = $ const. This is the energy equation and expresses the conservation of energy for a simple harmonic oscillator. (The constant is the total energy.) If we set $L = T - V$, where T is the kinetic energy and V is the potential energy, then the appropriate functional is $I = \int_{t_1}^{t_2} L\, dt$. L is called the *Lagrangian*; it will be studied in detail below.

CALCULUS OF VARIATIONS; EULER'S EQUATION

We shall be concerned with one-dimension (x) in the x, y plane for simplicity. We shall use Fermat's principle or the principle of least time to introduce the calculus of variations. Fermat's principle states that of all possible paths taken by a light ray in going from point A to point B (these points being fixed in the x, y plane), the actual path is the one that *takes the least time*. If t is time and s is the arc length along any curve $y = y(x)$ joining A and B, the time taken for the light ray to travel along the curve $y = y(x)$ is given by the integral

$$I(y) = \int_{t_A}^{t_B} dt = \int_A^B \left(\frac{dt}{dx}\right)\left(\frac{ds}{dx}\right) dx$$

$$= \int_A^B \frac{\sqrt{1 + y'^2}}{c}\, dx = I_B - I_A, \qquad y' = \frac{dy}{dx}, \qquad (9.1)$$

where $c = c(x, y(x))$ is the velocity of the light ray. The functional I is a function of y because, as implied above, the integrand involves an infinite set of paths passing through the fixed end points given by the

CHAPTER NINE

curves $y(x)$, the actual path being the one that minimizes I. [The paths are contained in $c = c(x, y)$.] It is clear that $I = I_B - I_A$. This represents the time taken for the light ray to go from A to B. How do we minimize the functional to find the actual curve $y = y(x)$? The answer leads to Euler's equation. It is derived in the next section.

Euler's Equation

The integrand in I is of the form $f(x, y, y')$. In the spirit of the calculus of variations, we generalize Fermat's principle and formulate the calculus of variations in one dimension as follows: Of the set of admissible curves $y(x)$ passing through the end points A and B (x_A, y_B), find the function $y = y(x)$ that gives the functional a *stationary value*. This means I is either a maximum or a minimum, though, as mentioned, most problems in physics require that I be a minimum. We therefore want the necessary condition that minimizes the integral

$$I(y) = \int_{x_A}^{x_B} F(x, y, y') \, dx. \tag{9.2}$$

(The reason for having y' in the integrand is that many problems in the calculus of variations can be formulated involving the slope of the curve.)

We assume that the curves $y(x)$ are continuously differentiable, so that they represent a family of smooth curves passing through the end points A, B. We also assume the same differentiable conditions on F. Let $y = \bar{y}(x)$ be any curve in this family such that the functional $I(\bar{y})$ is not a minimum, and let $y = y(x)$ be the curve in the family such that $I(y)$ is a minimum. This is called the *minimizing curve*. The curve represented by $\bar{y}$ is assumed to lie in the neighborhood of the minimizing curve. Let δy be the *variation* of $y(x)$. We write

$$\delta y(x) = \bar{y}(x) - y(x) = \varepsilon \eta(x), \tag{9.3}$$

where ε is a small positive constant that determines the amount of variation and $\eta(x)$ is an arbitrary variable that specifies the curve. Since all curves pass through the end points we must have

$$\eta(x_A) = \eta(x_B) = 0. \tag{9.4}$$

VARIATIONAL METHODS

We assume that the functional I is never negative. For $I(y)$ to be a minimum we must have

$$I(y + \varepsilon\eta) \geq I(y). \tag{9.5}$$

The left-hand member of this inequality is a continuously differentiable function of the parameter ε, so that a necessary condition for y to minimize I is

$$\left.\frac{dI(y + \varepsilon)}{d\varepsilon}\right|_{\varepsilon=0} = 0.$$

Differentiating eq. (9.2) with respect to ε yields

$$\left.\frac{dI(y + \varepsilon)}{d\varepsilon}\right|_{\varepsilon=0} = \int_A^B (\eta F_y - \eta' F_{y'}) \, dx = 0. \tag{9.6}$$

Integrating the second term in the integral by parts gives

$$\int_A^B F_{y'}\eta' \, dx = -\int_A^B \eta \left(\frac{dF_{y'}}{dx}\right) dx,$$

where we used the homogeneous end conditions $\eta_A = \eta_B = 0$. We then write eq. (9.6) as

$$\int_A^B (F_y - dF_{y'})\eta(x) \, dx = 0. \tag{9.7}$$

Equation (9.7) tells us that the integral vanishes for every value of $\eta(x)$ that satisfies the homogeneous end conditions. It follows that the integrand must vanish, yielding

$$F_y - \frac{dF_{y'}}{dx} = 0. \tag{9.8}$$

Equation (9.7) is called *Euler's equation*. It is the necessary condition for I to have an extremum (maximum or minimum). Note that it is not a sufficient condition for an extremum. This means that we are not assured that an extremum exists. But, if it does exist then Euler's equation is satisfied.

CHAPTER NINE

Example. What is the shortest distance between two points in a plane? This is a trivial example, since the path is obviously a straight line. However, let us see how it can be deduced from Euler's equation. An element of arc length in the plane is $ds = \sqrt{dx^2 + dy^2}$. The length of any curve $y(x)$ between points A and B is the functional I, where

$$I = \int_A^B ds = \int_{x_A}^{x_b} \sqrt{1 + \left(\frac{dy}{dx}\right)^2}\, dx.$$

The function F in Euler's equation is the integrand $F = \sqrt{1 + y'^2}$. It follows that

$$F_y = 0, \qquad F_{y'} = \frac{y'}{\sqrt{1 + y'^2}}.$$

Since $dF_{y'}/dx = 0$ we have

$$\frac{y'}{\sqrt{1 + y'^2}} = \text{const.}$$

This means that $y' = \text{const}$. The conclusion is that the curve $y(x)$ joining the points A and B in the x, y plane that minimizes the distance from A to B is the one with a constant slope, or a straight line.

Example. The simple harmonic oscillator revisited. In the previous example of the simple harmonic oscillator we introduced the function L defined by $L = T - V$. We called L the Lagrangian, since it was introduced by the great French mathematical physicist Lagrange. Recall that T is the kinetic energy and V the potential energy so that $T = \frac{1}{2}m\dot{x}^2$, $v = \frac{1}{2}kx^2$. $L = \frac{1}{2}m\dot{x}^2 - \frac{1}{2}kx^2$. Setting $F = L$, Euler's equation given by eq. (9.7) leads to the equation of the simple harmonic oscillator:

$$m\ddot{x} = -kx.$$

To prove this we need only change notation so that Euler's equation becomes

$$\frac{dL_{\dot{x}}}{dt} = L_x.$$

We have $L_{\dot{x}} = m\dot{x}$, $dL_{\dot{x}}/dt = m\ddot{x}$, and $L_x = -kx$. The term $L_{\dot{x}}$ is the linear momentum, the term $dL_{\dot{x}}/dt$ is the rate of change of the momentum or the inertial force, and the term L_x is the spring force.

Euler's equation in terms of L is

$$\frac{dL_{\dot{x}}}{dt} = L_x. \tag{9.9}$$

Equation (9.9) is called *Langrange's equation of motion* for the one-dimensional case. As mentioned above, the Lagrangian L is given by

$$L = T - V, \tag{9.10}$$

where T is the kinetic energy and V is the potential energy of the dynamical system. This will be treated in more detail below.

CONFIGURATION SPACE

In order to generalize the variational method to systems involving more than one coordinate and velocity, we introduce the concept of *configuration space*. To determine the motion of a dynamical system over a fixed time interval, we describe the instantaneous configuration of the system by the set of n independent coordinates called *generalized coordinates*, $\mathbf{q} = (q_1, q_2, \ldots, q_n)$. The configuration space is an n-dimensional hyperspace such that each point in the space is characterized by the n-vector $\mathbf{q}$ representing n generalized coordinates. $\mathbf{q}$ is called the *system point*. Starting with an initial time, as time increases the state of the dynamical system changes and the system point $\mathbf{q}$ traces a trajectory in configuration space, so that $\mathbf{q} = \mathbf{q}(t)$. This means that time is a parameter that monotonically increases from its initial to its final position, thus generating a trajectory in configuration space. Note that each point on the trajectory represents the configuration of the entire system (all the particles making up the system providing there are no constraints). We emphasize the fact that a trajectory in configuration space has no necessary connection with the path of the system in three-dimensional space.

CHAPTER NINE

KINETIC AND POTENTIAL ENERGIES

The calculus of variations is based on energy considerations. We start with a *conservative system*; the total energy of the system is a constant—the sum of the kinetic and potential energies. Our system is assumed to be free of constraints so that its instantaneous configuration is represented by a point on a trajectory in configuration space. To completely describe the motion of the system along the trajectory, we must introduce the *generalized velocity* $\dot{q} = (\dot{q}_1, \ldots, \dot{q}_n)$. For each q_j there corresponds a $\dot{q}_j$. From an energy point of view the mechanical properties of our system are determined by the kinetic energy T and the potential energy V. In general, T is assumed to be a function of the generalized coordinates and velocities. Specifically, we assume T to be a quadratic function of $\dot{q}$ of the form

$$T = \sum_{i,j} P_{ij}(q_1, \ldots, q_n) \dot{q}_i \dot{q}_j, \qquad (9.11)$$

where the coefficients P_{ij}s are known functions of $\mathbf{q}$. In many problems these coefficients are constants.

We assume V to be a quadratic in $\mathbf{q}$ of the form

$$V = \sum_{i,j} b_{ij} q_i q_j, \qquad (9.12)$$

where the b_{ij}s are given constants.

HAMILTON'S VARIATIONAL PRINCIPLE

Hamilton's principle is essentially a generalization of the fundamental extremum principle in the calculus of variations for a system in configuration space. To this end we wish to generalize Euler's equation (9.8) to a hyperspace of n dimensions so that the integrand $F = F(\mathbf{y}(x), \mathbf{y}'(x))$ where the vector $\mathbf{y} = (y_1(x), y_2(x), \ldots, y_n(x))$ and the vector $\mathbf{y}' = (y'_1(x), \ldots, y'_n(x))$. We replace the scalars y, $\bar{y}$, and η by n-dimensional vectors in eqs. (9.3)–(9.6), integrate by parts, and obtain the generalization of eq. (9.8), which is

$$F_{y_i} - \frac{dF_{y'_i}}{dx} = 0, \quad i = 1, 2, \ldots, n. \tag{9.13}$$

In order to formulate Hamilton's principle we let $F = T - V = L$, where L is the Lagrangian function. To correlate this with configuration space, we let the vector $\mathbf{y}$ correspond to the generalized coordinate vector $\mathbf{q} = (q_1, q_2, \ldots, q_n)$, let the vector $\mathbf{y}'$ correspond to the generalized velocity vector $\dot{\mathbf{q}} = (\dot{q}_1, \ldots, \dot{q}_n)$, and let x correspond to the time t. Then the functional I becomes

$$I = \int_{t_1}^{t_2} L(q_i, \dot{q}_i, t) \, dt. \tag{9.14}$$

Hamilton's principle can be stated as follows: *Of all possible paths that a dynamical system can take in configuration space, the actual path is the one that extremizes (minimizes) the functional given by eq.* (9.14). The necessary condition for the variation $\delta I = 0$ is the set of n equations given by eq. (9.13) (applied to configuration space by the above correspondences). We get

$$\frac{d}{dt} \frac{\partial L}{\partial \dot{q}_i} - \frac{\partial L}{\partial q_i} = 0, \quad i = 1, 2, \ldots, n \tag{9.15}$$

(written in old-fashioned notation). The set of n equations given by (9.14) is called *Lagrange's equations of motion*. This set of equations represents an alternate way of representing Newton's equations of motion for a system with n degrees of freedom. To see this, we note that the term $\partial L/\partial \dot{q}_i$ is the linear momentum of the part of the system having the ith generalized coordinate and velocity, and the term $\partial L/\partial q_i$ is the generalized force acting on the "ith part of the system." (The generalized force is the resultant force acting on the system in configuration space.) The system given by eq. (9.15) then tells us that the rate of change of the linear momentum is equal to the resultant external force acting on the system in configuration space.

We have just shown that Lagrange's equations of motion follow from Hamilton's variational principle. This principle of Hamilton has been described as "elegant," since in it is contained all the mechanics of conservation systems. It has the further merit of involving physical

CHAPTER NINE

quantities that can be defined without reference to a particular coordinate system. This means that Lagrange's equations are invariant with respect to the choice of coordinates for the system, thus giving much flexibility.

Whittaker [37, p. 246] presents a clear though rather involved proof that Hamilton's principle follows from Lagrange's equations of motion. He considers an actual and a neighboring path in configuration space (although he does not call it that). He shows that the functional $I = \int_{t_1}^{t_2} L(q, \dot{q}, t)\, dt$ taken along the actual path in configuration space has a stationary value (is an extremum) if he invokes Lagrange's equations of motion. Thus he proves Hamilton's principle if he assumes the validity of Lagrange's equations.

We now need to derive Lagrange's equations from Newton's equations of motion. To do this we invoke the *principle of virtual work*.

Principle of Virtual Work

The principle of virtual work was suggested by the Swiss mathematician James Bernoulli and developed further by the French mathematician D'Alembert. It is a method whereby the constraints acting on a system can be handled by a transformation from generalized to Cartesian coordinates. We can then make use of Newton's equations to derive Lagrange's equations.

In order to develop the principle of virtual work we must first introduce the concept of a *virtual displacement*. A virtual displacement of a system of n particles in three-dimensional space is defined as an arbitrary infinitesimal change of the particle coordinates $\delta \mathbf{r}$ consistent with the forces and constraints imposed on the system at a given time t. The displacement is called "virtual" to distinguish it from the actual displacement of the system occurring in a time interval dt during which the forces and constraints may be changing. Note that the radius vector $\mathbf{r}$ representing the instantaneous position of the system is not necessarily the same as the generalized coordinate vector $\mathbf{q}$.

As implied above, the operator δ operating on $\mathbf{r}$ means a virtual displacement of the system for a fixed time, to distinguish it from the operator d operating on $\mathbf{r}$, signifying an actual displacement of the system to a time dt later. From elementary mechanics, the virtual work W_i done by a force $\mathbf{F}_i$ acting on the ith particle in displacing it by an amount $\delta \mathbf{r}_i$ is $\delta W_i = \mathbf{F}_i \cdot \delta \mathbf{r}$ (taking the scalar product of the force and

VARIATIONAL METHODS

virtual displacement). If the system of particles is in equilibrium then $F_i = 0$, $i = 1, 2, \ldots, n$. Therefore, the condition for the system to be in equilibrium is that the virtual work done on the system shall vanish. This gives

$$\sum_{i=1}^{n} \mathbf{F}_i \cdot \delta \mathbf{r} = 0. \tag{9.16}$$

We now assume that the force on the ith particle is the sum of the applied force $\mathbf{F}'_i$ and a constraining force $\mathbf{f}_i$, so that the total force is given by

$$\mathbf{F}_i = \mathbf{F}'_i + \mathbf{f}_i. \tag{9.17}$$

Then the condition for the virtual work of the system in equilibrium to vanish is

$$\sum_i \mathbf{F}'_i \cdot d\mathbf{r} + \sum_i \mathbf{f}_i \cdot \delta \mathbf{r} = 0. \tag{9.18}$$

We now restrict our system to one where the virtual work of the forces of constraint is zero. This will not be true for a system having frictional forces. For the present we exclude this case. Then the condition for the system to be in equilibrium is that the virtual work of the applied forces acting on the system must vanish. This yields

$$\sum_i \mathbf{F}'_i \cdot \delta \mathbf{r} = 0. \tag{9.19}$$

The principle of virtual work, given by eq. (9.19), is based on a system in equilibrium. This is unsatisfactory for our purpose since we are interested in setting the stage for wave propagation problems, which involve dynamical systems. We therefore need a generalization of the principle of virtual work to cover dynamical systems. This is just what D'Alembert did. He ingeniously wrote the equations of motion in the following form:

$$F_i - \dot{p}_i = 0, \quad i = 1, 2, \ldots, n. \tag{9.20}$$

$p_i = m_i v_i$, where p_i is the momentum of the ith particle, m_i its mass, and v_i its velocity. The inertial force on the ith particle is $\dot{p}_i$. The right-hand side of eq. (9.20) is the zero force vector. This seemingly

CHAPTER NINE

innocuous device of putting all the forces on the left-hand side has the advantage of treating the inertial force as the external force. Therefore this equation tells us that the sum of the external and inertial forces on the particle (that is, all the forces) vanishes, giving the same equilibrium situation we had before. D'Alembert, by this simple algebraic trick, transformed a system of n particles from a problem in dynamics to an apparent problem in statics, so that the principle of virtual work can be applied to the system under the action of the $F - \dot{p}$ forces. This device is aptly called *D'Alembert's principle*. This principle can also be stated as follows: The system of n particles is in equilibrium under the action of the resultant of the external forces and the reversed (negative) inertial force. The principle of virtual work can now be written as

$$\sum_i (\mathbf{F}_i - \dot{\mathbf{p}}_i) \cdot \delta \mathbf{r} = 0. \tag{9.21}$$

This equation is the dynamical analogue of eq. (9.19). The external force $\mathbf{F}$ is assumed to contain no constraints.

Transformation to Generalized Coordinates

The radius vector $\mathbf{r}$ does not have independent coordinates due to the constraints on the system, but the generalized coordinates are linearly independent. We therefore seek a transformation from the Cartesian coordinate system $\mathbf{r}$ to the generalized coordinate system $\mathbf{q}$. This is given by

$$\mathbf{r}_i = \mathbf{r}_i(q_i, q_2, \ldots, q_n). \tag{9.22}$$

The virtual work on the dynamical system given by (9.21) is to be invariant with respect to this transformation so that the momentum in the generalized coordinate system is the generalized momentum corresponding to the generalized coordinate.

We consider a dynamical system of n particles. The velocity of the ith particle is $\mathbf{v}_i = d\mathbf{r}_i/dt$. We express v_i in terms of generalized coordinates by expanding according to the chain rule; we obtain

$$\mathbf{v}_i = \sum_{j=1}^n \frac{\partial \mathbf{r}_i}{\partial q_j} \dot{q}_j + \frac{\partial \mathbf{r}_i}{\partial t}. \tag{9.23}$$

VARIATIONAL METHODS

Similarly, the expansion of the virtual displacement $\delta \mathbf{r}_i$ of the ith particle is

$$\delta \mathbf{r}_i = \sum_j \frac{\partial \mathbf{r}_i}{\partial q_j} \delta q_j. \tag{9.24}$$

The virtual work δW of all the particles due to $\mathbf{F}$ is

$$\delta W = \sum_i \mathbf{F}_i \cdot \delta \mathbf{r}_i = \sum_{i,j} \mathbf{F}_i \cdot \frac{\partial \mathbf{r}_i}{\partial q_j} \delta q_j = \sum_j \mathbf{Q}_j \, \delta q_j, \tag{9.25}$$

where $\mathbf{Q}_j$ is the *generalized force* of the jth particle. It is defined by

$$\mathbf{Q}_j = \sum_i \mathbf{F}_i \cdot \frac{\partial \mathbf{r}_i}{\partial q_j}. \tag{9.26}$$

The generalized force $\mathbf{Q}$ of the system is the sum over all the particles of $\mathbf{Q}_i$, and is a function of the generalized coordinates and velocities.

We now put the principle of virtual work given by eq. (9.21) into generalized coordinates. To this end we manipulate the term

$$\sum_i \dot{\mathbf{p}}_i \cdot \delta \mathbf{r}_i = \sum_i m_i \ddot{\mathbf{r}}_i \cdot \frac{\partial \mathbf{r}_i}{\partial q_j} \delta q_j.$$

We express the right-hand side of this equation in the form

$$\sum_i m_i \ddot{\mathbf{r}}_i \cdot \frac{\partial \mathbf{r}_i}{\partial q_j} = \sum_i \left[\frac{d}{dt}\left(m_i \dot{\mathbf{r}}_i \cdot \frac{\partial \mathbf{r}_i}{\partial q_j} \right) - m_i \dot{\mathbf{r}}_i \cdot \frac{d}{dt}\left(\frac{\partial \mathbf{r}_i}{\partial q_j} \right) \right].$$

In the last term of the right-hand side we interchange the operators d/dt and $\partial/\partial q_j$ and obtain

$$\frac{d}{dt}\frac{\partial \mathbf{r}_i}{\partial q_j} = \sum_k \frac{\partial^2 \mathbf{r}_i}{\partial q_j \partial q_k} \dot{q}_k + \frac{\partial^2 \mathbf{r}_i}{\partial q_j \partial t} = \frac{\partial \mathbf{v}_i}{\partial q_j}.$$

From these expressions we obtain

$$\sum_i m_i \ddot{\mathbf{r}}_i \cdot \frac{\partial \mathbf{r}_i}{\partial q_j} = \sum_i \left[\frac{d}{dt}\left(m_i \mathbf{v}_i \cdot \frac{\partial \mathbf{v}_i}{\partial q_j} \right) - m_i \mathbf{v}_i \cdot \frac{\partial \mathbf{v}_i}{\partial q_j} \right].$$

CHAPTER NINE

This gives

$$\sum_i \dot{\mathbf{p}}_i \cdot \delta \mathbf{r}_i = \sum_j \left[\frac{d}{dt} \frac{\partial}{\partial \dot{q}_i} \left(\sum_i \frac{1}{2} m_i v_i^2 \right) - \frac{\partial}{\partial q_j} \left(\frac{1}{2} m_i v_i^2 \right) \right] \delta q_j.$$

The kinetic energy of the system $T = \frac{1}{2} \sum_i m_i v_i^2$, so that D'Alembert's principle (9.21) becomes

$$\sum_j \left(\frac{d}{dt} \frac{\partial T}{\partial \dot{q}_j} - \frac{\partial T}{\partial q_j} - Q_j \right) \delta q_j = 0. \qquad (9.27)$$

Since all the virtual displacements δq_j are independent, the coefficient of each δq_j must vanish, yielding

$$\frac{d}{dt} \frac{\partial T}{\partial \dot{q}_j} - \frac{\partial T}{\partial q_j} = Q_j, \qquad j = 1, 2, \ldots, n. \qquad (9.28)$$

The system of n equations given by (9.28) are the Lagrange equations for the kinetic energy T and the generalized force Q_j in terms of the generalized coordinates and velocities. (Note: There is one equation for each particle.)

For a conservative system the potential energy V is a quadratic function of $\mathbf{q}$. Therefore the external force $\mathbf{F}$ is derivable from V, so that for each particle we have $F_i = -\nabla_i V$, and the generalized force becomes

$$Q_j = \sum_i \mathbf{F}_i \cdot \frac{\partial \mathbf{r}_i}{\partial q_j} = -\sum_i \nabla_i V \cdot \frac{\partial \mathbf{r}_i}{\partial q_j} = -\frac{\partial V}{\partial q_j}. \qquad (9.29)$$

Recalling that T is a function of $\mathbf{q}$ and $\dot{\mathbf{q}}$ and V is a function of $\mathbf{q}$, and using the Lagrangian $L = T - V$, eq. (9.28) becomes

$$\frac{d}{dt} \frac{\partial L}{\partial \dot{q}_j} - \frac{\partial L}{\partial q_j} = 0, \qquad j = 1, 2, \ldots, n. \qquad (9.30)$$

These are the n Lagrange equations for the system in a conservative force field.

VARIATIONAL METHODS

RAYLEIGH'S DISSIPATION FUNCTION

Suppose the system of n particles is acted upon by frictional or viscous forces so that the external force is not derivable from a potential. The energy is dissipated in the form of heat. Even for this case we can still use Lagrange's equations provided we introduce a generalized potential $U = U(q, \dot{q})$, defined in terms of the generalized force by

$$Q_j = -\frac{\partial U}{\partial q_j} + \frac{d}{dt}\frac{\partial U}{\partial \dot{q}_j}. \tag{9.31}$$

Clearly, if U does not depend on the generalized velocity then the potential V exists such that $U(\mathbf{q}) = V(\mathbf{q})$. If the Lagrangian L is now defined by

$$L = T - U \tag{9.32}$$

(where U is substituted for V to take care of the velocity terms), then eq. (9.28) becomes

$$\frac{d}{dt}\frac{\partial (L + U)}{\partial \dot{q}_j} - \frac{\partial (L + U)}{\partial q_j} = Q_j. \tag{9.33}$$

Inserting the expression for Q_j in eq. (9.31) into eq. (9.32) yields Lagrange's equations (9.30).

If only some of the generalized forces acting on the system are derivable from the potentials U or V, Lagrange's equations can always be written as

$$\frac{d}{dt}\frac{\partial L}{\partial \dot{q}_j} - \frac{\partial L}{\partial q_j} = Q_j, \tag{9.34}$$

where L contains U or V, and Q_j represents the generalized force not derivable from the potentials U or V. Such a situation arises for the case of a frictional or viscous force, where it is experimentally shown that such a force is proportional to the particle velocity. Let the frictional force $\mathbf{f}$ on the ith particle be

$$f_i = -kv_i,$$

CHAPTER NINE

where $k = (k_x, k_y, k_z) > 0$ and the particle velocity $v_i = (u_i, v_i, w_i)$. Lord Rayleigh [29, vol. I, p. 102] showed that if this frictional force is added to the external force **F** in D'Alembert's principle of virtual work given by eq. (9.21) in such a manner as to retard the motion, then a scalar quadratic function of the velocity exists called the *Rayleigh dissipation function* $\mathscr{F}$ defined by

$$\mathscr{F} = \tfrac{1}{2}\sum_i \left(k_x u_i^2 + k_y v_i^2 + k_z w_i^2\right). \tag{9.35}$$

From this definition we obtain

$$\mathbf{f} = -\boldsymbol{\nabla}\mathscr{F}, \qquad \boldsymbol{\nabla} = \mathbf{i}\frac{\partial}{\partial u} + \mathbf{j}\frac{\partial}{\partial v} + \mathbf{k}\frac{\partial}{\partial w}, \tag{9.36}$$

where $(\mathbf{i},\mathbf{j},\mathbf{k})$ are the usual unit vectors. We may give a physical interpretation of the dissipation function by considering a differential amount of work dW done by the frictional force $\mathbf{f}$. This is given by

$$dW = -\mathbf{f}\cdot d\mathbf{r} = -\mathbf{f}\cdot\mathbf{v}\,dt = -\left(k_x u^2 + k_y v^2 + k_z w^2\right) = -2\mathscr{F},$$

which means that the dissipation function is half the rate of energy dissipated by friction.

Let Q_j be the jth component of the generalized force due to friction. This is given by

$$Q_j = \sum_i \mathbf{f}_i\cdot\frac{\partial \mathbf{r}_i}{\partial q_j} = -\sum_i \boldsymbol{\nabla}\mathscr{F}\cdot\frac{\partial \dot{\mathbf{r}}_i}{\partial \dot{q}_j}$$

$$= -\frac{\partial \mathscr{F}}{\partial \dot{q}_j}. \tag{9.37}$$

Lagrange's equations now become

$$\frac{d}{dt}\frac{\partial L}{\partial \dot{q}_j} - \frac{\partial L}{\partial q_j} = -\frac{\partial \mathscr{F}}{\partial \dot{q}_j}. \tag{9.38}$$

Setting $L = T - V$, Lagrange's equations become

$$\frac{d}{dt}\frac{\partial T}{\partial \dot{q}_j} + \frac{\partial V}{\partial q_j} = -\frac{\partial \mathscr{F}}{\partial \dot{q}_j}. \qquad (9.39)$$

The three quadratic functions can be expressed in terms of generalized coordinates and velocities as

$$T = \tfrac{1}{2}\sum_{i,j} a_{ij}\dot{q}_i\dot{q}_j,$$

$$V = \tfrac{1}{2}\sum_{i,j} b_{ij}q_i q_j,$$

$$\mathscr{F} = \tfrac{1}{2}\sum_{i,j} c_{ij}\dot{q}_i\dot{q}_j.$$

HAMILTON'S EQUATIONS OF MOTION

Recall that Lagrange's equations for a system of n particles consist of n second-order PDEs for $L(\mathbf{q}, \dot{\mathbf{q}})$, thus requiring n initial values of the generalized coordinates and n of the generalized velocities. These are the same requirements on the initial conditions demanded by Newton's equations of motion.

In formulating the equations of motion for a system of n particles, Hamilton used another approach. He made use of the fact that Lagrange defined the ith generalized momentum as $\partial L/\partial \dot{q}_i = p_i$. He used the n generalized coordinates $\mathbf{q}$ as one set of independent variables, but (because of the above-mentioned definition of the momentum) introduced the generalized momentum $\mathbf{p}$ as the other set. More specifically, to each q_i he introduced a p_i, which he called the *conjugate momentum* (conjugate to the corresponding q_i). Clearly, each $p_i = m_i\dot{q}_i$.

Consider a conservative force system. This means that there exists a potential V such that

$$\mathbf{F} = -\nabla V$$

for the external force $\mathbf{F}$. The vector form of the equations of motion for a conservative force field then becomes

$$\dot{\mathbf{p}} = -\nabla V. \qquad (9.40)$$

CHAPTER NINE

This is a first-order system of PDEs for **p** and V. But we cannot use this form of the transformed Lagrange's equations, since Hamilton looked for a first-order system of PDEs whose independent variables are **p** and **q**.

The approach Hamilton used is motivated by first taking the total time derivative of the Lagrangian L and then using Lagrange's equation of motion. We have

$$\frac{dL}{dt} = \sum_i \left(\frac{\partial L}{\partial q_i} \frac{dq_i}{dt} + \frac{\partial L}{\partial \dot{q}_i} \frac{d\dot{q}_i}{dt} \right).$$

Using Lagrange's equation of motion the above equation becomes

$$\frac{dL}{dt} = \sum_i \left(\frac{d}{dt} \frac{\partial L}{\partial \dot{q}_i} \frac{d\dot{q}_i}{dt} + \frac{\partial L}{\partial \dot{q}_i} \frac{d\dot{q}_i}{dt} \right)$$

$$= \sum_i \frac{d}{dt}\left(\dot{q}_i \frac{\partial L}{\partial \dot{q}_i} \right).$$

This yields

$$\frac{d}{dt}\left[\sum_i \dot{q}_i \frac{\partial L}{\partial \dot{q}_i} - L \right] = \frac{d}{dt}\left[\sum_i \dot{q}_i p_i - L \right] = 0. \tag{9.41}$$

Integrating eq. (9.41) yields the fact that the bracketed terms equal a constant. We shall call this constant H, the *Hamiltonian*, which is an important function as shown below. Thus

$$\sum_i \dot{q}_i p_i - L = H. \tag{9.42}$$

From this important equation we can show that H is the total energy of a conservative system so that $H = T + V$. Using the definition of L and recognizing that V is independent of $\dot{q}$ we get

$$p_i = \frac{\partial T}{\dot{q}_i}.$$

VARIATIONAL METHODS

Since T is a homogeneous quadratic function of the $\dot{q}_i$s, we get

$$\sum_i \dot{q}_i \frac{\partial T}{\partial \dot{q}_i} = 2T.$$

Equation (9.42) then becomes

$$2T - L = 2T - (T - V) = T + V = H,$$

so that H is indeed the total energy of a conservative system.

Instead of having H be a mere constant equal to the total energy, Hamilton went further and made an inspired decision: He let H be a dependent variable which is a differentiable function of $\mathbf{q}$ and $\mathbf{p}$ and also t. This decision allowed him to derive a set of two first-order PDEs involving the partial derivatives of H with respect to $\mathbf{p}$ and $\mathbf{q}$. We shall follow his line of reasoning by proceeding as follows: Again considering an n-particle system, we let $H = H(\mathbf{p},\mathbf{q},t)$, expand dH by the chain rule, and obtain

$$dH = \sum_i \frac{\partial H}{\partial q_i} dq_i + \sum_i \frac{\partial H}{\partial p_i} dp_i + \frac{\partial H}{\partial t} dt. \tag{9.43}$$

We now consider H defined by eq. (9.42) and form dH [recognizing that $L = L(\mathbf{q}, \dot{\mathbf{q}}, t)$]. We get

$$dH = \sum_i \dot{q}_i\, dp_i + \sum_i p_i\, d\dot{q}_i - \sum_i \frac{\partial L}{\partial q_i} dq_i - \sum_i \frac{\partial L}{\partial \dot{q}_i} d\dot{q}_i - \frac{\partial L}{\partial t}. \tag{9.44}$$

The coefficients of the $d\dot{q}_i$s are zero because $p_i = \partial L/\partial \dot{q}_i$. We also have

$$\frac{\partial L}{\partial q_i} = \dot{p}_i.$$

(since T is independent of $\mathbf{q}$). Therefore eq. (9.44) reduces to

$$dH = \sum_i \dot{q}_i\, dp_i - \sum_i \dot{p}_i\, dq_i - \frac{\partial L}{\partial t} dt. \tag{9.45}$$

CHAPTER NINE

Equating corresponding coefficients of eq. (9.45) with (9.43) yields the following system of $2n + 1$ first-order PDEs:

$$\frac{\partial H}{\partial p_i} = \dot{q}_i, \qquad \frac{\partial H}{\partial q_i} = -\dot{p}_i, \qquad i = 1, 2, \ldots, n, \qquad (9.46)$$

$$\frac{\partial H}{\partial t} = -\frac{\partial L}{\partial t}. \qquad (9.47)$$

The system of n equations given by (9.46) are called *Hamilton's canonical equations of motion*. As mentioned, they constitute a set of $2n$ first-order PDEs for $H(\mathbf{p}, \mathbf{q}, t)$, which replace the set of n second-order PDEs for $L(\mathbf{q}, \dot{\mathbf{q}}, t)$. Equation (9.47) relates H to L for the time-varying case.

Cyclic Coordinates

If the Lagrangian does not contain a particular generalized coordinate q, then that q is said to be a *cyclic coordinate*. Suppose q_i is a cyclic coordinate. Clearly we have

$$\frac{\partial L}{\partial q_i} = 0,$$

so that the Lagrange equation of motion for the cyclic coordinate becomes

$$\frac{d}{dt}\frac{\partial L}{\partial \dot{q}_i} = \dot{p}_i = 0,$$

which means that $p_i = $ const. This gives the theorem: *The generalized momentum conjugate to a cyclic coordinate is conserved.*

Looking at eq. (9.42), we see that $L = -H$ if q_i is cyclic. This tells us that we can use the Hamiltonian formulation for a cyclic coordinate. Moreover, since the momentum p_i conjugate to a cyclic coordinate yields $\dot{p}_i = 0$, so that $\partial H/\partial q_i = 0$, we get the following theorem: *A cyclic coordinate will be absent from the Hamiltonian H. Conversely, if a generalized coordinate does not occur in H then it is cyclic.*

VARIATIONAL METHODS

Cyclic coordinates lead to certain symmetry properties. For example, suppose a cyclic coordinate corresponds to a pure rotation of constant angular velocity about the x axis. Then the conjugate momentum is the angular momentum, which is conserved. If the system is spherically symmetric then all the components of the angular momentum are conserved. Here is another example: If a coordinate corresponding to the displacement of the center of gravity of the system in a certain direction is cyclic, then the conjugate linear momentum is conserved so that the system is invariant with respect to a pure translation in that direction.

If we expand dH/dt by the chain rule, we modify eq. (9.43) by dividing each term by dt. Inserting Hamilton's canonical equations (9.46) into this expansion and then using eq. (9.47) yields

$$\frac{dH}{dt} = \frac{\partial H}{\partial t} = -\frac{\partial L}{\partial t}.$$

Clearly, if L is independent of t then H is constant, and is equal to the total energy only if V is velocity independent. A dissipative force may exist that has a velocity-dependent potential so that H is constant but not the total energy.

Suppose the Lagrangian is a function of n generalized coordinates and velocities but one of the generalized coordinates is cyclic. There are still n Lagrange's equations of motion since there are n generalized velocities. This means that we must still solve an n-degree-of-freedom system even though we have the simplification that the momentum conjugate to the cyclic coordinate is constant. (We must still find n trajectories.) On the other hand, if we use the Hamilton formulation, the cyclic coordinate truly deserves its alternative description as an *ignorable coordinate*, since the Hamiltonian contains $n - 1$ degrees of freedom. Suppose q_n is cyclic. Then the conjugate momentum can be written as $p_n = \alpha$, where α is constant. H becomes

$$H = H(q_1, q_2, \ldots, q_{n-1}, p_1, p_2, \ldots, p_{n-1}, \alpha).$$

We see that, in contrast to the Lagrangian representation, the Hamiltonian actually describes an n-particle system involving $n - 1$ degrees of freedom if one of the coordinates is cyclic. This means that we need to solve for only $n - 1$ trajectories from Hamilton's equations. We completely ignore the cyclic coordinate and recognize that its conjugate

CHAPTER NINE

momentum α, which is a constant of integration, is given as an initial condition.

HAMILTON-JACOBI THEORY

In this section we discuss an important approach to the treatment of PDEs that occur in wave phenomena. Aside from this area, the Hamilton-Jacobi (H-J) theory is of great importance in adding to our fundamental knowledge of the general theory of PDEs.

The H-J theory was developed by Hamilton in the early part of the nineteenth century, primarily for problems in particle dynamics called "analytical dynamics." He went further afield by attempting to tie together the fields of optics and mechanics through the variational principle he developed in mechanics and its relation to Fermat's principle in optics. Hamilton therefore seized upon the analogy between the variational principles of geometric optics and analytical dynamics and used them as guides in the development of optical and dynamical theory. Then Hamilton's canonical equations of motion were extended by the German mathematician Jacobi to first-order PDEs, which are basic in continuum dynamics and wave propagation.

The mathematical models of analytical dynamics essentially consist of finding the properties of systems of ODEs with constraints, since systems of particles are involved. When we treat wave propagation in continuous media, PDEs are involved. The relationship between ODEs and PDEs was brought into sharp focus when we described hyperbolic PDEs by characteristic theory. The characteristic ODEs are involved in the solution of the hyperbolic PDEs.

Canonical Transformations

The basis of H-J theory is to obtain the appropriate canonical transformations into cyclic coordinates. We know that the conjugate momentum to a cyclic coordinate is conserved, so that we may take it as the initial momentum. As a motivation for the H-J theory we take a problem for which the solution of Hamilton's canonical equations is trivial. Let the Hamiltonian be conserved (a constant equal to the total energy). Now let all the generalized coordinates be cyclic. Then all the conjugate momenta are constant:

$$p_i = \alpha_i, \quad i = 1, 2, \ldots, n,$$

which we take as the initial conditions on the momenta. The Hamiltonian can be written as

$$H = H(\alpha_1, \alpha_2, \ldots, \alpha_n).$$

Consequently, Hamilton's equations for each $\dot{q}_i$ are

$$\dot{q}_i = \frac{\partial H}{\partial \alpha_i} = \gamma_i,$$

where the γ_is are functions of the α_is, so that they are also time-independent. The solution for **q** is

$$q_i = \gamma_i t + \beta_i, \quad i = 1, 2, \ldots, n,$$

where β_i is the constant of integration and represents the initial condition on the ith generalized coordinate.

Generalizing this example, we seek a canonical transformation from the generalized coordinates and momenta at any time t to a set of constants $(\mathbf{q}_0, \mathbf{p}_0)$, which are taken as the $2n$ initial conditions. The transformation must preserve the form of Hamilton's canonical equations given by (9.46) and (9.47). Such a canonical transformation is called a *contact transformation*; it clearly yields cyclic coordinates since the momentum vector $\boldsymbol{\alpha}$ is constant. The method of finding such a class of transformations is the basis of the H-J theory.

To this end we let Q_i, P_i be the initial values of q_i, p_i, respectively. The transformation equations take the form

$$\mathbf{Q} = \mathbf{Q}(\mathbf{q}, \mathbf{p}), \quad \mathbf{P} = \mathbf{P}(\mathbf{q}, \mathbf{p}). \tag{9.48}$$

This is a transformation of the original vectors (canonical coordinates) **q**, **p** into the transformed canonical coordinates represented by the initial vectors **Q**, **P**, where q_i is the ith component of **q**, etc. We solve for L in eq. (9.42) and apply Hamilton's principle, which can be put in the form

$$\delta \int_{t_A}^{t_B} [\dot{\mathbf{q}} \cdot \mathbf{p} - H(\mathbf{q}, \mathbf{p}, t)] \, dt = 0. \qquad (9.49)$$

As implied above, the canonical transformation given by eq. (9.48) must preserve Hamilton's principle. If we let $K(\mathbf{Q}, \mathbf{P}, t)$ be the Hamiltonian in transformed coordinates, Hamilton's principle involving K becomes

$$\delta \int_{t_A}^{t_B} [\dot{\mathbf{Q}} \cdot \mathbf{P} - K(\mathbf{Q}, \mathbf{P}, t)] \, dt = 0. \qquad (9.50)$$

Equation (9.50) is the image of (9.49) in the transformed coordinate system. Hamilton's canonical equations (9.46) apply to K with respect to the transformed canonical coordinates given by the vectors $\mathbf{P}, \mathbf{Q}$. We get

$$\frac{\partial K}{\partial P_i} = \dot{Q}_i, \qquad \frac{\partial K}{\partial Q_i} = -\dot{P}_i. \qquad (9.51)$$

The integrands in eqs. (9.49) and (9.50) must be connected by the following relationship:

$$\dot{\mathbf{q}} \cdot \mathbf{p} - H(\mathbf{q}, \mathbf{p}, t) = \dot{\mathbf{Q}} \cdot \mathbf{P} - K(\mathbf{Q}, \mathbf{P}, t) = \frac{dS}{dt}. \qquad (9.52)$$

Note that the right-hand side is not zero but is an arbitrary function, which we write as dS/dt such that

$$\delta \int_{t_A}^{t_B} S \, dt = 0.$$

S is analogous to the Lagrangian L. We shall see below that S is an important function playing a key role in the H-J equation (which we shall derive below). S is called *Hamilton's principal function*. S is a *generating function*, which allows us to transform from $(\mathbf{q}, \mathbf{p}, t)$ to $(\mathbf{Q}, \mathbf{P}, t)$ space. There are four possible relationships for S between the old and transformed coordinates: $S(\mathbf{q}, \mathbf{Q}, t)$, $S(\mathbf{p}, \mathbf{P}, t)$, $S(\mathbf{p}, \mathbf{Q}, t)$, and $S(\mathbf{q}, \mathbf{P}, t)$. Instead of investigating each one in detail, we shall choose the possibility $S = S(\mathbf{q}, \mathbf{P}, t)$. The reason for this choice is that we choose the initial momenta $\mathbf{P}$ and want to determine the spatial dependence of S on $\mathbf{q}$.

VARIATIONAL METHODS

Suppose we choose the vector $\mathbf{P} = \mathbf{0}$. Then $S = S(\mathbf{q}, t)$. We expand the total derivative and obtain

$$\frac{dS}{dt} = \frac{\partial S}{\partial q_i} d\dot{q}_i + \frac{\partial S}{\partial t}, \qquad (9.53)$$

where we sum over i from one to n using the Einstein summation convention. This equation is also valid in the more general case where $S = S(\mathbf{q}, \mathbf{P}, t)$ since $\mathbf{P}$ is independent of t. We can rewrite eq. (9.51) as

$$K(\mathbf{Q}, \mathbf{P}, t) - H(\mathbf{q}, \mathbf{p}, t) = \frac{dS}{dt} + \dot{\mathbf{Q}} \cdot \mathbf{P} - \dot{\mathbf{q}} \cdot \mathbf{p}. \qquad (9.51a)$$

We want to eliminate the initial canonical coordinates from the right-hand side. To this end we introduce the function $G(\mathbf{q}, \mathbf{P}, t)$ such that

$$S(\mathbf{q}, \mathbf{P}, t) = G(\mathbf{Q}, \mathbf{P}, t) - \mathbf{P} \cdot \mathbf{Q}. \qquad (9.54)$$

We obtain

$$S_t = G_t, \qquad \dot{S} = \dot{G} - P_i \dot{Q}_i.$$

We then get

$$K - H = -p_i \dot{q}_i + \frac{\partial G}{\partial q_i} \dot{q}_i + \frac{\partial S}{\partial t}. \qquad (9.55)$$

Since the components of the vector $\mathbf{q}$ are linearly independent coordinates, the coefficient of each $\dot{q}_i$ must vanish. This gives the following system:

$$\frac{\partial G}{\partial q_i} = \frac{\partial S}{\partial q_i} = p_i, \qquad (9.56)$$

$$K - H = \frac{\partial S}{\partial t}. \qquad (9.57)$$

There is no loss in generality in choosing the initial coordinates, $\mathbf{Q} = \mathbf{P} = \mathbf{0}$. We further demand that $K = 0$. We know that $H =$

CHAPTER NINE

$H(\mathbf{q}, \mathbf{p}, t)$. We therefore obtain the following PDE:

$$H\left(q_1, \ldots, q_n, \frac{\partial S}{\partial q_1}, \ldots, \frac{\partial S}{\partial q_n}, t\right) + \frac{\partial S}{\partial t} = 0. \qquad (9.58)$$

Equation (9.58) is the H-J equation. It is a first-order PDE for $S = S(\mathbf{q}, \mathbf{P}, t)$, which is, in general, nonlinear. We see that the H-J equation is a PDE for the $n + 1$ independent variables $(q_1, \ldots, q_n, t)$ for S. Consequently, a complete solution involves $n + 1$ constants. We observe that S does not appear in the H-J equation; only its first-order partial derivatives with respect to $\mathbf{q}$ and t appear. Therefore, if S is a solution to the H-J equation, then $S + \alpha_{n+1}$ is also a solution, where α_{n+1} is an additive constant. The complete solution of the H-J equation can then be written as

$$S = S(q_1, \ldots, q_n, \alpha_1, \ldots, \alpha_n, t) + \alpha_{n+1},$$

where

$$\alpha_i = P_i, \qquad i = 1, 2, \ldots, n.$$

We have thus chosen the n-component vector $\boldsymbol{\alpha}$ to be the initial momentum vector $\mathbf{P}$. The vector $\mathbf{p}$ can be determined from eq. (9.56) upon solving the H-J equation for S. The cyclic canonical coordinate conjugate to P_i is

$$Q_i = \frac{\partial S}{\partial \alpha_i} = \beta_i. \qquad (9.59)$$

If the Jacobian of the mapping from $(\mathbf{q}, \mathbf{p})$ to $(\mathbf{Q}, \mathbf{P})$ is not zero, then we obtain the trajectories $\mathbf{q} = \mathbf{q}(\boldsymbol{\alpha}, \boldsymbol{\beta}, t)$, which gives the canonical coordinates in terms of t and the $2n$ initial conditions expressed by the vectors $(\boldsymbol{\alpha}, \boldsymbol{\beta})$.

If we expand dS/dt and use eq. (9.56) we obtain

$$\frac{dS}{dt} = \frac{\partial S}{\partial q_i} \dot{q}_i + \frac{\partial S}{\partial t} = p_i \dot{q}_i - H = L. \qquad (9.60)$$

Integrating over t tells us that S is the functional that satisfies Hamilton's principle. But this is what we asserted when we first introduced S by saying "S is analogous to L." We now see that $\dot{S} = L$.

VARIATIONAL METHODS

Example. Simple harmonic oscillator. We consider a system consisting of a single mass m and a spring whose constant is k. This example is one dimensional: the canonical coordinates are (q, p). The Hamiltonian is

$$H = \frac{p^2}{2m} + \frac{kq^2}{2}.$$

Using eq. (9.56) we set $p = \partial S/\partial q$. The H-J equation (9.58) then becomes

$$\frac{1}{2m}\left(\frac{\partial S}{\partial q}\right)^2 + \frac{kq^2}{2} + \frac{\partial S}{\partial t} = 0,$$

where $S = S(q, \alpha, t)$, α being the initial momentum. Since we have a conservative system $H = E$, the total energy. From the nature of the H-J equation for this system we may separate out the time-dependent part of S by introducing the function $W = W(q, \alpha)$ and setting

$$S(q, \alpha, t) = W(q, \alpha) - Et.$$

We shall see why we introduced the term $-Et$ when we insert the above expression into the H-J equation, for it becomes

$$H = \frac{1}{2m}\left(\frac{\partial W}{\partial q}\right)^2 + \frac{kq^2}{2} = E,$$

which is the H-J equation for W. We see that $H = E$, the total energy. We solve this first-order PDE for W by an integration and obtain

$$W = \sqrt{mk}\int_{Q=\beta}^{q}\sqrt{\left(\frac{2E}{k}\right) - q^2}\, dq = S + Et.$$

We are not interested in W as such. However we note that $\partial W/\partial E = t$, which is what we want. We therefore obtain

$$\frac{\partial W}{\partial E} = \sqrt{\frac{m}{k}}\int \frac{dq}{\sqrt{\left(\frac{2E}{k}\right) - q^2}} = -\sqrt{\frac{m}{k}}\cos^{-1} q\sqrt{\frac{k}{2E}} + \tau = t,$$

369

where τ is a constant having the dimensions of time. Solving for q yields

$$q = \sqrt{\frac{2E}{k}} \cos \omega(t - \tau).$$

The natural frequency of the oscillator is $\omega = \sqrt{k/m}$ and $\omega\tau$ is the phase angle. The momentum conjugate to q is

$$p = -\sqrt{2Em} \sin \omega(t - \tau).$$

The cyclic coordinate $P = \alpha = \sqrt{2Em} \sin \omega\tau$. We can calculate L and show that S is the time integral of L. (We leave this to a problem.)

This example shows that we can integrate the H-J equation because S was separated into a time-independent part and a part proportional to time. This decomposition of S is always possible if H is independent of time.

EXTENSION OF W TO $2n$ DEGREES OF FREEDOM

We now go back to the more general case where the vectors $\mathbf{q}$ and $\mathbf{p}$ are n dimensional. We consider a conservative system where $H = E$ so that we can decompose S as follows:

$$S(q, \alpha, t) = W(q, \alpha) - Et, \tag{9.61}$$

where α is the usual initial momentum vector. Inserting eq. (9.61) into the H-J equation (9.58) yields

$$H\left(q_1, \ldots, q_n, \frac{\partial W}{\partial q_1}, \ldots, \frac{\partial W}{\partial q_n}\right) = E. \tag{9.62}$$

Using eq. (9.56) we see that

$$\frac{\partial W}{\partial q_i} = p_i. \tag{9.63}$$

W is called *Hamilton's characteristic function*. Equation (9.62) is the H-J equation for Hamilton's characteristic function.

VARIATIONAL METHODS

We wish to decompose W into the following sum of n functions, each term of which depends on a single component of $\mathbf{q}$:

$$W(q_1,\ldots,q_n,\alpha_1,\ldots,\alpha_n) = \sum_{i=1}^{n} W_i(q_i,\alpha_1,\ldots,\alpha_n). \qquad (9.64)$$

This decomposition is a separation of variables on W. If this can be done, then H can be decomposed as follows:

$$H\left(q_1,\ldots,q_n,\frac{\partial W}{\partial q_1},\ldots,\frac{\partial W}{\partial q_n}\right) = \sum_{i=1}^{n} H_i\left(q_i,\frac{\partial W_i}{\partial q_i}\right). \qquad (9.65)$$

This is a separation of variables on H.

Suppose all the components of $\mathbf{q}$ are cyclic except q_1. Then

$$\frac{\partial W}{\partial q_i} = p_i = \alpha_i, \qquad W_i = \alpha_i q_i, \qquad i \neq 1. \qquad (9.66)$$

It follows that

$$W = W_1 + \sum_{i=2}^{n} \alpha_i q_i,$$

which means that W is decomposed into the $(n-1)$-component cyclic vector $\mathbf{q}$ (where q_1 is noncyclic). The H-J equation for W_1 reduces to

$$H_1\left(q_1,\frac{\partial W_1}{\partial q_1},\alpha_2,\ldots,\alpha_n\right) = \gamma_1,$$

where γ_1 is a constant of integration.

Now let q_2 be the only noncyclic coordinate. This gives

$$W = W_1 + W_2 + \sum_{i=3}^{n} \alpha_i q_i.$$

The H-J equation for W_2 becomes

$$H_2\left(q_2,\frac{\partial W_2}{\partial q_2}\right) = \gamma_2.$$

CHAPTER NINE

We repeat this process until all the components of **q** are separable. We then add $H_1, H_2, \ldots, H_n$ and obtain eq. (9.65).

H-J Theory and Wave Propagation

We now apply the H-J theory to wave propagation problems. We again consider a conservative system so that $H = E$, the total energy. We consider the n-dimensional configuration space given by $\mathbf{q} = \mathbf{q}(q_1, \ldots, q_n)$. Hamilton's principal function S and the characteristic function W are imbedded in this space. Again we take $S = S(\mathbf{q}, \mathbf{P})$. For a conservative system, S and W are related by

$$S(\mathbf{q}, \mathbf{P}, t) = W(\mathbf{q}, \mathbf{P}) - Et. \tag{9.67}$$

Note that W is fixed in configuration space while S is a linear function of time, so that a surface of constant S moves according to the law given by eq. (9.67). Suppose that at some time t the surface of constant S corresponds to the surface $W = $ const. At time $t + dt$ the surface of constant S coincides with the surface for which $W = S + E\,dt$. The surface $S = $ const is interpreted as the wave front that propagates in configuration space. The outward-drawn normal at each point on the surface $S = $ const gives the direction of the wave specified by its wave or phase velocity c. $c = ds/dt$ where ds is the element of arc length normal to the surface. For a planar wave the surface $S = $ const is a plane, so that the phase velocity is in one direction at any given time. During the time interval dt the surface $S = $ const travels from W to a new surface given by $W + dW$. From the definition of dW and eq. (9.67), we get

$$\frac{dW}{dt} = \nabla W \cdot \mathbf{c} = E, \tag{9.68}$$

so that

$$c = \frac{ds}{dt} = \frac{E}{|\nabla W|}. \tag{9.69}$$

Example. A single particle without constraints. The configuration space reduces to three dimensions so that $\mathbf{q} = (x, y, z)$. The H-J equation for the characteristic function becomes

$$\frac{1}{2m}\left[(W_x)^2 + (W_y)^2 + (W_z)^2\right] = \frac{1}{2m}\nabla W \cdot \nabla W$$

$$= \frac{1}{2m}(\nabla W)^2 = E - V, \quad (9.70)$$

where V is the potential energy. Recall that $E - V = T$ for a conservative system. Using eqs. (9.69) and (9.70), we get the following expression for the phase velocity c:

$$c = \frac{E}{\sqrt{2mT}} = \frac{E}{p}. \quad (9.71)$$

Since the ith component of grad W is p_i, we have

$$\mathbf{p} = \nabla W. \quad (9.72)$$

Clearly, $\mathbf{p}$ is in the direction of the propagating wave; therefore grad W determines the direction of the propagating wave and is normal to the surface of constant S or W. A family of surfaces of constant W gives a set of trajectories of possible motion. As the particle moves along one of these trajectories, the surfaces of S generating the motion (which are the wave fronts) will also travel through this one-dimensional configuration space with the phase velocity c. Note from the reciprocal relation between c and $\mathbf{p}$ as shown by eq. (9.71) that as the particle speeds up the surfaces slow down, and vice versa.

The results given above for a single particle also hold for a system of n particles, only now these particles are imbedded in the n-dimensional configuration space. Instead of the trajectory of a single particle we deal with the path of the system point in configuration space. Recall that a system point is a point on the trajectory defined by the n-dimensional vector $\mathbf{q}$.

Application to Light Waves

The surfaces of constant S have been characterized as wave fronts because they propagate in space in the same manner as wave surfaces

CHAPTER NINE

of constant phase. We have computed the wave or phase velocity. However, we have not yet discussed the periodicity of the wave motion —the frequency and wavelength spectra of the waves associated with S. To this end we examine some of the properties of waves and use as our model light waves. Maxwell proved that electromagnetic waves have the same properties as light waves, being transverse waves traveling with the velocity of light. Only the frequency spectrum is different.

We start by considering a generic scalar potential ϕ in thee-dimensional space, such as the electromagnetic potential. The wave equation that ϕ must satisfy is

$$\nabla^2 \phi - \frac{n^2}{c_0^2} \phi_{tt} = 0, \quad \text{where } \nabla^2 = \frac{\partial^2}{\partial x^2} + \frac{\partial^2}{\partial y^2} + \frac{\partial^2}{\partial z^2}. \quad (9.73)$$

c_0 is the velocity of light in a vacuum and n is the index of refraction, which is defined as the ratio of c_0 to the velocity of light c in the optical medium. In general, n will be a function of space since c is also a function of space.

The potential ϕ in the wave equation (9.73) can be decomposed into the product of a function of $\mathbf{r}$ and a function of t. Since the function of t is oscillatory we may set

$$\phi = u(\mathbf{r}) e^{i k_0 c_0 t}. \quad (9.74)$$

Inserting eq. (9.74) into (9.73) yields the following PDE for $u(\mathbf{r})$:

$$\nabla^2 u + n^2 k_0^2 u = 0. \quad (9.75)$$

This is an elliptic PDE so that there are no real characteristics.

A note on the wave number: k_0 is the wave number in a vacuum where $n = 1$ and $c = c_0$, so that it is independent of $\mathbf{r}$. Let k be the wave number for $n = n(\mathbf{r})$. It follows that

$$k_0 = \frac{\omega}{c_0}, \quad k = \frac{\omega}{c} = \left(\frac{\omega}{c_0} \frac{c_0}{c} \right) = k_0 n, \quad \text{where } n = \frac{c_0}{c}.$$

We note that the frequency ω is assumed to be independent of n.

We first consider a plane wave solution of the wave equation. We may further decompose u into

$$u(\mathbf{r}) = v(\mathbf{r}) e^{i k_0 \boldsymbol{\nu} \cdot \mathbf{r}}, \quad (9.76)$$

VARIATIONAL METHODS

where $\boldsymbol{\nu}$ is the unit vector in the direction of the wave number. The exponent in the exponential term tells us that we have a plane wave (planar wave front) normal to the direction of $\mathbf{r}$, where $k_0\boldsymbol{\nu} \cdot \mathbf{r}$ is the projection of the wave number in the direction $\mathbf{r}$ of the propagating wave, which is a light ray normal to the wave front.

We now introduce the *optical path length* $\boldsymbol{\Psi}$ defined as a vector in the direction of the propagating wave. For a plane wave $\boldsymbol{\Psi}$ is equal to $\mathbf{r}$, but for a nonplanar wave where n is spatially dependent [$n = n(\mathbf{r})$], we replace the term $\mathbf{k}_0 \cdot \mathbf{r}$ by $\mathbf{k}_0 \cdot \boldsymbol{\Psi}(\mathbf{r})$, where the optical path length is spatially dependent. Then u becomes

$$u = e^{i\mathbf{k}_0 \cdot \boldsymbol{\Psi}} v(r). \tag{9.77}$$

Calculating $\nabla^2 \phi$ from eq. (9.74), the same equation for ϕ becomes

$$ik_0[2\,\boldsymbol{\nabla} A \cdot \boldsymbol{\nabla}\psi + \nabla^2\psi]\phi + \left[\nabla^2 A + (\nabla A)^2 - k_0^2(\nabla\psi)^2 + n^2 k_0^2\right]\phi = 0. \tag{9.78}$$

Since both A and ψ are present, this equation holds only if the two expressions in square brackets separately vanish. This gives

$$\nabla^2 A + (\nabla A)^2 + k_0^2\left[n^2 - (\nabla\psi)^2\right] = 0,$$
$$2\,\boldsymbol{\nabla} A \cdot \boldsymbol{\nabla}\psi + \nabla^2\psi = 0. \tag{9.79}$$

So far both equations of (9.79) are exact since no approximations have yet been made. We now make the approximation that the index of refraction n varies slowly in the sense that the distance (which is of the order of the wavelength) is small. This means that the wave number k_0 is taken to be large. This is the high-frequency case since $k_0 = n(\omega/c_0)$. This is the case we are considering in optics. We therefore divide the first equation of (9.79) by k_0 and let it become infinite. The first equation is then approximated by

$$(\nabla\psi)^2 = n^2. \tag{9.80}$$

Equation (9.80) is a very important equation in optics and in other wave phenomena of high frequency. It is called the *eikonal equation*. If n is constant then eq. (9.80) determines the wave front for a planar wave. If n varies in space then the wave front is still determined by eq. (9.80) but it is no longer planar.

CHAPTER NINE

Quantum Mechanics

We now digress and present a short exposition of the major points of quantum mechanics in order to have a background for a discussion of the relationship between the H-J theory and quantum mechanics. Quantum mechanics is concerned with the duality of particles and waves: the "dualistic theory of matter." Roughly speaking, this theory relates optics (which involves wave phenomena) to mechanics by way of the H-J theory (which is a particle approach to physics). But quantum mechanics is more than that, for it involves quantized energy levels, the uncertainty principle of Heisenberg, etc. These revolutionary ideas are not involved in either classical mechanics or physical optics.

The concept of quantized energy levels originated with Max Planck who, in 1900, formulated the quantum theory of black body radiation. Before Planck got on the scene physicists such as Lord Rayleigh and Sir James Jeans attempted unsuccessfully to give a theoretical explanation of the experimental fact that the intensity of the thermal radiation of a black body (a theoretically perfect radiator) per unit frequency at a given temperature rises from zero at very low frequencies to a maximum and then falls off to zero at very high frequencies. Rayleigh and Jeans used the equipartition theorem of classical statistical mechanics (which says that the mean kinetic energy of a system of particles in thermal equilibrium depends only on the temperature) to develop the classical law of black body radiation, called the Rayleigh-Jeans law. This law agreed with experiment at low frequencies but broke down at high frequencies, where it implied an ever-increasing energy density level leading to the *ultraviolet catastrophe*. Planck attempted to explain the distribution of energy in the black body spectrum by suggesting that the classical equipartition theory breaks down for the high-frequency spectrum. He therefore abandoned the classical approach, and then made the brilliant hypothesis that the vibrating particles should be restricted to certain quantized (discrete) energy levels. This hypothesis led to a theory that fitted the experimental data throughout the complete frequency range. It was a revolutionary theory in the sense that it replaced the classical theory and led to a complete revolution in the foundations of physics. It was incompatible with both Newton's mechanics and Maxwell's classical electromagnetic theory. Planck was very unhappy about this, for he was a conservative at heart, and tried fruitlessly to rescue the classical theory. Einstein showed that Planck's revolutionary hypothesis can be applied to the structure of matter. In particular, an atom can absorb or radiate energy only in a quantized manner, where the energy resultant U depends only on the frequency ν of the light being absorbed or radiated. This gives the fundamental law of quantum mechanics: $U = h\nu$, where h is *Planck's constant* ($h = 6.55 \times 10^{-27}$ erg sec). Einstein was awarded the Nobel prize for his contributions to the photoelectric effect, and not for his work on relativity theory.

In the early twentieth century Rutherford of England developed a theory of the atom which involved negative electrons circling the positive nucleus. In 1913 N. Bohr united Rutherford's nuclear concept of the atom with the quantized energy level hypothesis to formulate his famous theory of the structure and spectrum of the hydrogen atom (for which he earned a Nobel prize). Up to around 1923 there was a battle between the proponents of the corpuscular theory and the advocates of the wave theory of light, which led to a stalemate. In 1923 Louis de Broglie suggested that matter may share the dualistic characteristics of being corpuscular and having wavelike properties which give radiation effects. Heisenberg came on the scene around 1925 and stated his famous "uncertainty principle" that one cannot get an exact measurement of both the momentum and position of an atomic particle. This reformulated the theory of matter from an exact science (in classical physics) to a probabilistic one, where, for example, an electron circling the nucleus of an atom has no well-defined path but one that is governed by a probability function (the wave function). More generally, Heisenberg reformulated quantum theory. Schroedinger in 1926 developed a PDE based on de Broglie's approach, and both of them developed wave mechanics.

An Analogy between Geometric Optics and Classical Mechanics

With this background in the dualistic theory of radiation we must invent a suitable PDE for the wave function. This equation is assumed to have a form similar to that of the wave equation of optics, but must be so designated as to harmonize with the H-J theory (which is a theory in mechanics) in the limiting case where diffraction effects are negligible. We shall see below that the equation we are interested in is the Schroedinger equation for the wave function.

CHAPTER NINE

We now compare the optical path length ψ with Hamilton's characteristic function W by referring to eq. (9.70), which we rewrite as

$$(\nabla W) = \sqrt{2m(E-V)} = \sqrt{2mT} = p. \tag{9.80a}$$

Recall that eq. (9.70a) is the H-J equation for W. This is the case of classical mechanics. The *principle of least action* tells us that

$$\Delta \int_A^B \sqrt{2mT}\, ds = \Delta \int_A^B p\, ds = \Delta \int_A^B (\nabla W)\, ds = 0, \tag{9.81}$$

where the end points A and B are fixed. This principle tells us that the functional $\int_A^B p\, ds$ is an extremum. [See the appendix to this chapter for a derivation of eq. (9.81).] Recall that eq. (9.81) applies to the classical mechanics case. For the optics case we appeal to Fermat's principle of least time, which may be written as

$$\Delta \int_{t_A}^{t_B} dt = \Delta \int_A^B \left(\frac{dt}{ds}\right) ds = \Delta \int \frac{ds}{c} = \Delta \int \frac{n}{c_0}\, ds$$

$$= \Delta \int_A^B \frac{(\nabla \psi)}{c_0}\, ds = 0, \qquad n = \frac{c_0}{c}. \tag{9.82}$$

This important equation of optics tells us that the gradient of the optical path length obeys an extremum principle in the sense that it is the integrand of the functional involved in the principle of least time. This is analogous to the case in classical mechanics, where the gradient of Hamilton's characteristic function $\nabla W = p$ is the integrand in the functional that obeys the principle of least action.

The above treatment shows us that the H-J equation, which was derived for the classical mechanics case, has its analogy with the wave equation occurring in optics. We shall go further and show the relationship of the H-J equation with quantum mechanics.

We now go back to the H-J theory. We attempt to determine what the scalar wave equation would look like in the setting of the quantized energy levels of quantum mechanics for which the H-J equation represents the high-frequency end of the electromagnetic spectrum. To this end we use the similarity of the eikonal equation (9.80) with the H-J equation for Hamilton's characteristic function W given by eq. (9.70). This similarity tells us that the optical path length ψ is related to the

VARIATIONAL METHODS

characteristic function W. We shall consider the simplest relation by provisionally letting ψ be proportional to W. Recall that Hamilton's principal function S is related to W by $S = W - Et$, where E is the total energy of a conservative system. If the optical path length is assumed to be proportional to the characteristic function, it follows that Hamilton's principal function must be proportional to the total phase of the light wave, which is given by the imaginary part of the exponent in eq. (9.76). Now $k_0 = c_0 = \omega = \nu/2\pi$, so the exponent becomes

$$W - Et = \alpha k_0 [\psi_0(r) - c_0 t] = \alpha \left[k_0 \psi(r) - \frac{\nu}{2\pi} t \right], \quad (9.83)$$

where α is the proportionality constant. It follows that the total energy $E = \alpha(\nu/2\pi)$. The constant $\alpha = 2\pi h$, since the energy is quantized, so that

$$E = h\nu. \quad (9.84)$$

h is Planck's constant (which is a universal constant independent of the coordinate system). Equation (9.84) is of fundamental importance in quantum mechanics, since it is the basis for the modern theory of atomic and molecular structure. The energy depends only on the frequency in a discrete or quantized manner. For example, if an electron is excited by an energy light source of frequency ν then it emits a quantum of energy equal to $h\nu$.

The frequency, wavelength, and wave speed are connected by the relationship $c = \lambda \nu$. From the H-J equation given by (9.71), which connects the wave speed, total energy, and momentum, we obtain another fundamental relationship in quantum mechanics:

$$\lambda = \frac{h}{p}, \quad (9.85)$$

which connects the wavelength to the momentum. The optical wave equation for the potential ϕ is given by eq. (9.73). We separate variables by introducing the function $u = u(r)$. We have

$$\phi(r, t) = u(r) e^{i\omega t} = u(r) e^{2\pi i c t / \lambda}. \quad (9.86)$$

We have set $\omega = 2\pi c/\lambda$ in order to get the reduced wave equation below in terms of λ rather than ω. Inserting eq. (9.86) into (9.73) gives

$$\nabla^2 u + \left(\frac{4\pi^2}{\lambda^2} \right) u = 0. \quad (9.87)$$

CHAPTER NINE

Equation (9.87) is called the *reduced wave equation* corresponding to the optical wave equation. Since the H-J theory yields the formula $p = \sqrt{2m(E-V)}$ [from eqs. (9.70), (9.72), and (9.85)] we get

$$\lambda = \frac{h}{\sqrt{2m(E-V)}}. \tag{9.88}$$

Inserting this expression for the wavelength into eq. (9.76), the reduced wave equation becomes

$$\nabla^2 u + \left(\frac{8\pi^2 m}{h^2}\right)(E-V)u = 0. \tag{9.89}$$

Equation (9.89) is the Schroedinger equation of quantum mechanics. The function u is called the amplitude or space factor of the wave function, and is a fundamental variable in quantum mechanics.

A note on the relationship between the wavelength and momentum expressed by eq. (9.85): This equation holds for problems that have atomic dimensions. For example, an oxygen molecule with a speed corresponding to the mean thermal energy of 300 K has a wavelength of approximately 1.5×10^{-8} cm, which is a dimension of the order of magnitude of an atomic diameter and is in the X-ray region. With this wavelength, diffraction effects play a major role. On the other hand, let us apply eq. (9.85) to a body of macroscopic dimensions, such as a golf ball, which weights 47 g. Let it travel with a speed of 1 m/sec. A simple calculation shows that the wavelength is $\lambda = 1.41 \times 10^{-20}$ m. This wavelength is so small that one cannot demonstrate diffraction effects.

In summary, we have used the H-J theory of classical mechanics, which gives $\nabla W = p = \sqrt{2m(E-V)}$ and $c = E/p$, to manipulate the optical wave equation into the Schroedinger equation of quantum mechanics, using the fundamental theory of quantum mechanics that $E = h\nu$.

ASYMPTOTIC THEORY OF WAVE PROPAGATION

There is a relationship between the classical mechanics of the Hamilton-Jacobi theory and wave phenomena in optics. The technique to be used here is based on the method of *asymptotic expansions* for large

VARIATIONAL METHODS

values of the wave number or frequency. Optical phenomenon is the limiting case as the wave number or frequency becomes very large. There are two wave numbers, k_0 and k. They are related as follows: $k_0 = \omega/c_0$ is the wave number if the index of refraction $n = 1$. k is the wave number for the index of refraction $n = n(\mathbf{r})$. It follows that

$$k = \frac{\omega}{c} = \left(\frac{\omega}{c_0}\right)\left(\frac{c_0}{c}\right) = k_0 n$$

using the definition of the refractive index n, as shown above. Note that ω is independent of n. For convenience we expand in inverse powers of the wave number k_0. Our starting point is the wave equation (9.73) that the scalar potential ϕ must satisfy. We are interested in the asymptotic behavior of $u(\mathbf{r})$ as $k_0 \to \infty$, which means very large frequencies. To this end we separate the wave number k_0 times the optical path length or eikonal ψ in the term $e^{ik\psi}$ from the spatial part of the reduced wave equation and expand in inverse powers of ik_0. We thereby obtain the following expansion for u:

$$u \sim e^{ik_0\psi} \sum_{n=0}^{\infty} \frac{v_n(r)}{(ik_0)^n}. \tag{9.90}$$

It is clear that we expand in inverse powers of ik_0 because ik_0 appears in the separated exponential term. Equation (9.90) is called the asymptotic expansion of $u(\mathbf{r})$. This means that the ratio of u to the series in the right-hand side tends to unity as k_0 becomes infinite. The term $v_n(\mathbf{r})$ is called the nth expansion coefficient. We shall obtain a set of recursion relations wherein each $v_n(\mathbf{r})$ is given as a function of v_n and v_{n-1}. Then solving for v_0 we can recursively obtain $v_1, v_2, \ldots$.

The procedure is a straightforward one: We insert the asymptotic series (9.90) into the reduced wave equation (9.87) and equate coefficients of like powers of $(ik_0)^{-n}$. It is convenient to rewrite eq. (9.90) as

$$u = e^{ik_0\Psi} v, \tag{9.91}$$

where

$$v = \sum_{n=0}^{\infty} \frac{v_n}{(ik_0)^n}. \tag{9.92}$$

We now insert (9.91) into the reduced wave equation (9.87). In order to calculate the Laplacian $\nabla^2 u$ we need to apply the grad or ∇ operator

CHAPTER NINE

twice to eq. (9.91). We obtain

$$\nabla u = e^{ik\Psi}[\nabla v + ikv\,\nabla\Psi],$$

$$\nabla^2 u = e^{ik\Psi}[\nabla^2 v + ik\,\nabla\Psi \cdot (\nabla v + ikv\,\nabla\Psi) + ik\nabla(v\,\nabla\Psi)]$$
$$= e^{ik\Psi}[\nabla^2 v + ikv(\nabla^2\Psi) + 2ik(\nabla\Psi)\cdot(\nabla v) - k^2 v(\nabla\Psi)^2].$$

Inserting the above expression for $\nabla^2 u$ into (9.87) yields

$$\nabla^2 v + ik[2(\nabla\Psi)\cdot(\nabla v) + v(\nabla^2\Psi)] + k^2 v[1 - (\nabla\Psi)^2] = 0. \quad (9.93)$$

If we insert the asymptotic expansion for v given by (9.92) into (9.93) we get

$$\sum_{n=0}^{\infty}(ik^{-n})\Big\{\nabla^2 v_n + ik[2(\nabla\Psi)\cdot(\nabla v_n) + v_n\,\nabla^2\Psi]$$
$$+ k^2 v_n[1 - (\nabla\Psi)^2]\Big\} = 0. \quad (9.94)$$

In order for eq. (9.94) to hold, the coefficient of each inverse power of ik must be set equal to zero. This yields

$$\nabla^2 v_n + 2(\nabla\Psi)\cdot(\nabla v_{n+1}) + v_{n+1}\nabla^2\Psi + v_{n+2}[1 - (\nabla\Psi)^2] = 0,$$

$$v_n = 0 \quad \text{for } n < 0. \quad (9.95)$$

Setting $n = -2$ in eq. (9.95) yields

$$v_0[1 - (\nabla\Psi)^2] = 0.$$

Since we demand that $v_0 \neq 0$ we must have

$$(\nabla\Psi)^2 = 1. \quad (9.96)$$

Equation (9.96) is the eikonal equation for a constant index of refraction $n = 1$. Recall eq. (8.80), which was obtained not by the asymptotic method but by an exact method [putting eq. (9.76) into the wave equation (9.73)]. Also recall that Ψ is the eikonal or optical path length. Using eq. (9.96) and replacing $n + 1$ by n (for convenience), eq. (9.95) becomes

$$v_n \nabla^2\Psi + 2(\nabla\Psi)\cdot(\nabla v_n) = -\nabla^2 v_{n-1}, \quad n = 0, 1, 2, \ldots, v_{-1} = 0. \quad (9.97)$$

VARIATIONAL METHODS

Since $n = 1$ we have plane waves. Therefore surfaces of constant Ψ are planar wave fronts, and $\nabla\Psi$ is normal to each point on the wave front. Consider the term $(\nabla\Psi) \cdot (\nabla v_n)$ in eq. (9.97). For simplicity set $v_n = f$. Consider a line stemming from a point on the wave front and running normal to the wave front. This line is a light ray. Let s be the arc length along the ray. Then, expanding the total derivative of f with respect to s, we get

$$\frac{df}{ds} \equiv f = f_x x' + f_y y' + f_z z' = (\nabla f) \cdot \left(\frac{d\mathbf{r}}{ds}\right).$$

df/ds is the total derivative of f and is tangent to the light ray. This follows from the fact that $\nabla f \cdot \mathbf{r}'$ is the projection of ∇f in the direction tangent to the light ray, where $\mathbf{r}' = d\mathbf{r}/ds$ is the unit tangent vector. The vector $\nabla\Psi$ is normal to the wave front or is in the direction of the light ray. It follows that

$$(\nabla\Psi) \cdot (\nabla f) \equiv (\nabla\Psi) \cdot (\nabla v_n) = \frac{dv_n}{ds}. \tag{9.98}$$

Coming back to eq. (9.97): we see that this represents a recursive system where v_n is to be determined given v_{n-1}. Going back to $n = 0$, we must know v_0. The solution of eq. (9.97) for v_n is easily seen to be

$$v_n(s) = v_n(s_0) e^{-(1/2)\int_{s_0}^s \nabla^2 \psi \, ds} - \frac{1}{2} \int_{s_0}^s e^{-(1/2)\int_\tau^s \nabla^2 \psi \, ds'} \nabla^2 v_{n-1}(\tau) \, d\tau.$$
(9.99)

The first term on the right-hand side of eq. (9.99) is the complementary solution. The second term is a particular integral that depends on the previous iterate v_{n-1}. Setting $n = 0$ (recognizing that $v_{-1} = 0$) we get

$$v_0(s) = v_0(s_0) e^{-(1/2)\int_{s_0}^s \nabla^2 \psi \, ds}. \tag{9.100}$$

The rays orthogonal to the surfaces $\psi = \text{const}$ are straight lines. This is seen from eq. (9.96), where $(\nabla\psi)^2 = \nabla\psi \cdot \nabla\psi = (d\psi/ds)^2 = 1$.

Recall that Ψ depends on the wave front we are concerned with. Having obtained v_n for $n = 1, 2, \ldots$ (given v_0), we use eq. (9.90) to obtain u.

The applications of this method to cylindrical waves and diffraction problems are given in [18] and [19]. The above analysis is essentially due to J. B. Keller [18], [19].

CHAPTER NINE

APPENDIX: THE PRINCIPLE OF LEAST ACTION

We shall call Hamilton's principle the δ variation and the principle of least action (to be described) the Δ variation. The δ variation corresponds to virtual displacements where time is fixed and the coordinates varied according to the constraints imposed upon the system. The varied path for the δ variation need not correspond to the actual path, since the virtual displacement does not necessarily correspond to an actual displacement occurring in the course of the motion. This is especially true when the constraints depend on time. Moreover, the Hamiltonian may not be conserved during the motion. In contrast, the Δ variation will be seen to depend on displacements that involve a change in dt, so that the varied path is obtained by a succession of displacements each including a different time change dt. We can then require that the varied path be consistent with the physical motion. For example, if H is conserved on the actual path it must also be conserved on the varied Δ path. The time of transit for the trajectory might not be the same as the actual path, since the system point may have to speed up or slow down in order to keep the Hamiltonian constant.

In order to keep track of the varied paths for the Δ variation, we assign a parameter α to each curve. Since the variations include the variation in time associated with each system point, it follows that t is a function of α. The variation in the ith generalized coordinate is $q_i = q_i(t, \alpha)$. Δq depends on the total derivative of q, so that we may write

$$\Delta q \rightarrow \left(\frac{dq}{d\alpha}\right) d\alpha = \left(\frac{\partial q}{\partial \alpha} + \frac{dt}{d\alpha}\dot{q}\right) d\alpha$$
$$= \delta q + \dot{q}\,\Delta t, \tag{9A.1}$$

since the term $\delta q \rightarrow (\partial q/\partial \alpha)\,d\alpha$, and $\dot{q}\,d\alpha$ represents the change in t due to the Δ variation and can thus be designated as Δt. Therefore (9A.1) may be written as

$$\Delta q = \delta q + \dot{q}\,\Delta t. \tag{9A.2}$$

This relation holds for any function $f(\mathbf{q}, t)$ so that

$$\Delta f = \delta f + \dot{f}\,\Delta t. \tag{9A.3}$$

VARIATIONAL METHODS

The expansion for Δf in terms of the components of q and t becomes

$$\Delta f = \sum_i \frac{\partial f}{\partial q_i} \Delta q_i + \frac{\partial f}{\partial t} \Delta t$$

$$= \sum_i \frac{\partial f}{\partial q_i} \delta q_i + \left(\sum_i \frac{\partial f}{\partial q_i} \dot{q}_i + \frac{\partial f}{\partial t} \right) \Delta t. \quad (9A.4)$$

In order to motivate the concept of the *action integral*, we take a look at the term $\sum_i p_i \dot{q}_i$ in eq. (9.42). We have $\sum_i p_i \dot{q}_i = H + L = E + T - V = (T + V) + T - V = 2T$. The integral over time (fixed end points) of the expression $\sum_i p_i \dot{q}_i$ is called the action integral. Let A be the action integral, then the definition becomes

$$A = \int_{t_1}^{t_2} \sum_i p_i \dot{q}_i \, dt = \int_{t_1}^{t_2} 2T \, dt = \int_A^B 2T \left(\frac{dt}{ds} \right) ds = \int_A^b 2\frac{T}{\dot{q}} \, ds = \int_A^B p \, ds, \quad (9A.5)$$

where ds is an element of arc length in configuration space. The last equation of (9A.5) tells us that the action integral is equal to the spatial integral of the momentum p from $s = A$ to B (the fixed end points).

The principle of least action states that, in a system for which H is constant, the Δ variation of the action integral must be zero, which means that A has a stationary value. We have

$$\Delta \int_{t=t_1}^{t_2} \sum_i p_i \dot{q}_i \, dt = \Delta \int_{t_1}^{t_2} 2T \, dt = \Delta \int_A^B p \, ds. \quad (9A.6)$$

We shall prove that A has a stationary value. In the proof it will be convenient to use the time integral rather than the spatial integral. We observe that the principle of least action states that the spatial integral of the momentum over a trajectory from $s = A$ to B (the functional) must be stationary.

We now prove that the Δ variation of the action integral A is zero, which gives a stationary value for the functional A. We have

$$\Delta A = \Delta \int_{t_1}^{t_2} \sum_i p_i \dot{q}_i \, dt = \Delta \int_{t_1}^{t_2} (L + H) \, dt$$

$$= \Delta \int_{t_1}^{t_2} L \, dt + H(\Delta t_2 - \Delta t_1). \quad (9A.7)$$

CHAPTER NINE

The last step follows from the conservation of energy. The limits of the integration are also subject to the Δ variation. We set $\int L\, dt = I(t)$. Using eq. (9A.4) we write

$$\Delta \int_{t_1}^{t_2} L\, dt = \Delta I(t_2) - \Delta I(t_1)$$

$$= \delta I(t_2) - \delta I(t_1) + \dot{I}(t_2)\Delta t_2 - \dot{I}(t_1)\Delta t_1, \qquad \dot{I} \equiv \frac{dI}{dt},$$

and finally get

$$\Delta \int_{t_1}^{t_2} L\, dt = \delta \int_{t_1}^{t_2} L\, dt + L[\Delta t_2 - \Delta t_1]. \tag{9A.8}$$

We might think that the first term on the right vanishes because of Hamilton's principle. Hamilton's principle requires that the variation of δq_i be zero at the end points, whereas in the Δ process Δq_i becomes zero at the end points, not Δt. $\delta \int_{t_1}^{t_2} L\, dt$ becomes

$$\delta \int_{t_1}^{t_2} L\, dt = \int_{t_1}^{t_2} \delta L\, dt = \int_{t_1}^{t_2} \sum_i \left(\frac{\partial L}{\partial q_i} \delta q_i + \frac{\partial L}{\partial \dot{q}_i} \delta \dot{q}_i \right) dt$$

$$= \sum_i \int_{t_1}^{t_2} \left[\frac{d}{dt}\left(\frac{\partial L}{\partial \dot{q}_i}\right) \delta q_i + \frac{\partial L}{\partial \dot{q}_i} \frac{d}{dt} \delta q_i \right] dt$$

$$= \sum_i \int_{t_1}^{t_2} \frac{d}{dt}\left(\frac{\partial L}{\partial \dot{q}_i} \delta q_i \right) dt,$$

where we used Lagrange's equations of motion. Using eq. (9A.2), $\delta \int_{t_1}^{t_2} L\, dt$ becomes

$$\delta \int_{t_1}^{t_2} L\, dt = \sum_i \int_{t_1}^{t_2} \frac{d}{dt}\left(\frac{\partial L}{\partial \dot{q}_i} \Delta q_i - \frac{\partial L}{\partial \dot{q}_i} \dot{q}_i \Delta t \right) dt$$

$$= \sum_i \left(\frac{\partial L}{\partial \dot{q}_i} \Delta q_i - \frac{\partial L}{\partial \dot{q}_i} \dot{q}_i \Delta t \right)\Big|_{t_2}^{t_1},$$

$$= -\sum_i p_i \dot{q}_i \Delta t \Big|_{t_2}^{t_1}, \tag{9A.9}$$

VARIATIONAL METHODS

Using eqs. (9A.8) and (9A.9), (9A.7) becomes

$$\Delta A = \Delta \int_{t_1}^{t_2} L\, dt + H(\Delta t_2 - \Delta t_1) = \delta \int_{t_1}^{t_2} L\, dt + (L + H)(\Delta t_2 - \Delta t_1)$$

$$= -\sum_i p_i \dot{q}_i \, \Delta t \Big|_{t_1}^{t_2} + (L + H)(\Delta t_2 - \Delta t_1)$$

$$= -\sum_i p_i \dot{q}_i \, \Delta t \Big|_{t_1}^{t_2} + \sum_i p_i \dot{q}_i \, Dt \Big|_{t_1}^{t_2} = 0. \qquad (9\text{A}.10)$$

This completes the proof of the principle of least action.

Problems

1. Form a surface of revolution by taking a curve $y = y(x)$ between two fixed points and revolving it around the y axis. Find the total area in the form $I = \int_{P_1}^{P_2} f(x, dv/dx)\, dx$. Minimize this functional and show that the minimum surface area is the catenary

$$y = a\, \text{arc cosh}\, \frac{x}{a} + b \quad \text{or} \quad x = a \cosh \frac{y - b}{a}.$$

Find the constants a and b such that the minimizing curve goes through the end points (x_1, y_1), (x_2, y_2).

2. Find the curve joining two points along which a particle falls from rest from the higher to the lower point under the influence of gravity (no friction). This is called the *brachistochrone problem*. *Hint:* the integrand is

$$f = \sqrt{\frac{1 + dy/dx}{2gy}},$$

where g is the acceleration of gravity.

3. Show that the geodesics of a spherical surface are great circles (circles whose centers are at the center of the sphere). A geodesic is the shortest distance between two points on the surface.

CHAPTER NINE

4. The transverse oscillations of a stretched string can be approximated by an equally spaced distribution of mass-spring elements coupled in series.

(a) Show that if the spacing tends to zero (yielding an infinite number of elements) the Lagrangian approaches in the limit

$$L = \tfrac{1}{2}\int \left[\mu(u_t)^2 - T(u_x)^2 \right] dx,$$

where μ is the mass per unit length of the string and T is the constant tension.

(b) Show that the solution of Lagrange's equation of motion yields the wave equation for the oscillating string.

(c) Obtain the Lagrangian for the continuous string by finding the kinetic and potential energies due to the transverse motion. The potential energy can be obtained from the work done in stretching the string in the course of its oscillations.

5. Find the trajectory of a projectile under the influence of gravity with no friction, and the dependence of the coordinates on time, by the H-J method.

6. Show directly that the transformation

$$Q = \log\left(\frac{1}{q}\sin p\right), \qquad P = q \cot p,$$

is canonical.

Bibliography

The subject of wave propagation in various media is multidisciplinary in nature, spanning different aspects of applied mathematics and physics. Therefore this bibliography contains references to various fields of applied mathematics, as well as to the physical sciences of fluid and solid mechanics and water waves. A selected sampling of the pertinent standard works in these fields is given rather than references to numerous papers, which would only tend to confuse the nonspecialist reader.

1. Birkhoff, G. and S. MacLane. *A Survey of Modern Algebra*, 4th ed. Macmillan, 1977.
2. Bland, D. R. *The Theory of Linear Elasticity*. Pergamon Press, 1960.
3. Bliss, G. A. *Lectures on the Calculus of Variations*. University of Chicago Press, 1947.
4. Chester, Clive R. *Techniques in Partial Differential Equations*. McGraw-Hill, 1971.
5. Churchill, R. V. *Complex Variables and Applications*, 2nd ed. McGraw-Hill, 1960.
6. Churchill, R. V. *Fourier Series and Boundary Value Problems*, 2nd ed. McGraw-Hill, 1963.
7. Churchill, R. V. *Operational Mathematics*, 3rd ed. McGraw-Hill, 1972.
8. Courant, R. and D. Hilbert. *Methods of Mathematical Physics Vol. II: Partial Differential Equations*. Interscience, 1962.
9. Courant, R. and K. O. Friedrichs. *Supersonic Flow and Shock Waves*. Interscience, 1948.
10. Davis, J. L. *Introduction to Dynamics of Continuous Media*. McGraw-Hill, 1987.
11. Davis, J. L. *Wave Propagation in Solids and Fluids*. Springer-Verlag, 1988.
12. Ewing, W. M., W. S. Jardetsky, and F. Press. *Elastic Waves in Layered Media*. McGraw-Hill, 1957.
13. Feynman, R. P., R. B. Leighton, and M. Sands. *The Feynman Lectures on Physics*. Addison-Wesley, 1964.
14. Goldstein, H. *Classical Mechanics*. Addison-Wesley, 1959.
15. Gurtin, M. E. *The Linear Theory of Elasticity*, Vol. 6a/2 of *Handbuch der Physik* (in English). Springer-Verlag, 1965.
16. Hadamard, J. *Lectures on Cauchy's Problem*. Dover, 1952.

BIBLIOGRAPHY

17. Hildebrand, F. B. *Advanced Calculus for Applications*, 2nd ed. Prentice-Hall, 1976.
18. Keller, J. B. "A Geometric Theory of Diffraction," in *Calculus of Variations and its Applications, Proceedings of the Symposia on Applied Mathematics*, vol. 8, pp. 27–52. American Mathematical Society, 1958.
19. Keller, J. B., R. M. Lewis, and B. D. Seckler, "Asymptotic Solution of Some Diffraction Problems," *Communications in Pure and Applied Mathematics* 9:207–265, 1956.
20. Kolsky, H. *Stress Waves in Solids*. Dover, 1963.
21. Lamb, H. *Hydrodynamics*, 1st American ed. Dover, 1945.
22. Landau, L. D. and E. M. Lifshitz. *Fluid Mechanics*. Pergamon Press, 1959.
23. Landau, L. D. and E. M. Lifshitz. *Theory of Elasticity*. Pergamon Press, 1959.
24. Love, A. E. H. *A Treatise on the Mathematical Theory of Elasticity*. Dover, 1944.
25. Milne-Thomson, L. M. *Theoretical Hydrodynamics*, 2nd ed. Macmillan, 1950.
26. Murnaghan, F. D. *Finite Deformation of an Elastic Solid*. Wiley and Sons, 1951.
27. Nowacki, W. *Thermoelasticity*. Addison-Wesley, 1962.
28. Prager, W. *Introduction to Mechanics of Continua*. Ginn and Co., 1961.
29. Rayleigh, J. W. S. *The Theory of Sound*, 2 vols., 1st American ed. Dover, 1945.
30. Sears, F. W. and M. W. Zemansky. *University Physics*, 2nd ed. Addison-Wesley, 1962.
31. Shapiro, A. *The Dynamics and Thermodynamics of Compressible Flow*. Ronald Press, 1958.
32. Sokolnikof, I. S. *Mathematical Theory of Elasticity*, 2nd ed. McGraw-Hill, 1956.
33. Sokolnikof, I. S. *Tensor Analysis*, 2nd ed. Wiley, 1964.
34. Sommerfeld, A. *Mechanics of Deformable Bodies*. Academic Press, 1950.
35. Stoker, J. J. *Water Waves*. Interscience, 1957.
36. Timoshenko, S. and J. N. Goodier. *Theory of Elasticity*. McGraw-Hill, 1957.
37. Whittaker, E. T. *A Treatise on the Analytical Dynamics of Particles and Rigid Bodies*. Dover, 1944.

Index

Acoustic impedance, 155
Adiabatic condition, 17
Adjoint operators, 75
Amplitude modulated wave as envelope of high frequency waves, 37

Bernoulli's law, 159f.
Bessel's equation, 82, 133
Boundary layer, 190f.
Brachistochrome problem, 387
Bulk modulus, 16

Calculus of variations
Canonical form of a second-order PDE, 70f.
Canonical transformations, 364f.
Cauchy initial value problem, 44f., 102f.
Cauchy-Riemann equations, 55, 303
Caustic or focal curves, 83
Characteristic coordinates, 15, 71, 89
Characteristic equations (ODEs)
—Classification of, 59
—First-order PDEs, 55f.
—Geometric interpretation for two-dimensional steady flow, 166
—Hodograph plane, 165f.: Adiabatic gas, 167
—Nonlinear PDEs, 65, 66
—Parametric form, 43, 52
—Second-order PDEs, 58, 63
—Wave equation, 86, 87
Characteristic line element, 64
Characteristics, method of, 43f.
—Linear PDEs, 57f.
—One-dimensional compressible flow, 151f.
—Parametric representation, 49f., 65
—Second-order PDEs, 61f.
—Two-dimensional steady flow, 161
Characteristic surfaces, 63, 229
Complex modulus and compliance, 258
Compressible fluids
—One-dimensional, 145f.
—Two-dimensional steady flow, 157
Compression ratio, 215
Configuration space, 349
Conservation laws (*see* Eulerian and Laplacian representation)
Critical sound speed, 159
Curvilinear orthogonal coordinates, 237f.
Cyclic coordinates, 362f.

D'Alembert's paradox, 307f.
D'Alembert's solution of the wave equation, 97f.
Deformation, 214
Depth of penetration for viscous wave over a plate, 195
Dimensionless parameters (*see* Reynold's law)
Direction field, 65
Directional derivatives, 42f., 86, 152
Discrete wave propagating systems, 3f.
Dispersion and group velocity, 36f.
Dispersive waves, 125
Dissipation function (Rayleigh), 357f.
Divergence of stress tensor, 188
Divergence theorem, 73f., 150
Domain of dependence and range of Influence, 101f.
Doppler effect, 33f.
Doppler shift (*see* Doppler effect)
Duhamel-Neumann law, 282f.

Eigenfunctions, 93

391

INDEX

Eigenvalues, 93
—Negative eigenvalues, 127
Eikonal equation, 375
Electrical transmission line, 122f.
Electromechanical analogy between electrical circuits and VE material, 271f.
Energy dissipation in a viscous fluid, 191f.
Enthalpy, specific, 160
Entropy, specific, 160
Epicycloid, 167f.
Euler's equations
—Calculus of variations, 345f.
—Inviscid fluid, vector form, 160
—Two-dimensional steady flow, 158
—Water waves, 299f.
Eulerian representation
—One-dimensional compressible flow, 149: Conservation of mass, 149f.; Conservation of momentum, 150f.
—Viscous fluid, 185f.: Conservation of mass, 185; Conservation of momentum, 186f.
—Water waves, 299f.

Fermat's principle, 344
Finite bar, 113f.
Finite system of mass-springs, 4
Fluid dynamics (*see* Wave propagation- In fluids)
Friction, internal, 251

Generalized coordinate, conjugate to generalized momentum, 359
Generalized coordinate transformation, 354f.
Generalized force, 355
Generalized momentum, conjugate to generalized coordinate, 359
Generalized velocity, 350
Green's identity, 74
Group velocity (*see* dispersion and group velocity)

Hamiltonian, 360
Hamilton-Jacoby theory, 364f.
H-J theory and wave propagation, 372f.
—Application to light waves, 373f.

Hamilton's equations of motion, 359f.
Hamilton's principle function, 366
Hamilton's variational principle, 350f.
Harmonics and modes of vibrations, 28, 33
Harp, vibrations of a plucked harp string, 96
Heaviside step function, 112
Helmholtz equation (*see* Wave equation- Reduced wave equation)
Hodograph transformation, 163f.
—Hodograph plane, 165
Hooke's law, 222
Hugoniot relation for an adiabatic gas, 180f.

Ideal or perfect gas, 17
Impedance, characteristic, 235
Integral surfaces, 64, 65f.
Interference phenomena, 21f.
—Constructive and destructive interference, 23, 24
Internal energy, specific, 160
Inviscid fluids, 145f.
Isothermal condition, 17

Jacobian, 119, 163f, 215

Kinetic energy, 350

Lagrangian function, 351
Lagrangian representation
—One-dimensional compressible gas, 146: Conservation of mass, 146; Conservation of momentum, 147; Energy equation, 148
—Water waves, 314f.
Lagrange's equations of motion, 351, 357, 358
Lame constants, 222
Laminar flow, 183
Laplace transform method, 105f.
—Application to the wave equation, 111
Laplacian, 126
Least action, principle of, 384f.
Least time, principle of, 344, 345
Limit speed, 161

Memory function, 261

INDEX

Minimizing curve, 346
Monge axis, 64
Monge cone, 67

Navier equations, 223f.
—Generalization for a VE material, 265f.
—Generalized for thermoelasitic media, 293f.
Navier-Stokes equation, 189
—Boundary conditions on N-S equations, 189, 190
—Linearized form, 189

Optics, geometric analogy to classical mechanics, leading to Schroedinger's equation, 377f.
Orthogonal relations among Bessel functions, 134
Oscillating body of arbitrary shape, 196f.
Oseen approximation, 208f.

Partial differential equations (PDEs)
—Canonical form of a second-order PDE (*see* Canonical form of a second-order PDE)
—Characteristic equations for (*see* Characteristic equations for first-order PDEs)
—Classification of, 59f.
—Equivalent to two first order PDEs, 51
—First-order PDEs, 55f.
—Geometric nature of PDEs, 42
—Quasilinear: Geometric interpretation of, 63f.
—Second-order linear PDEs treated by by characteristic theory, 57f.
—Types of, 41f.
Physics of propagating waves, 3f.
Planck's constant, 379
Plane elastic waves, 227f.
Plane harmonic waves, 287f.
Poiseuille flow, 199f.
Poisson's ratio, 222
Potential energy, 350
Prandtl's relation, 181
Progressing wave, 9

Quantum mechanics, 376f.

Radially symmetric waves, 240f.
Reduced wave equation (*see* Wave equation- Reduced wave equation)
Reflection of light waves, 25f.
—Construction to show minimizing principle, 27
Reflection of waves in a string, 27f.
Regressing wave, 9
Relatively undistorted waves, 124
Reynold's law, 197f.
Reynold's number, 197
Riemann invariants, 154
Riemann's method, 73f.
—Riemann's test function, 75

Self-adjoint operator, 75
Separation of variables, 92f.
Shock wave phenomena, 169f.
—Advancing shock front, 173
—Conservation laws across shock front, 174f.: Conservation of mass, 175; Conservation of momentum, 175; Conservation of energy, 175, 176; Entropy jump across shock front, 176
—Jump conditions, 176f.
—Momentum flux across shock front, 178
—Qualitative concepts, analogy with particle motion, 173
—Receding shock front, 173
—Shock conditions, 177f.
—Shock line, 171
—Shock strength, 172
—Shock transition, 182
—Shock tube, 171f.
—Stationary shock front, 173
—Surfaces of discontinuity, 170
Similarity conditions (*see* Reynold's law)
Simple harmonic motion, 20
Sinusoidal waves, 15, 19f.
Sound velocity, Laplace's expression, 18
Specific heats, 18
Spherically symmetric waves, 232
Standing waves, 25, 26, 319f.
Stokes' flow, 201f.
Stokes' law, 208
Strain as a function of displacement, 217f.
Strain energy density function, 221

393

INDEX

Strain rate tensor, 206
Strain tensor, 215f.
Stream function, 301
Streamlines, 159
Stress relaxation experiment, 254
Stress tensor, 219f.
Stress waves
—Elastic solids, 213f.
—Inviscid fluids, 145f.
—Thermoelastic media, 282f.
—VE solids, 250f.
—Viscous fluids, 193f.
Stress-strain relations, 220f.
String (*see* Vibrating string)
Superposition principle, 19

Telegraph equation, 123
Thermoelastic media
—Three dimensions, 135f.
—Three-dimensional thermal waves, 291f.:
  Tidal waves, 337f.
Transmission coefficients, 122
Transverse oscillations of string, 9f.
—Transverse wave velocity of string, 10f.
Traveling waves in general, 11f.
Two-dimensional steady flow, 157f.
Two-dimensional steady flow, 157

Variational methods, 344f.
Velocity potential, 159
—Supersonic, 163
Vibrating string
—As combined initial value and boundary value problem, 90f.
—Speed of transverse wave of a string, 10, 11
—Transverse oscillations, 9f.
Virtual work, principle of, 352f.
Viscoelastic solids, 250f.
—Constitutive equations, three-dimensional for VE material, 264f.
—Continuous distribution function: Maxwell model, 261f.; Voigt model, 263
—Continuous Maxwell model, 260
—Continuous Voigt model, 263
—Creep experiment, 256f.

—Discrete viscoelastic models, 252f.:
  Maxwell model, 253, 254; Voigt or Kelvin model, 255, 256
—Stress-relaxation experiment, 254, 255
Viscosity, elementary treatment, 183f.
Viscosity coefficient
—Dynamic, 183
—Kinematic, 184
Viscous flow around body of arbitrary
Viscous fluids, 183f.
Vortex motion (Kelvin), 309

Wave equation
—Bar, one-dimensional, 5f.
—Equivalent to two first order PDEs, 51f.
—General solution for one-dimensional wave equation, 15
—Longitudinal waves in a bar, 234, 235: In a finite bar, 235, 236
—Nonhomogeneous, 116f.: In characteristic coordinates, 117
—One-dimensional, 8, 85f.: Factorization of, 65, 86
—Reduced wave equations, 126, 230
—Speed of progressing and regressing waves, 9
—Three-dimensions: Huygen's principle, 138f.; Spherical symmetry, 135f.
—Two-dimensions, 125f.: Membrane; Circular, 131f.; Rectangular, 127f.; Square boundary, 129f.
Wave propagation
—Asymptotic theory, 380f.
—In fluids, 145f.
—Invariance of wave form, 14
—Media of different velocities, 120f.
—Nonlinear, 67f.: Characteristic equations for, 68, 69
—Progressing and regressing waves, 23
—Sound wave propagation, 16, 29f.: Oscillating air columns, 30f.
—Standing waves, 26, 319f.
—Surface waves (Rayleigh waves), 243f.
—VE media (one-dimensional wave propagation), 266f.
—Viscous fluid, 193f.

INDEX

Water waves, 297f.
—Approximation theories, 327f.: Shallow-water theory, 330f.; Small amplitude waves, 330
—Boundary conditions, 322f.
—Canal, straight, 311f.
—Complex variable treatment, 302f.
—Gravity waves (small amplitude), 311
—Instability, 325f.
—Perturbation methods (*see* Approximation theories)
—Spherical symmetry, 135f.
—Standing waves, 319f.
—Stretching parameter, 333
—Surface wave problem, 324, 325
—Tidal waves, 337f.: Dynamical theory, 339f.; Equilibrium theory, 337f.
—Two-dimensional flow, 300f.
—Waves of finite depth, 321f.

Young's modulus, 7